Department of the Environment

Acidic Emissions Abatement Processes

Manual of Acidic Emission Abatement Technologies

VOLUME 2: OIL-FIRED SYSTEMS

London: HMSO

© Crown copyright 1991
First published 1991

ISBN 0 11 752214 7

Preface

This study of abatement technologies applicable at source to air pollution arising from the combustion of fossil fuels was commissioned by the Department of the Environment and undertaken by The Fellowship of Engineering. The intention has been to produce a comprehensive survey of abatement technologies giving details of processes, plant and equipment, input materials required and by-products arising and to present pollution emission data in the form of emission factors suitable for calculation of the effects of emission inventories.

The advantages and disadvantages, both technical and economic, of the many processes are stated wherever practicable, but these are necessarily dependent to a significant degree on the particular site, grade of fuel and size and type of emitter considered; in particular, costs should be regarded as indicative only. To facilitate comparative calculation, datum cases before the application of abatement processes are defined in detail and the format of each volume is standardised insofar as is practicable.

So far as is known, it is the first time that such an extensive survey has been undertaken in the UK. Data was collected over a period of more than a year. The work proceeded by questionnaire and inquiry of the many producers and licensors of processes as well as by the collection and analysis of published data. While the information in the four volumes is fairly comprehensive there are inevitably some gaps, and it should not be considered exhaustive. While the accuracy of all the data presented cannot be guaranteed, wherever possible references are given to enable specific items to be checked at source as necessary.

The work represented in this Volume was completed during the period 1986 to 1987 to provide data for the Department of the Environment and others in consideration and implementation of the European Community Directives on Air Pollution which were then under discussion. The emphasis is, therefore, primarily on the application of abatement technologies to existing large plants. Acidic emission abatement by energy saving means including combined heat and power and combined cycle technology was considered separately.

Contents

	Page
Acknowledgements	vii
Study Group	ix
Acronyms	xi
Units and Conversion Factors	xiii
Executive Summary	xv

1. Introduction   1

    1.1 Background and Purpose of the Manual   3
    1.2 Acid Emission Inventories   3
    1.3 Scope   4
    1.4 General Assumptions for Datum Combustion Systems   5
    1.5 Appraisal of Processes   5
    1.6 Notes for the Guidance of the Reader   7

2. Flue Gas Desulphurisation Processes   9

    2.1 Classification of the Processes   11
    2.2 Outline FGD Process Description   11
    2.3 General Appraisal of Processes   31
    2.4 Processes for Detailed Study   55
    2.5 Evaluation of Selected FGD Processes   56
    2.6 Costs for FGD Systems   87

3. Nitrogen Oxides Abatement Processes (Combustion Techniques)   91

    3.1 Classification of Combustion Techniques   93
    3.2 Outline Descriptions   93
    3.3 General Appraisal of Each Technique   98
    3.4 Comparison of Techniques   100
    3.5 Costs   103

4. Nitrogen Oxides Abatement Processes (Flue Gas Treatment)   105

    4.1 Classification of the Processes   107
    4.2 Outline Descriptions   107
    4.3 General Appraisal of Processes   113
    4.4 Processes for Detailed Study   121
    4.5 Evaluation of Selected $DeNO_x$ FGT Processes   122
    4.6 Costs   133

5. Combined SO$_2$-NO$_x$ Abatement Processes — 135

  5.1 Classification of the Processes — 137
  5.2 Outline Descriptions of Combined SO$_2$–NO$_x$ Abatement Processes — 137
  5.3 General Appraisal of Processes — 153
  5.4 Processes for Detailed Study — 162
  5.5 Evaluation of Selected Combined Abatement Processes — 164

Appendix 1 : Bibliography — 173

Appendix 2 : Vendor Information — 195

Appendix 3 : Cost Estimates and Procedures — 201

Appendix 4 : Index — 213

Acknowledgements

We are grateful to the following Companies, Institutions and Consultants who have provided information for incorporation in this report:

UNITED KINGDOM

Airoil-Flaregas Ltd, West Drayton, Middx.
Babcock Power Ltd, London
Beijer Institute, York
Berridge Environmental Laboratories Ltd, Chelmsford, Essex
British Coal Corporation (Coal Research Establishment), Cheltenham
British Gas, London
British Gypsum, Nottingham
Calor, Slough, Bucks.
Central Electricity Generating Board (Generation, Development &
    Construction Division), Barnwood, Glos.
Central Electricity Research Laboratories (CEGB), Leatherhead,
    Surrey
Chem Systems, London
Department of the Environment, London
Dunphy Oil & Gas Burners Ltd, Rochdale, Lancs.
E.T.S.U., Harwell
Fellowship of Engineering, London
Hamworthy Combustion Ltd, Poole, Dorset
International Energy Agency (Coal Research), London
John Zink Co. Ltd, St. Albans, Herts.
Johnson Matthey, Royston, Herts.
Lodge Cottrell, Birmingham
Lurgi (UK) Ltd, London
Metra Consulting, London
Nu-Way Ltd, Droitwich, Worcs.
Peabody Holmes Ltd, Maidstone, Kent
Pennwalt Ltd, Camberley, Surrey
Saacke Ltd, Portsmouth, Hants.
Stordy Combustion Engineering Ltd, Wolverhampton, West Midlands
Warren Spring Laboratory, Stevenage, Herts.

DENMARK

Niro Atomizer AS., Soeborg, Copenhagen

FEDERAL REPUBLIC OF GERMANY

Deutsche Babcock Anlagen, Krefeld
G. Bischoff GmbH, Essen
Rheinisch–Westfalisches Elektrizitatswerk, Essen
L & C Steinmuller GmbH, Gummersbach
Thyssen Engineering, Essen
Umweltbundesamt, Berlin
VGB, Essen

FINLAND

Bioneer Oy, Hämeenlinna
Tampella, Tampere

FRANCE

CITEPA, Paris
Organisation for Economic Co-operation & Development, Paris
Societe Foster Wheeler Francaise, Paris

JAPAN

Chiyoda Chemical Engineering & Construction Co. Ltd, Yokohama
Ishikawajima-Harima Heavy Industries Co. Ltd., Tokyo
Mitsui Miike Engineering Corporation, Tokyo
Sumitomo Heavy Industries, Tokyo

THE NETHERLANDS

Concawe, The Hague
De Jong Coen BV, Schiedam
ESTS, Ijmuiden
Foster Wheeler International Corporation, The Hague

SWEDEN

AB Aroskraft, Vasteras
Fläkt Industri AB, Vaxjo
National Swedish Environmental Protection Board, Solna

USA

Andersen 2000 Inc., Atlanta, GA.
Battelle, Columbus, Ohio
DB Gas Cleaning Corp., Orinda, CA.
Davy McKee, Lakeland, FA.
Electric Power Research Institute, Palo Alto, CA.
Foster Wheeler Development Corporation, Livingston, NJ.
General Electric Environmental Services Inc., Lebanon, PA.
Otto H. York, Parsippany, NJ.
Peabody Process Systems, Norwalk, CT.
R.E. Sommerlad (Consultant), Cranford, NJ.
Tennessee Valley Authority, Chattanooga, Tenn.

Study Group

The following assisted in the preparation of this manual:

FELLOWSHIP OF ENGINEERING

Steering Group : Sir Frederick Page (Chairman)
Mr. M. Kneale (Project Manager and Nominated Officer for the Fellowship of Engineering)

Dr. F. Steele Professor J.F. Davidson
Mr. J.R. Appleton Professor S. Eilon
Mr. J.G. Dawson Professor I. Fells
Mr. G.A. Lee Mr. R.J. Kingsley
Professor G.F.I. Roberts Mr. V.J. Osola
Dr. J. Gibson Mr. K.R. Vernon
Dr. J.H. Chesters Dr. D. Train
Dr. A.J. Apling (Nominated Officer of the Department of the Environment)
Mr. J. Murlis (Department of the Environment)

Coal Task Group : Dr. D.R. Cope
Professor J.F. Davidson
Dr. J. Gibson (Chairman)

Oil Task Group : Mr. P. Brackley
Professor I. Fells
Mr. G.A. Lee (Chairman)
Mr. J. Solbett

Gas Task Group : Dr. C.G. James
Professor G.F.I. Roberts (Chairman)
Mr. P. Scott
Dr. F.E. Shephard
Dr. W.A. Simmonds
Professor A. Williams

Mobile Sources Group : Professor G.P. Blair
Mr. J.G. Dawson (Chairman)
Mr. A. Silverleaf

FOSTER WHEELER POWER PRODUCTS

: Dr. R. Fletcher
Mr. K. Johnson
Mr. H. Luaw
Mr. D. McSherry
Mr. H.T. Wilson (Programme Manager)

HOY ASSOCIATES (UK)	:	Mr. D.W. Gill
		Mr. H.R. Hoy (Director)
		Mr. A.G. Roberts
		Mr. J.E. Stantan
		Mr. D.M. Wilkins
WARREN SPRING LABORATORIES	:	Dr. M. Williams
		Mr. J. Potter
		Dr. J.H. Weaving

Acronyms

AAF	American Air Filters
AFBC	atmospheric fluid-bed combustion
B&W	Babcock and Wilcox
BBF	biased burner firing
BF	Bergbau Forschung
BHK	Babcock-Hitachi K.K. (Japan)
BOOS	burners out of service
C-E	Combustion Engineering (USA)
CEA	Combustion Engineering Associates
CEC	Commission of the European Communities
CEGB	Central Electricity Generating Board
CFBC	circulating fluidised-bed combustion
COD	chemical oxygen demand
CONCAWE	Conservation of Clean Air and Water–Europe (Oil Companies' International Study Group)
DBA	dibasic acid
DMB	distributed mixing burner
DRB	dual register burners
DSAA	down-fired sequential air addition
ECE	Economic Commission for Europe (UN)
EDTA	ethylenediamine tetra-acetic acid
EGR	exhaust gas recirculation
EHE	external heat exchanger
EPA	Environmental Protection Agency (USA)
EPDC	Electric Power Development Co. Ltd. (Japan)
EPRI	Electric Power Research Institute (USA)
ESP	electrostatic precipitators
FBC	fluidised-bed combustion
FBHE	fluidised bed heat exchanger
FGD	flue gas desulphurisation
FGR	flue gas recirculation
FRG	Federal Republic of Germany
FW	Foster Wheeler
GEESI	General Electric Environmental Services Inc. (USA)
ID	induced draft
IEA	International Energy Agency
IFNR	in-furnace NO_x reduction
IFP	Institut Francais du Petrole
IHI	Ishikawajima-Harima Heavy Industries (Japan)
IIP	Institut fur Industriebetriebsehre und Industrielle Produktion (University of Karlsruhe, FRG)
JBR	jet bubbling reactor (Chiyoda)
KHI	Kawasaki Heavy Industries (Japan)
KVC	Kawasaki volume combustion system
LEA	low excess air
LHV	lower heating value (net calorific value)
LIMB	limestone injection/multi-stage burner
LNB	low-NO_x burner
LNCFS	low-NO_x concentric firing system
LVHR	low volume heating rate

MACT	Mitsubishi Advanced Combustion Technology
MCR	maximum continuous rating
MHI	Mitsubishi Heavy Industries (Japan)
MIT	Massachusetts Institute of Technology (USA)
NATO-CCMS	North Atlantic Treaty Organisation–Committee on the Challenge of Modern Society
n.a.	not available
NCB	National Coal Board (British Coal)
NCR	non-selective catalytic reduction
OECD	Organisation for Economic Co-operation and Development
OFA	overfire air (injection)
OSC	off-stoichiometric combustion
PCF	primary combustion furnace
PENSYS	Pittsburgh Environmental Systems Inc. (USA)
PETC	Pittsburgh Energy Technology Centre (USA)
pf	pulverised fuel
PFBC	pressurised fluidised-bed combustion
PM	pollution minimum (burner)
PVC	polyvinyl chloride
RAP	reduced air preheat
SCR	selective catalytic reduction
SDA	spray dryer absorber
SGR	separate gas recirculation
SNR	selective non-catalytic reduction
TCA	turbulent contact absorber
TET	turbine entry temperature
TVA	Tennessee Valley Authority (USA)
UFI	upper fuel injection
UOP	Universal Oil Products
VDI	Verein Deutscher Ingenieure (Dusseldorf, FRG)

Units and Conversion Factors

Thermal   *Equivalent Units*

GJt	gigajoule	0.95×10^6 Btu
kWt	kilowatt	3,412 Btu/h
MWt	megawatt	3.4×10^6 Btu/h
TWth	terawatt hour	3.4×10^{12} Btu
MJ/kg	megajoules per kilogramme	430 Btu/lb
therm	therm	1.0×10^5 Btu

Electrical

GJe	gigajoule	278 kWh
kWe	kilowatt	1.34 hp
MWe	megawatt	1341 hp
MWh	megawatt hour	3.4×10^6 Btu
kW/MWt	kilowatt (consumption) per megawatt thermal (capacity)	–
kW/MWe	kilowatt (consumption) per megawatt electrical (capacity)	0.1%

Volumetric Flowrates

Nm³/h	normal cubic metre per hour, i.e. at 0°C and 1 atmosphere	0.59 ft³/min (0°C, 1 atm)
m³/h	actual cubic metre per hour	0.59 ft³/min (at same temperature and pressure)
Nm³/s	normal cubic metre per second	2119 ft³/min (0°C, 1 atm)
l/Nm³	litre of liquid per normal cubic metre of gas (liquid/gas ratio)	6.23×10^{-3} gallon/ft³
m³/MWh	cubic metre (water or effluent) per megawatt hour (electrical output)	–
l/GJt	litre (water or effluent) per gigajoule (thermal output)	0.23 gallon/10^6 Btu

Concentration

mg/Nm³*	milligramme (pollutant) per normal cubic metre (flue gas)	0.35 vppm (SO_2) 0.487 vppm (NO_2) 0.614 vppm (HCl)
mg/m³	milligramme (pollutant) per actual cubic metre (flue gas)	–
g/GJ	gramme (pollutant) per gigajoule (thermal input)	2.33×10^{-3} lb/10^6 Btu
ng/J	nanogrammes per joule	1.0 g/GJ
lb/10^6 Btu	pound per million Btu	430 g/GJ

ppm*	part (pollutant) per million (parts of fluid). Typically by volume for gas (see also vppm); by weight for liquids	0.00001%
vppm	part per million by volume	2.86 mg (SO_2)/Nm^3 2.05 mg (NO_2)/Nm^3 1.63 mg (HCl)/Nm^3

*Although SO_2 and HCl concentrations are normally quoted in units of mg/Nm^3, it is common practice to express NO_x concentrations in ppm (vol) or vppm. This is because continuous NO_x monitors operate on a molar basis which relates directly to vppm units, and is independent of temperature, pressure and NO/NO_2 ratio. Separate detection of NO and NO_2 is therefore not required.

Mass

t	tonne (metric ton)	2205 lb
ton	US (short) ton	2000 lb
g/l	gramme per litre	0.010 lb/gallon
kg/MWh	kilogramme (reagent) per megawatt hour (electrical output)	–

Efficiency Factor

GJe/tonne	gigajoule (electrical output) per tonne of fuel fired	–
GJt/tonne	gigajoule (thermal output) per tonne of fuel fired	–

Emission Factor

kg/tonne	kilogramme (elemental pollutant) per tonne of fuel fired	–

Pressure

mbar	millibar	0.0145 1lbf/in^2 (psi)
mm H_2O	millimetre of water gauge	0.10 mbar
atm	atmospheric pressure	14.7 psig; 1.013 bar
Pa	pascal	0.01 mbar; 0.1 mm H_2O

Miscellaneous

mill	one thousandth of US dollar	$ 0.001 (0.1 cent)
ha	hectare	2.47 acres 10,000 m^2
micron	micrometre (particle size)	0.000039 inch
m/s	metre per second	3.28 ft/s

Manual of Acidic Emission Abatement Technologies
Executive Summary
(Oil)

0.1 Background and Purpose of the Study

The Fellowship of Engineering has been commissioned by the Department of the Environment to prepare a factual Manual describing the essence of individual technologies or processes for abatement of acidic emissions from combustion of fossil fuel in mobile and stationary sources; the present study is concerned with emissions from stationary combustion systems. The objective is to consider the abatement technologies or processes currently available commercially, and those still undergoing development but showing promise of commercial adoption in the UK by the end of the century.

The aim has been to prepare the Manual in a form suitable for publication and for subsequent updating. The anticipated readership includes air quality professionals requiring information on current or likely future commercially developed technologies, and organisations needing a data base for assessing emission control economics. The Manual for stationary sources deals in separate volumes with combustion systems fired with coal, oil and gas; this Volume is concerned with oil combustion systems.

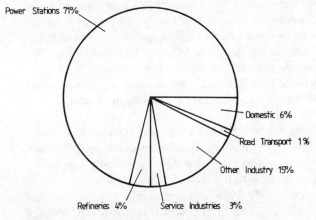

Figure 0.1 UK Man-made Sources of SO_2 Emission (1985) Total SO_2 Inventory 3.58 million tonnes

0.2 The Acidic Emissions

The emissions most important as precursors of acid deposition are sulphur dioxide, SO_2, and nitrogen oxides, NO_x. Estimates of the total UK inventory of man-made emissions for 1985 have been prepared by Warren Spring Laboratory (WSL). The estimates for SO_2 emissions are presented in Figure 0.1. It is seen that power stations are overwhelmingly the principal source, with industry (refineries, service and other industry) accounting for most of the rest.

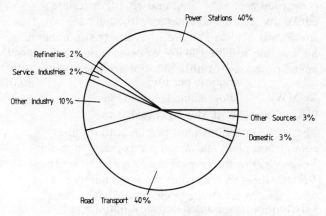

Figure 0.2 UK Man-made Sources of NO_x Emission (1985) Total NO_x Inventory (as NO_2) 1.84 million tonnes

The WSL estimates for the total NO_x emissions inventory are presented in Figure 0.2. When allowance is made for the high proportion of NO_x emissions originating from road transport, the remaining sources of emissions are found to be of similar relative significance for both SO_2 and NO_x, i.e. predominantly from power stations with a smaller but important contribution from industry. With regard to the latter, WSL estimates of NO_x emissions from the combustion of various fuels in boilers generating up to 230 tonne/h steam are shown in Figure 0.3. The main contribution is from boilers

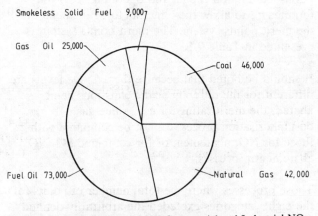

Figure 0.3 WSL Estimate of UK Commercial and Industrial NO_x Emissions (1982) Estimated Emissions, tonnes (as NO_2) from Various Fuels Total Estimated (as NO_2) 195,000 tonnes

fired with fuel oil, followed by coal, natural gas, gas oil and smokeless solid fuel. Together, however, these contribute a small proportion of the total, just over 10%.

Most of the UK emissions of HCl originate from the combustion of UK coal, which is unusually rich in chlorine, releases HCl on combustion, providing most of the UK HCl emissions; only a minor contribution comes from the incineration of industrial waste containing PVC. The emissions are an order of magnitude smaller than those of SO_2 from power stations. Emissions of hydrogen fluoride, HF, are not generally regarded as significant.

The acidic emissions with which the overall study is concerned are: SO_2, NO_x, and HCl. Particulate emissions are also considered; although not specifically acidic in nature, they are environmentally undesirable and are capable of being reduced by many of the processes for abatement of acidic emissions.

0.3 Content of the Manual

The acidic emissions most pertinent to oil-fired combustion plant and which are considered in this Volume (2) of the manual are SO_2 and NO_x. The Volume deals in turn with: the large number of processes for controlling emissions of SO_2. Next, combustion technologies giving reduced NO_x emissions are discussed and then flue gas treatment processes for NO_x removal. Finally, combined flue gas treatment processes for simultaneous removal of both SO_2 and NO_x are considered.

The approach is to classify and describe the available processes or technologies, and to present appraisals based on published information. In order to rank processes, the appraisals have been roughly quantified by awarding merit points for: information availability; process simplicity; extent and nature of operating experience; requirements for power, reagents and end-product disposal; and process applicability.

Published data on capital and operating costs were too meagre to allow these factors to be included in the merit points system. The merit points system is presented in Table 0.1.

It should be noted that because the average levels are different for the different acidic emissions being abated, the merit rating for e.g. a flue gas desulphurisation process cannot be compared with those for NO_x abatement or for combined SO_2/NO_x abatement processes.

Those processes where the total number of points for the eight categories exceeded the arbitrarily defined limit of 10 were regarded as being potentially applicable commercially in the UK, and their

Table 0.1 Merit Points System

Scale of Merit Points	
0	Poorer than average
1	Average
2	Better than average
3	Outstandingly better than average

Categories for which points are awarded	
1	Information available
2	Process simplicity
3	Operating experience
4	Operating difficulty
5	Energy consumption
6	Reagent requirements
7	Ease of end-product disposal
8	Process applicability

application to either one or two typical oil-fired combustion systems has been evaluated in detail.

In each of the detailed evaluations, the processes are described and their status and operating experience indicated, together with process variations that have been adopted, and development potential. Process requirements for the specific applications are evaluated, the by-products and effluents are identified and quantified, and then the emission and efficiency factors, and (where available data allow) capital and operating costs, are presented. Finally the process advantages and drawbacks are listed. The emissions (kg/h) and emission factors (kg/tonne fuel) are expressed throughout the Manual in terms of the mass of element (sulphur, nitrogen or chlorine), and not of the acidic compound (sulphur dioxide, nitrogen oxides or hydrogen chloride).

0.4 Datum Oil-Fired Combustion Systems

For the evaluation of the various processes and technologies, two datum combustion systems have been defined to cover the range of oil-fired boilers contributing mainly to SO_2 and NO_x emissions. The systems are: (1) Large industrial oil-fired boiler generating 450 tonne/h steam; and (2) Small factory oil-fired boiler generating 25 tonne/h steam.

Large oil-fired boilers for power stations (e.g. 500 MWe units) have not been considered in the present edition of the Manual as they make only a small contribution to the overall acidic emissions inventory. In 1985–6, for example, the declared net capability of oil-fired power stations in England and Wales was 7120 MWe with an average load factor of 11.1% [CEGB Statistical Yearbook]. This amounted to about 4.5% of the power output from coal-fired stations.

The assumptions made for the datum systems, and the performance and emissions to be expected from them (without emissions abatement) are summarised in Table 0.2.

Table 0.2 Assumptions for Datum Oil-Fired Combustion Systems Data for Systems Without Acidic Emissions Abatement

Case 1: Large industrial oil-fired boiler generating 450 tonne/h steam
Case 2: Small factory oil-fired boiler generating 25 tonne/h steam

Datum System No.	1	2
Output, tonne/h steam	450	25
Steam pressure, bar	100	10
Steam temperature, °C	480	180
Useful Energy Output, GJt/h	1468	67.8
Oil Composition (% by weight)		
Moisture	0.1	0.1
Ash	0.04	0.04
Carbon	86.3	86.3
Hydrogen	11.0	11.0
Sulphur	2.6	2.6
Chlorine, ppm	30	30
Gross CV, MJ/kg	42.4	42.4
Boiler Type	Water Tube	Twin Fire Tube
Oil fired, tonne/h	39.4	1.88
MWt (gross)	464	22.2
Excess air, %	3	10
Flue gas flow, Thousand Nm³/h		
Wet	456	23.2
Dry	408	20.9
Water vapour	48	2.3
Gas temperature, °C	150	230
Actual gas volume, Thousand m³/h	707	42.8
Emissions, mg/m³ dry gas		
SO_2	5010	4675
NO_x (as NO_2)	1025	615
HCl	3	3
Particulates	39	36
Elements in acidic emissions, kg/h		
Sulphur	1023	48.9
Nitrogen	127	3.92
Chlorine	1.2	0.06
Particulates	15.7	0.8

0.5 Flue Gas Desulphurisation

Flue gas desulphurisation (FGD) is the most widely practised of acidic emission abatement processes and technologies, and of the large number of processes that have been developed, twenty-seven have been appraised. Six of the processes had more than 10 merit points and were selected for detailed evaluation; they are described briefly below, and their appraisals are summarised in Table 0.3.

Sea water scrubbing (Process Code No. S11.1): Gas is scrubbed with sea or river water; the effluent, containing sulphites and sulphates, is pumped to sea.

Limestone or lime slurry scrubbing (Process Code No. S21.1): Gas is scrubbed with limestone or lime slurry; forced atmospheric oxidation of the effluent slurry produces gypsum.

Lime spray dryer absorption (SDA–Process Code No. S22.1): Reaction of hot gas with lime slurry in a spray dryer absorber produces solids containing absorbed SO_2 as sulphite and sulphate.

Wellman-Lord process (Process Code No. S31.1): Gas is scrubbed with sodium sulphite solution, forming sodium bisulphite; the absorbent is regenerated thermally, with evolution of SO_2 for sale or conversion to sulphuric acid or elemental sulphur.

Magnesia scrubbing (Process Code No. S41.1): Gas is scrubbed with magnesia slurry; the absorbent is regenerated by calcination with carbon, releasing SO_2 for conversion to sulphuric acid.

Active carbon adsorption (Process Code No. S51.1): SO_2 is adsorbed by active carbon, which is regenerated thermally, releasing SO_2 and CO_2. The SO_2 can be sold or converted to sulphuric acid or elemental sulphur.

The processes evaluated in detail were those considered to have potential for application in the UK. Only the lime slurry spray dryer absorption process (13 merit points) was evaluated for both Systems 1 and 2, as it was considered to be the only process applicable to both operating scales.

Two other basic types of process with non-regenerable reagents were selected for Combustion System 1: limestone or lime slurry scrubbing (12 merit points) and sea water scrubbing (11 merit points). Because of the overwhelming preponderance of FGD plants that are based on limestone or lime slurry scrubbing, and the wide range of designs that have been developed, three representative variants of the limestone scrubbing process were evaluated: the IHI limestone/gypsum, Chiyoda Thoroughbred 121 limestone-gypsum, and Saarberg-Hölter-Lurgi (S-H-L) lime-gypsum processes.

Regenerable reagent processes, which because of their complexity are generally suitable only for the largest operating scale (e.g. 500 MWe power station boilers), could also be considered for a smaller scale (System 1) if adoption of FGD led to creation of an absorbent regeneration industry. Three such FGD processes (all with 11 merit points) were therefore evaluated for System 1: Wellman-Lord, magnesia slurry scrubbing and active carbon adsorption.

There are therefore eight individual processes evaluated in detail. In each instance, a sulphur capture was assumed that was within the range of capability of the process, with a more rigorous performance for System 1 than for System 2. The resulting emission factors for sulphur, together with those for nitrogen, chlorine and particulates, are summarised in Table 0.4 for the applications to the two Datum Combustion Systems and for 100% FGD plant availability; the processes are listed in the Table in descending order of merit.

An alternative to flue gas desulphurisation is to burn oil that has been desulphurised before combustion. This option, which is particularly attractive for the small user (e.g. System 2) is not considered in the Manual, but is illustrated in Table 0.4 by the emissions from combustion of gas oil of negligible ash content and of sulphur content 0.44%.

Efficiency factors, together with capital costs, and operating costs excluding the annualised capital cost element, are summarised in Table 0.5; however, because insufficient data were available at the time the processes were being evaluated, the costs have not been estimated for sea water scrubbing and active carbon adsorption (System 1) and for lime spray dryer absorption (System 2).

Table 0.3 Appraisal of Flue Gas Desulphurisation Processes

1	Code No.	S22.1	S21.1	S11.1	S31.1	S41.1	S51.1
2	Reagent or Process	Lime Spray Dryer	Limestone or Lime Slurry	Sea Water	Wellman-Lord	Magnesia Slurry	Active Carbon Adsorption
	Appraisal						
3	Merit Points	13	12	11	11	11	11
4	Merits	1,2*5,8	1,2,3*,8	2*,6	1,3,6,7,8	1,6,7,8	2,5,6,7,8
5	Drawbacks	4	4	–	4,5	3	3,4
	Status						
6	No. of Units	94	478	14	35	8	10
7	MWe Range	3–1400	1–1150	10–230	6–1550	120–360	0.3–370
8	Total MWe	17,000	126,000	1310	7425	1485	602
9	% S. Capture	70–95	50–99	80–99	90–99	90–98	80–98
	Detailed Evaluation						
10	System No.	1,2	1**	1	1	1	1

Notes
Line 1: Process Code No. denoting position of basic process in the classification.
Line 2: Principal reagent, or name of process.
Line 3: Total number of merit points (see Table 0.1 for scale of merit points).
Line 4: Categories with above average merit rating (see Table 0.1); asterisk indicates outstandingly better than average (i.e. 3 merit points).
Line 5: Categories with below average merit rating (see Table 0.1).
Line 6: Number of FGD units erected or on order worldwide
Line 7 and 8: Range of combustion plant sizes, and total capacity installed or on order, in equivalent MWe; 1 MWe equivalent = 1 MWe of power produced or 3000 Nm³/h of gas treated or 2.5 tonne/h of steam produced.
Line 9: Range of sulphur capture performance reported.
Line 10: Datum combustion systems for which application of the process has been evaluated in detail.

**Three variants of this process evaluated in detail.

Table 0.4 Emission Factors for FGD Processes

Code No.	FGD Process	Merit Points*	Emission Factors, kg/tonne oil fired			
			Sulphur	Nitrogen	Chlorine	Particulates
Datum System 1: Large industrial boiler–450 tonne/h steam						
–	None	–	26.0	3.23	0.03	0.40
–	Gas Oil	–	4.4	n.d.	n.d.	negl.
S22.1	Lime SDA	13	5.07	3.16	0.03	0.15
S21.1a	IHI	12	3.32	2.92	0.03	0.15
S21.1b	Chiyoda	12	3.33	3.18	0.03	0.15
S21.1c	S-H-L	12	3.34	3.19	0.03	0.15
S1.1	Sea water	11	3.36	3.21	0.03	0.15
S31.1	Wellman-Lord	11	3.32	3.17	0.03	0.15
S41.1	Magnesia	11	3.08	2.94	0.03	0.14
S51.1	Active carbon	11	3.32	2.92	0.03	0.15
Datum System 2: Small factory boiler–25 tonne/h steam						
–	None	–	26.0	2.09	0.03	0.40
–	Gas oil	–	4.4	n.d.	n.d.	negl.
S22.1	Lime SDA	13	5.4	2.04	0.03	0.16

n.d. No data
* See Table 0.1

Table 0.5 Efficiency Factors and Costs for FGD Processes

Code No.	FGD Process	Merit Points*	Efficiency Factor GJt/tonne oil	Capital Cost, £million		Operating Cost** £/kWt-year
				New Build	Retrofit	
Datum System 1: Large industrial boiler–450 tonne/h steam						
–	None	–	37.3			
S22.1	Lime SDA	13	36.5	18.3	24.0	47.61
S21.1a	IHI	12	36.6	18.9	30.0	36.17
S21.1b	Chiyoda	12	36.8	14.6	21.7	27.17
S21.1c	S-H-L	12	36.9	20.7	30.3	34.99
S11.1	Sea water	11	37.1	n.d.	n.d.	n.d.
S31.1	Wellman-Lord	11	34.8	31.6	41.4	54.71
S41.1	Magnesia	11	34.0	30.8	44.9	39.52
S51.1	Active carbon	11	36.6	n.d.	n.d.	n.d.
Datum System 2: Small factory boiler–25 tonne/h steam						
–	None	–	36.0	–	–	–
S22.1	Lime SDA	13	35.3	n.d.	n.d.	n.d.

n.d. No data–costs not calculated
* See Table 0.1
** Operating costs, £/equivalent kWt-year for 5694 full-load operating hours per year, assumed equal for new build and retrofit applications.
To convert from £/kWt-year to p/kWt-hour, multiply by 0.0176.

0.6 NO_x Abatement–Combustion Techniques

Combustion techniques for flue gas denitrification are, at present, capable of NO_x reductions of up to 50% individually, and up to 80% where combinations of techniques are utilised: see Table 0.6. (It should be noted that where NO_x reduction figures are reported for techniques such as low-NO_x burners, LNB, a comparison has been made with more conventional designs of burner).

The study appraised eleven combustion techniques for NO_x abatement which were classified into three broad categories: burner design, furnace design/modification and furnace operation.

Dual Register Burner (DRB): A burner with a primary gas annulus surrounding the core pipe housing the oil atomiser. Recirculated flue gas is mixed with the secondary air supply and directly supplies the primary gas annulus.

Tangential Firing System: A premixed oil burner which divides the flame into two combustion zones, fuel-rich and fuel-lean respectively. This is one possible configuration of many.

Axial-staged Recirculating Burner: A burner in which the combustion air is mixed with the fuel oil in two stages. Flue gas is recirculated to divide and delay combustion and to lower combustion temperature.

Reburning (Fuel Staging): A small part of the fuel is injected above the main combustion zone, producing a secondary, reducing combustion zone. Combustion air is added further downstream to complete combustion.

Unlike flue gas treatment processes for NO_x abatement, FGD and combined SO_2/NO_x abatement processes, a merit point system was not applied to $DeNO_x$ combustion techniques. However, a comparison of the leading techniques is summarised in Tables 0.7 to 0.9 inclusive.

Table 0.6 Estimates of Removal Efficiencies for Combustion Modifications

Technique	Reduction in NO_x Emission (%)
Low-NO_x Burners (LNB)	10–40
Reburning	30–50
Off-Stoichiometric Combustion (OSC) or Staged Combustion	10–40
Low Volume Heating Rate (LVHR)	30–40
Flue Gas Recirculation (FGR)	10–50
LVHR + LNB	37–64
LVHR + LNB + FGR	43–82
LVHR + LNB + OSC	43–78
LVHR + LNB + Reburning	56–82
LVHR + LNB + FGR + Reburning	60–91

0.7 NO_x Abatement-Flue Gas Treatment

Flue gas treatment (FGT) processes for denitrification are, at present, economically competitive with combustion modifications only where high NO_x reductions of 70–90% are required. The study appraised six basic types of process, and evaluated two of them in detail. These two process types, which are outlined below, had more than 10 merit points and were regarded as potentially applicable in the UK; their appraisals are summarised in Table 0.10. Only one other process is commercially available at present–dry adsorption (8 points); this was developed as an FGD process but is now offered as a combined SO_2-NO_x process. It should be noted that the number of points in the merit point systems adopted for FGD and combined SO_2-NO_x processes are not strictly comparable with those for NO_x processes.

Selective Catalytic Reduction (SCR): NO_x in the flue gas is reacted with ammonia gas in the presence of a catalyst, producing nitrogen and water which are discharged with the treated gas stream.

Selective Non-catalytic Reduction (SNR): NO_x in the flue gas is reacted with ammonia gas at high temperature, e.g. in the upper section of the boiler. No catalyst is necessary at these temperatures and the products, nitrogen and water, are discharged with the treated gas stream.

The two processes were evaluated for Datum Combustion System 1 only. For the smallest operating scale dealt with in this Volume (Datum System 2), it is anticipated that the lower NO_x reduction efficiencies required to meet acceptable NO_x emission levels can be readily, and more economically, attained by combustion modifications alone.

For the detailed evaluation of each of the two selected processes, a NO_x reduction level was assumed that was

Table 0.7 Comparison of Burners

Burner Type	Dual-Register Burner	Tangential Firing System	Axial-Staged Recirculation Burner	Reburning (Fuel Staging)
Technique	Secondary air introduced in stages and is mixed with recirculated flue gas.	Pre-mixed burner which divides flame into two combustion zones: one fuel-rich and the other fuel-lean.	Combustion air is mixed with the fuel oil in two stages, recirculated flue gas being used to divide the combustion zone.	A small part of the fuel is injected above the main combustion zone, and additional air is added downstream to complete combustion.
Applicability*	N,R	N,R	N,R	N,R
Operating Experience	50% NO_x reduction has been achieved–relative to conventional burner.	NO_x concentration in flue gas remained below 160 mg/m³ over a wide range of loads in tests at 265 MWe plant.	5 units have been equipped. NO_x removal efficiencies of up to 45% for light oil and up to 35% for heavy oil have been achieved.	Applied to several oil-fired boilers of up to 600 MWe in Japan. NO_x removal efficiency is about 50%.

* N = New units, R = Retrofits

Table 0.8 Comparison of Furnace Designs or Modifications

Furnace Design or Modification	Flue Gas Recirculation	Staged Combustion	Reduced Heat in Furnace
Technique	Low combustion temperature. Low oxygen concentration.	Two stage combustion	Low volumetric heating rate. (LVHR)
NO_x level achievable	Up to 15% reduction	10–35% reduction	20–25% reduction
Applicability*	N,R	N,R	N
Problems	Cost. Flame expansion. Energy penalty.	Incomplete combustion. Flame extension. Tube erosion.	Possible boiler de-rating.

* N = New units, R = Retrofits

Table 0.9 Comparison of Furnace Operation Techniques

Furnace Operation	Low Excess Air	Biased Burner Firing	Burners Out Of Service	De-rating
Technique	Low excess air combustion, i.e. reduced oxygen concentration.	Upper row burners operate fuel-lean. Lower row operates fuel-rich.	Upper row burners are out of service, injecting only air.	Reduced heat load in furnace, thereby lowering combustion temperature.
NO_x level achievable (% reduction)	40	30–40	30–40	20–30
Applicability*	R	R	R	N,R
Problems	Unburnt carbon. CO emissions. Fouling and corrosion.		Derating of unit.	

* N = New units, R = Retrofits

Table 0.10 Appraisal of NO_x Abatement FGT Processes

Process (& Code)	Merit Rating	Merits	Drawbacks	No of Units	Total MWe Equiv.	% NO_x Removal	Applied to
Column	1	2	3	4	5	6	7
SCR (N41)	13	1,3,5,7,8	–	163+	24,850	53–90	1
SNR (N42)	11	2,5,7,8	4	c.83	n.d.	40–85	1

Notes

Col. 1: See Table 0.1 for scale of merit points.
Col. 2: Categories with above average merit rating (see Table 1).
Col. 3: Categories with below average merit rating (see Table 1).
Col. 4: Number of units erected or on order (all fuels).
Col. 5: 1 MWe Equivalent = 1 MWe of power produced or 3000 Nm^3/h of gas treated. (n.d.–insufficient data for SNR, but have been fitted to utility boilers)
Col. 7: Datum system for which application of process has been evaluated in detail.

within the range of capability of the process giving due regard to anticipated outlet NO_x (and ammonia) concentrations. The resulting emission factors for nitrogen, and efficiency factors, are summarised in Table 0.11. It was assumed that SO_2, HCl and particulate concentrations in the flue gas would be unaffected by these processes: minor side-reactions with ammonia and particulate deposition were considered to be negligible. There was insufficient information to enable costs for either process to be presented, although the reader is referred to Section 4.6 where limited comparative $deNO_x$ cost data is presented.

Table 0.11 Summarised Results of Detailed Evaluations of NO_x Abatement FGT

$DeNO_x$ Process (& Code)	Merit Points*	Efficiency Factor GJt/tonne oil	Emission Factor (kg nitrogen per tonne oil)
Datum System 1: Large industrial boiler–450 tonne/h steam			
None	–	37.3	3.23
SCR (N41)	13	37.0	0.63
SNR (N42)	11	37.2	0.94

*See Table 0.1

0.8 Combined Sulphur Dioxide–Nitrogen Oxides Abatement Processes

In addition to new processes developed for the purpose, several of the FGD processes and flue gas denitrification processes have been modified to enable both SO_2 and NO_x to be removed from the flue gas simultaneously. The study appraised fifteen processes, and evaluated two of them in detail.

As with the processes for desulphurisation and for denitrification of flue gas, combined abatement processes with more than 10 merit points were regarded as being potentially commercially applicable in the UK. This criterion was met by only one process–Active Carbon/Selective Catalytic Reduction (12 merit points), and this was evaluated for Datum System 1, assuming the creation of an absorbent regeneration industry as already mentioned.

Like its FGD counterpart, the Lime Spray Dryer Absorption process has attractions for small as well as for large scale operation, but because of lack of information available and of operating experience, its rating is only 9 merit points. However, because of its promise for future applications to all operating scales, it was evaluated, in as much detail as the available information allowed, for both Datum Systems 1 and 2. Descriptions of the two processes selected are outlined below, and their appraisals are summarised in Table 0.12.

Table 0.12 Appraisal of Combined Abatement Processes

1	Code No.	S11.1	S31.1
2	Reagent or Process	Active Carbon/ Selective Catalytic Reduction	Lime Spray Dryer Absorber
	Appraisal		
3	Merit Points	12	9
4	Merits	2,5,6,7,8	2,5,8
5	Drawbacks	3	3,4
	Status		
6	No. of Units	7	2
7	MWe Range	0.7–370	7–32
8	Total MWe	606	39
	Performance		
9	% S Capture	90–98	90–95
10	% NO_x Abatement	60–90	20–60
	Detailed Evaluation		
11	System No.	1	1,2

Notes:

Line 1: Process Code No. denoting position of basic process in the classification.
Line 2: Principal reagent, or name of process.
Line 3: Total number of merit points (see Table 0.1 for scale of merit points).
Line 4: Categories with above average merit rating (see Table 0.1).
Line 5: Categories with below average merit rating (see Table 0.1).
Line 6: Number of combined abatement units erected or on order worldwide.
Lines 7 and 8: Range of combustion plant sizes, and total capacity installed or on order, in equivalent MWe; 1 MWe equivalent = 1 MWe of power produced or 3000 Nm^3/h of gas treated or 3 MWt.
Line 9: Range of sulphur capture performance reported.
Line 10: Range of NO_x abatement performance reported.
Line 11: Datum combustion system for which application of the process has been evaluated in detail.

Active carbon/Selective catalytic reduction (Process Code No. NS11.1): SO_2 and NO_2 are adsorbed in the 1st-stage reactor by carbon fed from the 2nd-stage reactor. Nitric oxide is catalytically reduced to elemental nitrogen by ammonia in the 2nd-stage reactor. Carbon from the 1st stage is regenerated thermally (producing SO_2, CO_2 and elemental nitrogen) and recycled to the 2nd stage; the SO_2 evolved can be sold, or converted to sulphuric acid or elemental sulphur.

Lime Spray Dryer Absorber (SDA–Process Code No. NS31.1): SO_2 and NO_x in hot gas are absorbed in an SDA by lime slurry containing sodium hydroxide, forming solids containing calcium sulphite, sulphate, nitrite and nitrate.

In making the detailed evaluations of the two processes selected, extents of SO_2 and NO_x abatement were assumed that were within the ranges of capability of the processes, with a more rigorous performance for Datum System 1 than for System 2. The resulting emission factors for sulphur and nitrogen, together with those for chlorine and particulates, are summarised in Table 0.13. Efficiency factors are also presented in the Table. There was insufficient information available at the time the evaluations were made to enable costs for either of the Combined Abatement processes to be estimated.

Table 0.13 Summarised Results of Detailed Evaluations of Combined Abatement Processes

Code	Combined Process	Merit	Efficiency GJt per tonne oil	Emission Factors, kg/tonne oil fired			
				Sulphur	Nitrogen	Chlorine	Partic's*
Datum System 1: Large industrial boiler–450 tonne/h steam							
–	None	–	37.3	26.0	3.23	0.03	0.40
NS11.1	Active C/SCR	12	36.9	2.06	1.56	0.03	0.15
NS31.1	Lime SDA	9	36.5	5.07	1.54	0.03	0.15
Datum System 2: Small factory boiler–25 tonne/h steam							
–	None	–	36.0	26.0	2.09	0.03	0.40
NS31.1	Lime SDA	9	35.1	5.44	1.30	0.03	0.16

* Total merit points awarded (see Table 0.1)

1. Introduction

1.1 Background and Purpose of the Manual

1.2 Acid Emission Inventories

1.3 Scope

1.4 General Assumptions for Datum Combustion Systems

1.5 Appraisal of Processes

1.6 Notes for the Guidance of the Reader

1. Introduction

1.1 Background and Purpose of the Manual

This Manual has been compiled by the Fellowship of Engineering in fulfilment of a commission from the Department of the Environment to prepare a Manual describing, comparing and appraising processes for abatement of acidic emissions from fossil fuel combustion systems. It is anticipated that the readership of the Manual will include those air quality professionals wishing to have information on the acidic emissions abatement technologies available, and organisations involved in assessing the economics of emission control.

The aim has been to include only those abatement technologies applied during or after combustion; acid emission abatement processes dependent on removal of pollutant elements (sulphur, nitrogen and halogens) from the fuel before combustion have therefore been excluded, but it is hoped to include them in a later edition of the Manual. Among the technologies thus excluded are: oil desulphurisation (of particular interest for small oil-fired combustion systems); and coal gasification (which has an important potential application in gasification-combined cycle power generation systems).

In order to provide a comprehensive data base, the Manual contains:

- Brief descriptions of each type of process or technology commercially available, or showing promise of becoming commercially available by the end of the century

- Information showing the status and applicability of the processes or technologies

- Accounts of operating experience and performance

- Detailed descriptions and cost data for application of selected processes and technologies to a range of coal-fired, oil-fired and gas-fired combustion systems.

In order to make it more manageable for the user, the Manual has been prepared in three volumes covering stationary combustion systems, one for each of the fuels coal, oil and gas; and a fourth volume for mobile sources. This Volume is concerned with stationary oil-fired combustion systems.

In preparing this first issue of the Manual for stationary combustion systems, the Fellowship of Engineering retained Foster Wheeler Power Products Limited with Hoy Associates Limited who acted as a collaborating sub-contractor. The direction of the work by the Fellowship was organised through an overall Steering Group, and specialised Task Groups, one for each of the three fuels–coal, oil and gas. The composition of the Steering and Task Groups is given in the Acknowledgements.

1.2 Acidic Emission Inventories

The main precursors of acid deposition are sulphur dioxide–SO_2, and nitrogen oxides–NO_x (primarily nitric oxide, NO),–emitted into the atmosphere from natural and man-made sources. In north-west Europe, man-made sulphur emissions contribute over 80% of total sulphur emissions into the atmosphere. Although the available data on NO_x emissions are less extensive than those for sulphur emissions, it is estimated that for north-west Europe, man-made sources of NO_x contribute between 75% and 93% of total NO_x emissions [38].

Warren Spring Laboratory (WSL) have prepared estimates of UK emissions of SO_2 and NO_x, and the quantities for 1985, and sources, are presented in Figures 1.1 and 1.2 [40]. The principal source of SO_2 emission is seen to have been power stations (71%); in normal circumstances, however, less than 5% of the UK electric power consumption is generated in

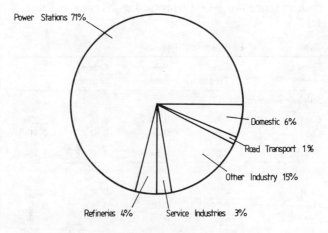

Figure 1.1 UK Man-made Sources of SO_2 Emission (1985)
Total SO_2 Inventory 3.58 million tonnes

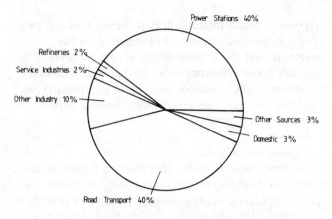
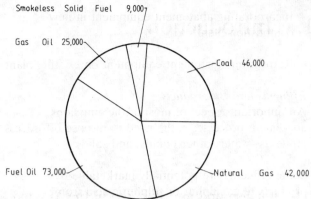

Figure 1.2 UK Man-made Sources of NO$_x$ Emission (1985) Total NO$_x$ Inventory (as NO$_2$) 1.84 million tonnes
Figure 1.3 WSL Estimate of UK Commercial and Industrial NO$_x$ Emissions (1982) Estimated Emissions, tonnes (as NO$_2$) from Various Fuels Total Estimated (as NO$_2$) 195,000 tonnes

oil-fired stations. Responsibility for the remaining 29% (about 1 million tonnes) of the 3.58 million tonnes UK SO$_2$ inventory was shared by oil and coal.

Similarly for NO$_x$, power stations and road transport accounted for 80% of the emissions, so that stationary oil combustion systems contributed less than 20% overall (i.e. less than 370,000 tonnes) to the NO$_x$ inventory of 1.84 million tonnes as NO$_2$. When stationary sources of emission alone are taken into consideration it is seen that the pattern for NO$_x$ emissions is very similar to that for SO$_2$, with power stations contributing 67%, refineries, service industries and other industry 23%, and domestic sources 5% of stationary-source emissions.

In 1982, WSL measured NO$_x$ emissions from 72 selected commercial and industrial coal-, oil- and gas-fired boilers of up to 230 tonne/h steam capacity. These data were used [41] as the basis for estimating UK emissions from a limited number of sources, as shown in Figure 1.3. Of the 195,000 tonnes of NO$_x$ (expressed as NO$_2$), 98,000 tonnes (about 50%) originated from combustion of gas oil and fuel oil.

On the basis of these data it can be estimated that annual emissions from stationary oil-fired systems are of the order of 0.71 million tonnes SO$_2$ and 0.22 million tonnes of NO$_x$ (expressed as NO$_2$); see Table 1.1.

Table 1.1 Acid Emission Inventories

Source	Road Transport	Power Stations	Other Sources	Total
Sulphur Dioxide				
Coal	–	2.43	0.36	2.77
Oil	0.04	0.11	0.64	0.81
Total	0.04	2.54	1.00	3.58
Nitrogen Oxides				
Coal	–	0.70	0.10	0.80
Oil	0.74	0.04	0.18	0.96
Gas	–	–	0.08	0.08
Total	0.74	0.74	0.36	1.84

Most of the UK emissions of HCl originate from the combustion of UK coal, which is unusually rich in chlorine; minor sources are principally the incineration of industrial waste containing PVC, and combustion of oil. The emissions of HCl are an order of magnitude smaller than those of SO$_2$ from power stations. Emissions of hydrogen fluoride are much lower and are not generally regarded as significant, and those of the other acid halides–hydrogen bromide and iodide–are lower still.

1.3 Scope

Abatement Processes and Technologies
The study in this Volume covers processes and technologies applicable to oil-fired systems, for:

– Flue gas desulphurisation

– Nitrogen oxides abatement

– Simultaneous sulphur and nitrogen oxides removal

Applications
To arrive at realistic evaluations, consideration has been given to the application of selected emissions abatement processes to representative types and sizes of oil-fired combustion plant (defined as the 'datum oil-fired combustion systems' in Section 1.4).

The datum combustion systems considered do not include large oil-fired power station boilers, e.g. for 500 MWe generator sets. This is because, although the installed oil-fired generating capacity of the CEGB is 7120 MWe, with about 1920 MWe in dual-fuel stations, the actual power sent out from these stations at present averages only 926 MWe and represents 3.6% of the UK power production [CEGB Statistical Yearbook]. The position is likely to change in exceptional circumstances (as it did during the recent mineworkers' strike) or if the cost of oil becomes competitive with that of coal.

The evaluations take into account the alternatives of:

- Incorporating abatement equipment in new combustion plant

- Retrofitting abatement equipment on existing plant

Effluents and By-products
An important aspect of most acidic emissions abatement processes is the need to dispose of process effluents–water, other liquids, and solids.

Some effluents are potentially marketable by-products, e.g. sulphur or sulphuric acid from regenerable flue gas desulphurisation processes, but credit for their sale will be possible only when a market exists, or if no unacceptable social penalty (e.g. destruction of an existing industry) has to be paid for marketing the product.

Reduction in Combustion Plant Efficiency
Many of the acidic emissions abatement techniques and processes result in a diminution in the thermal efficiency of the combustion plant. In practice, this will give rise to the need to burn more fuel, and in some instances to instal additional combustion plant, to produce the desired output of useful energy. This factor has to be borne in mind when considering the impact of installing abatement equipment on national pollutant inventories, or on the degree of abatement needed.

1.4 General Assumptions for Datum Combustion Systems

Two datum oil-fired combustion systems are considered in this Volume and defined in this Section.

Although separate evaluations are made for each system for retrofitting emissions abatement equipment and for new builds incorporating emissions abatement, it is considered that the assumptions made here do not have to be changed to accommodate these alternatives. This is because any differences between the thermal efficiencies of the combustion system in old plant (retrofitted with abatement equipment) and new builds (incorporating abatement equipment) will have only minor effects on the gas flows and compositions entering the emissions abatement equipment.

The datum combustion systems selected to represent the major large oil users in the UK, are defined as follows:

System 1: Large industrial oil-fired (heavy fuel oil) water-tube boiler: 450 tonne/h steam at 100 bar pressure and 480°C temperature; excess air level 3%; boiler efficiency 88%. Particulates emission (without acidic emissions abatement equipment) 39 mg/Nm^3 dry gas. Gas temperature leaving boiler plant to stack 150°C.

System 2: Small factory oil-fired (heavy fuel oil) twin fire-tube boiler: 25 tonne/h of steam at 10 bar pressure and 180°C (saturated); excess air level 10%; overall boiler efficiency 85%. Particulates emission (without acidic emissions abatement equipment) 36 mg/Nm^3 dry gas. Gas temperature leaving boiler plant to stack 230°C.

Assumed oil compositions
The compositions quoted in Table 1.2 represent those for heavy fuel oil and gas oil currently available, but with the decline in production of North Sea petroleum and increasing dependence on Middle East sources, the composition of oil–in particular the sulphur content of heavy fuel oil–will change to an extent that cannot be predicted.

Oil desulphurisation
An alternative to adoption of the flue gas desulphurisation processes described in Section 2 is to burn oil that has been desulphurised before combustion. This option, which is particularly attractive for the small user, e.g. System 2, has not been considered in the present edition of this Volume. However, the composition of gas oil, which has a low sulphur content, and the sulphur oxides emission resulting from its use, has been given in Tables 1.2 to 1.4 as an indication of the benefits accruing from the use of partially desulphurised oil.

Calculated datum flue gas compositions
Compositions and acidic emission levels (assuming zero abatement), estimated from the definitions of the datum combustion systems and from the assumed oil compositions, are presented in Table 1.3. The SO_2 and HCl contents represent the total inputs of sulphur and chlorine in the oil. The NO_x contents have been estimated from [37].

Datum efficiency and emission factors
Efficiency and emission factors (assuming zero abatement), calculated from the definitions of the datum combustion systems and from the assumed oil compositions, are presented in Table 1.4.

1.5 Appraisal of Processes

A procedure has been adopted to quantify the appraisals made in this Manual of processes for flue gas desulphurisation (Section 2.3), flue gas denitrification (Section 4.3) and combined SO_2-NO_x abatement (Section 5.3). In this rough appraisal procedure, 'merit points' have been awarded to each process for a number of features. The points have been awarded according to the scale:

0 – below average merit
1 – average merit
2 – above average merit
3 – outstandingly above average merit

The features to which these points have been assigned are as follows:

1. Information available: the amount of information available

2. Process simplicity

3. Operating experience: the extent of operating experience

4. Operating difficulty: availability, reliability

5. Loss of power: the equivalent percentage reduction in power sent out as a result of installing the process

   2 = Less than 2.5%
   1 = 2.5 to 7.5%
   0 = More than 7.5%

6. Reagent requirements: the general rule is

   2 = regenerative reagent processes
   1 = throw-away reagent processes

   but there are some exceptions.

7. Ease of end-product disposal

8. Process applicability: the general rule is

   2 = processes suitable for retrofitting (unless application is restricted by e.g. geographical considerations)
   1 = other processes

In some instances, lack of information has made it necessary to assume the number of merit points for a feature; the assumptions made were deduced from characteristics that are known of the process concerned. Process capital and operating costs could not be included in the list of features for merit points, as in most instances published cost data were unavailable or too meagre to be of value, and it was considered unsafe to assume merit ratings for these features.

The merit points system described above is based on comparisons with the average level for each of the eight features. However, because the average levels are different for the different acidic emissions being abated, the merit ratings for a flue gas desulphisation process, for example, cannot be compared with those for flue gas denitrification or for combined abatement processes.

A process that was average in all of the eight features would, of course, have a total of 8 merit points. Very few processes achieved 1 or more points for all of the eight features. On the other hand, several processes were above average in a sufficient number of features to achieve totals exceeding 8 merit points. It was arbitrarily decided that a process with over 10 merit points was potentially applicable commercially in the UK, and its application to either one or two of the datum oil-fired combustion systems has been evaluated in detail (Sections 2.5, 4.5 or 5.5).

Table 1.2 Assumed Compositions of Oil

	Heavy Fuel Oil (HFO)	Gas Oil (GO)
Composition, by weight:		
Moisture, %	0.1	0.01
Ash, %	0.04	Negl.
Carbon, %	86.3	–
Hydrogen, %	11.0	–
Sulphur, %	2.6	0.44
Chlorine, ppm	30	–
Gross calorific value, MJ/kg	42.4	45.5
Net calorific value, MJ/kg	40.4	42.7

Table 1.3 Calculated Flue Gas Composition and Acidic Emissions

System No.	1		2		2 (Alternative)	
Excess air, %	3		10		10	
Fuel	HFO		HFO		GO	
Flue gas composition by volume:	Wet	Dry	Wet	Dry	Wet	Dry
H_2O %	10.6	–	9.9	–		
CO_2 %	13.9	15.5	13.0	14.5		
O_2 %	0.5	0.6	1.8	2.0		
N_2 %	74.8	83.7	75.1	83.3		
SO_2 ppm	1567	1752	1472	1634	265	296
NO_x ppm	447	500	270	300		
HCl ppm	2	2	1	2		
Emissions, mg/Nm³ (dry gas; assuming no abatement):						
SO_2		5010		4675		850
NO_x as NO_2		1025		615		
HCl		3		3		
Particulates		39		36		

Table 1.4 Calculated Efficiency and Emission Factors

System No.	1	2
Oil heat input, MWt (gross)	464	22.2
Oil fired, tonne/h	39.4	1.88
Useful energy output, GJt/h	1468	67.8
Efficiency factor, GJt/tonne	37.3	36.0
Ashes output, kg/h	15.7	0.8
Flue gas, Thousand Nm³/h		
Wet	456	23.2
Dry	408	20.9
Water vapour	48	2.3
Gas temperature, °C	150	230
Actual gas volume, Thousand m³/h	707	42.8
Elements in acidic emissions, kg/h (assuming no abatement)		
Sulphur	1023	48.9(a)
Nitrogen	127	3.92
Chlorine	1.2	0.06
Emission factors, kg/tonne oil (assuming no abatement)		
Sulphur	26	26(b)
Nitrogen	3.23	2.09
Chlorine	0.03	0.03

(a) For gas oil, 7.7 kg/h
(b) For gas oil, 4.4 kg/tonne

1.6 Notes for the Guidance of the Reader

Plant size
Plant size is expressed in a number of ways in this Manual, depending upon the particular application for the combustion system. In order to make comparisons of operating scale, the following conversion factors may be used:

1 MWe of power generated is roughly equivalent to:

- 3.6 GJe/h of power generated (exact conversion)
- 3 MWt heat input to the combustion system
- 10.8 GJt/h of heat input to the combustion system
- 3000 Nm^3/h gas treated
- 2.5 tonne/h of steam generated by the boiler

In those cases where the original text expresses plant size in M^3/h or MW without precise definition, and it is uncertain which units apply (e.g. MWt or MWe; Nm^3 or m^3), no further qualification of the quoted figures has been made.

Descriptions of Equipment
In this Manual, simplified block diagrams accompany the Outline Descriptions of processes, and the descriptions of plant in detailed evaluations of process applications are illustrated by simplified flow diagrams. A convention has been adopted that plant items shown on the diagrams are referred to in the text with their names given capital initials (e.g. Absorber, Heat Exchanger) and items not shown on the diagrams are with lower case initials (e.g. crystalliser).

Applicability of processes to coal and oil
All of the flue gas desulphurisation, denitrification and combined abatement processes considered in this Volume are applicable to coal- and oil-fired combustion systems. However, for coal-fired systems, many of the processes need to incorporate a gas pre-scrubbing stage to remove particulates and acid halides as well as to cool the gas. In general, the Outline Descriptions of the basic process types have been described (Sections 2.2, 4.2 and 5.2) as for application to coal-fired combustion systems, and in some instances it is also necessary to refer to the electrostatic precipitator (ESP), and to disposal of fly-ash. When the same processes are applied to oil-fired combustion systems, where the particulates and acid halides concentrations are negligible, prescrubbing and ESPs are not needed and the gas can be cooled for example by evaporative spray cooling. The acidic abatement process is in other respects the same for coal- and oil-fired systems.

In view of this, in the appraisal of the processes in this Volume (Sections 2.3, 4.3 and 5.3) actual applications have been quoted from the literature whatever the fuel, and whatever the type or purpose of the combustion system.

Emission abatement levels assumed
In the detailed evaluations of selected emission abatement processes (Sections 2.5, 4.5 and 5.5) assumptions have been made regarding the concentrations of pollutants in the gas leaving the abatement plant. These assumptions have been arbitrary, but in all instances the emissions abatement assumed has been within the range of capability of the process concerned, with the most rigorous requirement for Datum System 1. In all instances the particulates emission was assumed to be reduced to 15 mg/Nm^3.

Emissions and emission factors
In the detailed evaluations of techniques and processes (and in Table 1.4) the emissions in kg/h, and emission factors in kg/tonne of fuel, are expressed as kg of the element (sulphur, nitrogen and chlorine) per hour or per tonne of fuel. To convert to kg of sulphur dioxide, nitric oxide or nitrogen dioxide, and hydrogen chloride, multiply the figures by the factors below.

To convert:	Multiply by
Sulphur to sulphur dioxide (SO$_2$)	1.998
Nitrogen to nitric oxide (NO)	2.142
Nitrogen to nitrogen dioxide (NO$_2$)	3.284
Chlorine to hydrogen chloride (HCl)	1.028

Unavailable information
At the time the Manual was being prepared, some of the information needed was unavailable. In order to preserve the format for the Manual, however, Section headings etc. have been retained even where no material could be included. It is hoped that much of this information will become available for inclusion in a later edition of the Manual; in the meantime, the absence of available information is indicated by three asterisks thus: ***

Effect of Plant Availability
In the detailed evaluations of processes in Sections 2.5, 4.5 and 5.5, an indication is given of the effect on annual average emissions and emission factors of the abatement plant being by-passed in the event of its being shut down for any reason. The effect is, of course, to increase the emissions and emission factors for those emissions that would otherwise be abated. For an abatement plant availability of A%, the average annual emissions and emission factors are increased, from values represented by E, to the higher values E' given by:

$$E' = E.A/100 + E_o(1 - A/100)$$

where E_o represents the corresponding emission or emission factor without abatement (given in Section 1.4). For illustration purposes, average annual emissions and emission factors for abatement plant availabilities of 100% and 90% are given for all of the processes evaluated.

2. Flue Gas Desulphurisation Processes

2.1 Classification of the Processes

2.2 Outline FGD Process Description

2.3 General Appraisal of Processes

2.4 Processes for Detailed Study

2.5 Evaluation of Selected FGD Processes

2.6 Costs for FGD Systems

2. Flue Gas Desulphurisation Processes

2.1 Classification of the Processes

The general classification of flue gas desulphurisation (FGD) processes adopted in this Volume is presented in Figure 2.1. The processes are divided into seven categories according to the scheme:

- First division: distinction between wet and dry processes.

- Second division: distinction between non-regenerable and regenerable reagents; however, if regeneration of the sorbent is based on transfer of the captured sulphur to combination with another reagent, the process is classified as non-regenerable.

- Third division: depends on the method of application of the reagent:
  - for wet reagents, whether the reagent is solution or slurry.
  - for dry reagents, whether the reagent is applied in the flue gas or (for non-regenerable processes only) in the furnace.

- Fourth division:
  - for non-regenerable wet processes, distinction between those producing a wet end-product, or a dry end-product (excluding wet end-products that are dewatered before sale or disposal).
  - for non-regenerable dry processes, distinction between reagent injection processes or absorption in a reactor.
  - for regenerable processes: distinction between those using thermal or chemical regeneration methods.

There are: four main wet reagent categories (Categories S10 to S40) and three main dry reagent categories (Categories S50 to S70), with the fourth division in each of these main categories denoted S11, S12, S21, S22, etc. The 'S' prefix denotes sulphur abatement technology.

Some of the processes yielding a dry end-product involve spray-drying of solutions or slurries as an integral feature of the process. Such processes are frequently described in the literature as 'dry', 'semi-dry' or 'wet-dry' processes. These terms are regarded as misleading for such processes, which are referred to in this Volume as solution–based or slurry–based (as appropriate) absorption processes with dry end products. In the Volume, the terms 'wet' and 'dry' are used to describe the state of the reagent contacted with the gas; the terms 'semi-dry' and 'wet-dry' are not used.

Process Code Numbers
A number of basic process types occur within each process category. Basic process types are assigned a Code Number comprising the relevant Category Number followed by a Type Number; e.g. the limestone slurry scrubbing processes fall into Category S21 and have been assigned the Code No. S21.1.

The Code Numbers assigned to the basic process types are shown in Table 2.1.

2.2 Outline FGD Process Description

The system used to classify flue gas desulphurisation processes is illustrated in Figure 2.1; the codes of processes dealt with in this Section are listed in Table 2.1.

CATEGORY S11

Process Code S11.1–Sea Water Scrubbing Process

Outline of Process: A simplified block diagram of the process is presented in Figure 2.2. Gas from the Boiler is cooled in a Heat Exchanger (H.E.) and by injection of sea water in the Cooler, and is then scrubbed in the Absorber with sea water. The cleaned gas is reheated in the H.E. and exhausted to the stack. The water leaving the Absorber contains sulphurous and sulphuric acid, together with hydrogen halides, trace elements and particulates, and is discharged to sea. In some circumstances the effluent can be neutralised with e.g. chalk, but it is then usually necessary to ensure that the concentration of the reaction product is below the solubility limit.

Chemistry of Process [122]: Sea water contains carbonate, bicarbonate, borate, phosphate and arsenate ions which exert a buffering effect, maintaining a pH of about 8.3 and assisting the absorption of SO_2. Sulphur oxides in the gas entering

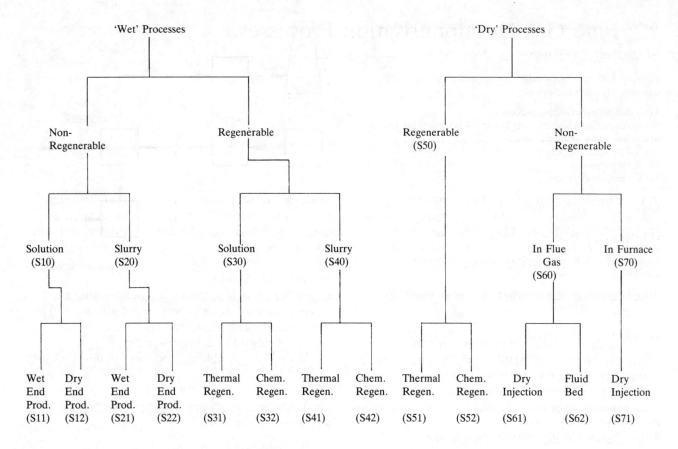

Figure 2.1 Classification of Flue Gas Desulphurisation Processes

the Absorber dissolve in the water to form sulphurous acid, H_2SO_3, and sulphuric acid, H_2SO_4. Sulphurous acid oxidises fairly readily, and some of it therefore reacts with excess oxygen in the gas to form sulphuric acid. These reactions are represented by:

Absorption: $SO_2 + H_2O = H_2SO_3$
                   $SO_3 + H_2O = H_2SO_4$

Oxidation: $2\ H_2SO_3 + O_2 = 2\ H_2SO_4$

The water also absorbs other acidic gases, e.g. acid halides, and washes out particulates from the flue gas. Because sulphurous, sulphuric and halogen acids are themselves water pollutants, and the discharge may also possibly contain toxic trace elements leached from the particulates, the process can be applied only where the effluent can be discharged harmlessly, e.g. to the sea; this explains the use of sea water as the scrubbing agent. The process has also been applied using the River Thames as a source of scrubber water and for receiving the discharge [156]. It was here that neutralisation was required:

Neutralisation:
   $H_2SO_4 + CaCO_3 = CaSO_4 + CO_2 + H_2O$

Process Code S11.2 – Alkali scrubbing process

Outline of Process: This basic process includes scrubbing with slurries of alkaline ash, from which alkali compounds are leached and react with sulphur oxides in solution. A simplified block diagram of the process is presented in Figure 2.3. Gas is cooled in a gas-gas Heat Exchanger (H.E.) or by injection of water, and is then scrubbed in the Absorber with a solution of alkali: caustic soda (NaOH); soda ash (impure sodium carbonate, Na_2CO_3); or a slurry of alkaline ash. The purified gas is reheated in the H.E. and/or by combustion of liquid or gaseous fuel, and exhausted to stack. The spent solution is oxidised with air to form sodium sulphate which is then crystallised out from solution; in some circumstances (where water pollution is not a consideration) the sodium sulphate can be disposed of in solution.

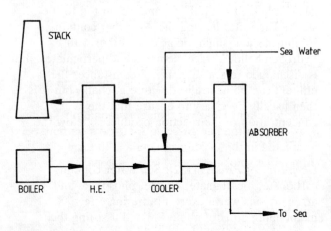

Figure 2.2 Block Diagram of Sea Water Scrubbing FGD Process

Table 2.1 Classification of flue gas desulphurisation processes

(a) Processes using wet reagents

Category S10: Non-regenerable solution-based wet processes
Category S11: Wet end-product

S11.1 Sea water scrubbing process
S11.2 Alkali (including alkaline fly-ash slurry) scrubbing process
S11.3 Ammonia scrubbing process
S11.4 Sulphuric acid scrubbing process
S11.5 Dowa process
S11.6 Dual alkali process

Category S12: Dry end-product

S12.1 Alkali scrubbing/spray drying process
S12.2 Walther process

Category S20: Non-regenerable slurry-based wet processes
Category S21: Wet end-product

S21.1 Limestone or lime (including high-calcium fly-ash) slurry scrubbing process

Category S22: Dry end-product

S22.1 Lime slurry scrubbing/spray drying process

Category S30: Regenerable solution-based wet reagent processes
Category S31: Thermal regeneration

S31.1 Wellman-Lord process
S31.2 Fläkt-Boliden process
S31.3 Catalytic Inc./IFP or ammonia scrubbing process

Category S32: Chemical regeneration

S32.1 Citrate process
S32.2 Conosox process
S32.3 Ispra Mark 13A process
S32.4 Aqueous carbonate process

Category S40: Regenerable slurry-based wet reagent processes
Category S41: Thermal regeneration

S41.1 Magnesia slurry scrubbing process

Category S42: Chemical regeneration

S42.1 Sulf-X process

(b) Processes using dry reagents

Category S50: Regenerable dry reagent processes
Category S51: Thermal regeneration

S51.1 Active carbon adsorption process

Category S52: Chemical regeneration

S52.1 Wet active carbon adsorption process
S52.2 Copper oxide process
S52.3 Catalytic oxidation process

Category S60: Non-regenerable dry reagent applied to flue gas
Category S61: Dry injection

S61.1 Hydrated lime injection process
S61.2 Alkali injection process

Category S62: Dry reactor

S62.1 Lurgi CFB lime absorber process

Category S70: Non-regenerable dry reagent applied in furnace
Category S71: Dry injection

S71.1 Sorbent direct injection process (limestone, lime, sodium bicarbonate or soda ash direct injection to furnace)

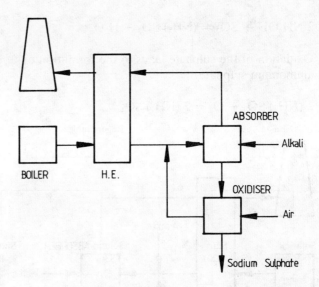

Figure 2.3 Block Diagram of Alkali Scrubbing FGD Process

Chemistry of Process: Sulphur dioxide reacts in the Absorber with the alkali to form sodium sulphite, Na_2SO_3:

$$2\ NaOH + SO_2 = Na_2SO_3 + H_2O$$

$$Na_2CO_3 + SO_2 = Na_2SO_3 + CO_2$$

Some of the sodium sulphite is oxidised by excess oxygen in the gas, and the process is completed in the Oxidiser, using air, to form sodium sulphate, Na_2SO_4:

$$2\ Na_2SO_3 + O_2 = 2\ Na_2SO_4$$

Process Code S11.3–Ammonia Scrubbing Process

Outline of Process: A simplified block diagram of the process (which is not to be confused with the regenerable ammonia scrubbing process outlined under Process Code S31.3) is presented in Figure 2.4. Gas is cooled in a gas-gas Heat Exchanger (H.E.) or by injection of water, and is then scrubbed in the Absorber with a solution of ammonium hydroxide (NH_4OH) obtained by injecting ammonia gas into the Absorber. The purified gas is reheated in the H.E. or by combustion of liquid or gaseous fuel, and exhausted to stack. The spent solution containing ammonium sulphite, $(NH_4)_2SO_3$, is filtered and the sulphite can be recovered for use in the manufacture of caprolactam. However, some ammonium sulphite is oxidised in the flue gas, and the oxidation may be completed in the Oxidiser, using air, to form ammonium sulphate, $(NH_4)_2SO_4$, as a marketable by-product.

Chemistry of Process: In the Absorber, sulphur dioxide reacts with the ammonium hydroxide to form ammonium sulphite:

$$2\ NH_4OH + SO_2 = (NH_4)_2SO_3 + H_2O$$

Oxidation of the sulphate leads to the formation of ammonium sulphate:

$$2\ (NH_4)_2SO_3 + O_2 = 2\ (NH_4)_2SO_4$$

Catalytic oxidation:
$$2\ FeSO_4 + SO_2 + O_2 = Fe_2(SO_4)_3$$
$$Fe_2(SO_4)_3 + SO_2 + 2\ H_2O = 2\ FeSO_4 + 2\ H_2SO_4$$

The sulphuric acid is neutralised with limestone, $CaCO_3$, to form gypsum, $CaSO_4.2H_2O$, which crystallises out from the solution:

$$H_2SO_4 + CaCO_3 + H_2O = CaSO_4.2H_2O + CO_2$$

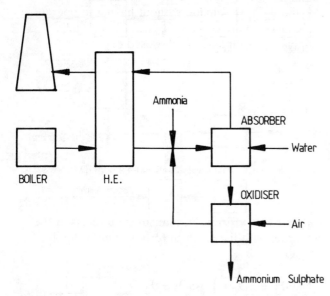

Figure 2.4 Block Diagram of Ammonia Scrubbing FGD Process

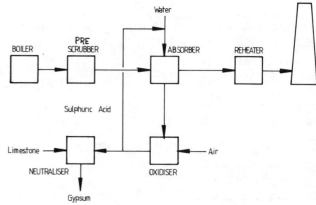

Figure 2.5 Block Diagram of Sulphuric Acid Scrubbing FGD Process

Process Code S11.4 – Sulphuric Acid Scrubbing Process

Outline of Process [172]: A simplified block diagram of the process is presented in Figure 2.5. Gas from the Boiler is passed through a Prescrubber to remove acid halides and particulates, and then into the Absorber where it is scrubbed with a recirculated stream of dilute sulphuric acid, which absorbs the SO_2. The gas then passes via a Reheater to the stack. The acid is oxidised with air in the Oxidiser to ensure conversion of absorbed SO_2 to sulphuric acid. A side stream of the acid is pumped to the Neutraliser, where it is converted to calcium sulphate di-hydrate (gypsum) which is crystallised out for sale, for the manufacture of plaster board, or safe disposal. The liquor from the gypsum separation process is pumped to a waste water plant.

Chemistry of Process [172]: The absorption and oxidation of SO_2 to give sulphuric acid is favoured by the use of the minimum practicable sulphuric acid concentration, but this concentration can be increased to levels making it easier for the subsequent production of gypsum if the absorption and oxidation is catalysed by the presence of ferrous sulphate, $FeSO_4$:

Absorption: $SO_2 + H_2O = H_2SO_3$

Oxidation: $2\ H_2SO_3 + O_2 = 2\ H_2SO_4$

Process Code S11.5 – Dowa Process

Outline of Process [95]: A simplified block diagram of the process is presented in Figure 2.6. Gas leaving the Boiler passes to an Absorber, where it is scrubbed with a solution of basic aluminium sulphate, which absorbs SO_2. The gas then flows via a Reheater to the stack. The solution from the Absorber is treated with air in the oxidiser, where aluminium sulphite formed in the Absorber is oxidised to the sulphate. The solution then passes to the Neutraliser, where it is treated with limestone, which precipitates gypsum, regenerating the basic aluminium sulphate. The gypsum is separated from the solution in a Thickener and Filter. The concentration of unwanted soluble compounds (e.g. chlorides and magnesium salts) is controlled by treating a side stream of the solution with an excess of limestone in the Alumina Recovery tank. This precipitates aluminium hydroxide; the precipitate is concentrated in a Thickener and returned to the system via a Redissolver tank where it is mixed with water. The gypsum joins that formed in the Neutraliser and is subsequently removed.

Although the actual sorbent, basic aluminium sulphate, is regenerated in the Neutraliser, the Dowa process is not classed as a regenerable process because the captured sulphur is transferred to another reagent.

Chemistry of Process [95]: In the Absorber, SO_2 reacts with basic aluminium sulphate solution, $Al_2(SO_4)_3.Al_2O_3$, to form the sulphate-sulphite:

$$Al_2(SO_4)_3.Al_2O_3 + 3\ SO_2 = Al_2(SO_4)_3.Al_2(SO_3)_3$$

The reaction occurs at a pH of 3.0–3.5. In the Oxidiser, the sulphite is oxidised by air to the sulphate. Reaction is rapid, as the SO_2 is present in the solution in the ionic state:

$$2\ Al_2(SO_4)_3.Al_2(SO_3)_3 + 3\ O_2 = 4\ Al_2(SO_4)_3$$

Treatment with limestone in the Neutraliser then regenerates the basic aluminium sulphate, precipitates calcium sulphate dihydrate (gypsum, $CaSO_4.2H_2O$) and evolves carbon dioxide:

$$2\ Al_2(SO_4)_3 + 3\ CaCO_3 + 6\ H_2O$$
$$= Al_2(SO_3)_3.Al_2O_3 + 3\ CO_2 + 3\ CaSO_4.2H_2O$$

If the treatment is carried out with a large excess of limestone, the reaction products are aluminium hydroxide, $Al_2(OH)_3$, which is precipitated at high pH, and gypsum:

$$Al_2(SO_4)_3 + 3\ CaCO_3 + 9\ H_2O$$
$$= 2\ Al(OH)_3 + 3\ CaSO_4.2H_2O + 3\ CO_2$$

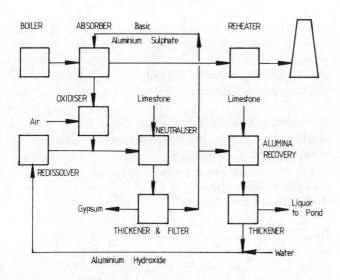

Figure 2.6 Block Diagram of Dowa FGD Process

Process Code S11.6 – Dual (or Double) Alkali Process

Outline of Process [95]: A simplified block diagram of the process is presented in Figure 2.7. The gas passes through a Prescrubber to remove particulates and acid halides, and is then treated in the Absorber with sodium sulphite solution to absorb sulphur dioxide. After reheating, the gas is exhausted to stack. The alkali solution contains sodium bisulphite and some sulphate. It is passed to the Regenerator, where it is regenerated with hydrated lime or limestone to produce solid calcium sulphite. This is separated from the regenerated sodium sulphite which is returned to the Absorber. A slip-stream of the unregenerated alkali solution is treated with either sulphuric acid (to reduce the pH) or alkali (to increase the pH) before treatment with lime; this is to allow calcium sulphate to be precipitated separately from sulphite without having to operate at very low sulphite concentrations. The calcium sulphite and sulphate are disposed of.

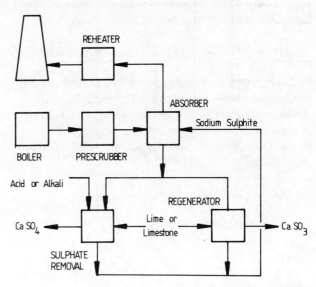

Figure 2.7 Block Diagram of Dual Alkali FGD Process

Although the actual sorbent, calcium sulphite, is regenerated in the Regenerator, the Dual Alkali process is not classed as a regenerable process because the captured sulphur is transferred to another reagent.

Chemistry of Process [122]: In the Absorber, sodium sulphite, Na_2SO_3, reacts with SO_2 to form sodium bisulphite, $NaHSO_3$:

$$Na_2SO_3 + SO_2 + H_2O = 2\ NaHSO_3$$

The ingoing solution has a pH of about 9, but the bisulphite solution leaving the Absorber has a pH of only 6.

The sulphite is regenerated in the Regenerator with hydrated lime, $Ca(OH)_2$, or limestone, $CaCO_3$; the insoluble calcium sulphite, $CaSO_3$, is separated from the solution, which is returned to the Absorber:

$$2\ NaHSO_3 + Ca(OH)_2 = Na_2SO_3 + CaSO_3 + 2\ H_2O$$

$$2\ NaHSO_3 + CaCO_3$$
$$= Na_2SO_3 + CaSO_3 + H_2O + CO_2$$

Some sulphate is formed in the Absorber by reaction with excess oxygen in the gas:

$$2\ Na_2SO_3 + O_2 = 2\ Na_2SO_4$$

The sodium sulphate can be converted to gypsum (calcium sulphate dihydrate, $CaSO_4.2H_2O$) by treatment, with hydrated lime:

$$Na_2SO_4 + Ca(OH)_2 + 2\ H_2O = CaSO_4.2H_2O + 2\ NaOH$$

but co-precipitation of the gypsum with the calcium sulphite formed in the normal regeneration reaction occurs only when the sulphite concentration is very low. One solution is to add sulphuric acid:

$$Na_2SO_4 + 2\ CaSO_3 + H_2SO_4 = 2\ CaSO_4 + 2\ NaHSO_3$$

but more sulphuric acid is needed to reduce the pH to the value of 2–3 needed for the precipitation. Another solution is to increase the pH to 13–14 by adding alkali, whereupon calcium sulphate is precipitated by lime, and the sodium sulphite is regenerated:

$$Na_2SO_4 + Ca(OH)_2 = CaSO_4 + 2\ NaOH$$

$$NaOH + NaHSO_3 = Na_2SO_3 + H_2O$$

CATEGORY S12

Process Code S12.1 – Alkali Scrubbing/Spray Drying Process

Outline of Process [122]: A simplified block diagram of the process is presented in Figure 2.8. Gas from the Boiler, at a temperature of 120–160°C, enters the top of a Spray Dryer into which a solution of alkali (soda ash or caustic soda) is sprayed. The gas passes down the Spray Dryer co-current with the spray droplets; the sulphur oxides and halogen acids in the gas react with the alkali, and water is evaporated from the droplets to produce a dry product which also contains the particulates present in the ingoing gas. Coarse particles of product are collected at the base of the Spray Dryer, and fine particles are separated from the gas in a Baghouse. The gas leaves the system at 60–80°C, and is exhausted via a Reheater (if required) to the stack. Part of the dry end product is recycled to the Feed Tank to increase the conversion of alkali. An elaboration of the process regenerates the alkali (see under Process Code S32.4).

The dry end product contains sodium sulphite, sulphate and halides, together with some sodium carbonate. It is usual to recycle part of the dry end product to the Feed Tank to increase the alkali utilisation. The acidic components of the gas are absorbed with very high efficiency, and reaction is faster than with hydrated lime (see Process Code S21.1) so that lower liquid/gas flow ratios can be used.

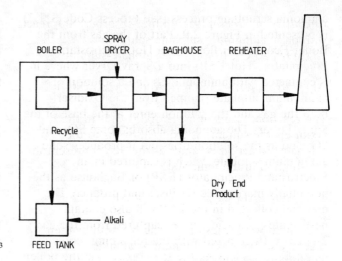

Figure 2.8 Block Diagram of Alkali Scrubbing/Spray Drying FGD Process

Chemistry of Process [122]: The sulphur oxides, SO_2 and SO_3, and acid halides, HF, HCl and HBr (represented below as 'HHa') react with the alkali solution as follows:

With caustic soda (NaOH)

$$2\ NaOH + SO_2 = Na_2SO_3 + H_2O$$
(forming sodium sulphite)

$$2\ NaOH + SO_3 = Na_2SO_4 + H_2O$$
(forming sodium sulphate)

$$NaOH + HHa = NaHa + H_2O$$
(forming sodium halides)

With soda ash (sodium carbonate, Na_2CO_3) the reactions are similar, but with liberation of CO_2 in place of H_2O:

$$Na_2CO_3 + SO_2 = Na_2SO_3 + CO_2$$
(forming sodium sulphite)

$$Na_2CO_3 + SO_3 = Na_2SO_4 + CO_2$$
(forming sodium sulphate)

$$Na_2CO_3 + 2\ HHa = 2\ NaHa + CO_2 + H_2O$$
(forming sodium halides)

Reaction is rapid whilst the droplets are wet, and the gas temperature in the Spray Dryer is maintained slightly above the saturation temperature (e.g. within 20°C) to delay the drying out of the droplets sufficiently for efficient absorption.

Process Code S12.2 – Walther Process

Outline of Process [Buckau-Walther Group promotional literature]: A simplified block diagram of the process, which is not to be confused with the ammonia scrubbing process (see Process Code S11.3) is presented in Figure 2.9. Part of the gas from the Boiler Economiser flows via a Hot Electrostatic Precipitator (Hot ESP) into a Spray Dryer where it is contacted with ammonia gas and a solution of ammonium sulphate pumped from the Oxidiser; both the gas and the solution enter at the base of the Spray Dryer. The ammonia absorbs some SO_2 and SO_3 and as the solution dries out it produces solid ammonium sulphate which is captured in an Electrostatic Precipitator (ESP) or Baghouse as the potentially marketable fertiliser end-product. The material collected in the ESP will also contain particulates and acid halides captured from the gas. The gas is then mixed with the remaining (cooler, unpurified) gas leaving the Economiser via the boiler Air Heater and ESP. The reunited gas stream is cooled in a Heat Exchanger (H.E.) then flows to two spray towers – a Scrubber and Rescrubber – denoted S1 and S2 in the block diagram. The gas passes up the spray towers counter-current to a flow of water which removes unreacted ammonia from the gas, and absorbs more SO_2. The purified gas is reheated in the H.E. and exhausted to stack. The solution leaving the Scrubber is oxidised with air in an Oxidiser and is then passed to the Spray Dryer.

Chemistry of Process [Buckau-Walther Group promotional literature]: The ammonia solution absorbs SO_2, SO_3 and some CO_2 from the gas in the Scrubber (S1), Rescrubber (S2) and Spray Dryer to form ammonium sulphite, sulphate and carbonate:

Sulphite formation:
$$2 NH_4OH + SO_2 = (NH_4)_2SO_3 + H_2O$$

Sulphate formation:
$$2 NH_4OH + SO_3 = (NH_4)_2SO_4 + H_2O$$

Carbonate formation:
$$2 NH_4OH + CO_2 = (NH_4)_2CO_3 + H_2O$$

In the Scrubber (S1) and Rescrubber (S2), much of the SO_2 absorption results from reaction with the ammonium carbonate solution:

Carbonate neutralisation:
$$(NH_4)_2CO_3 + SO_2 = (NH_4)_2SO_3 + CO_2$$

Sulphite is oxidised to sulphate by reaction with atmospheric oxygen in the Oxidiser, and the sulphate dried in the Spray Dryer and removed as a solid in the ESP or Baghouse, is potentially marketable as a fertiliser:

Oxidation reaction:
$$2 (NH_4)_2SO_3 + O_2 = 2 (NH_4)_2SO_4$$

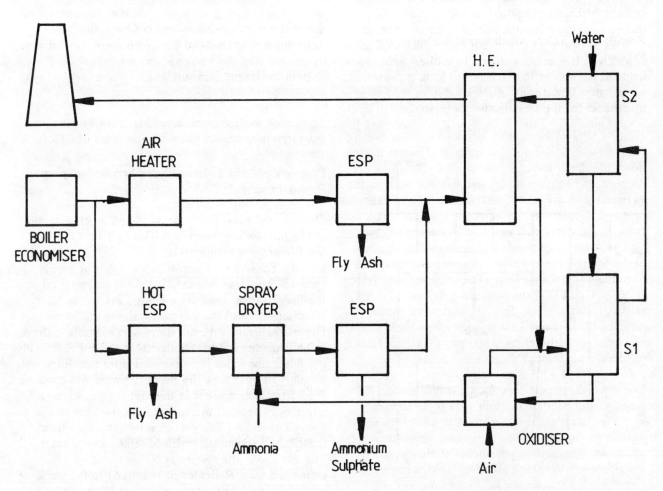

Figure 2.9 Block Diagram of Walther FGD Process

At the high temperature conditions in the Spray Dryer (i.e. above about 60°C) the carbonate and sulphite decompose quantitatively, reversing the formation reactions. The formation of carbonate is minimised by controlling the pH of the solution at about 7.

The material collected in the ESP will also contain particulates and acid halides captured from the gas.

CATEGORY S21

Process Code S21.1 – Limestone or Lime Slurry Scrubbing Process

Outline of process [173]: This basic process type includes scrubbing with slurries containing high-calcium fly-ash, and with slurries of magnesium hydroxide. A simplified block diagram of the process is presented in Figure 2.10. Gas from the Boiler is cooled in a gas-gas Heat Exchanger (H.E.) or by water injection. If the gas contains acid halides, or if it contains particulates and a pure gypsum by-product is to be produced, it is first scrubbed with water in the Prescrubber. The gas is then scrubbed in the Absorber with a slurry containing limestone or hydrated lime to remove SO_2. The purified gas is reheated in the H.E. and/or by combustion of oil or gaseous fuel, and exhausted to stack. The spent slurry, containing calcium sulphite hemihydrate, is difficult to deal with as it is not easily thickened or dewatered. It is either pumped to settling lagoons, or oxidised with air in the Oxidiser to form gypsum which, after dewatering, is disposed of or sold as a by-product for the manufacture of plasterboard.

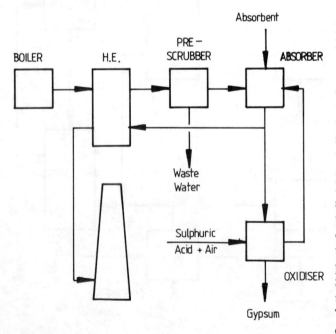

Figure 2.10 Block Diagram of Limestone/Lime Slurry FGD Process

Chemistry of process [122]: In the Absorber, sulphur dioxide reacts with limestone, $CaCO_3$, or hydrated lime, $Ca(OH)_2$, to form calcium sulphite hemihydrate, $CaSO_3.1/2H_2O$:

$2\ SO_2 + 2\ CaCO_3 + H_2O$
$= 2\ CaSO_3.1/2H_2O + 2\ CO_2$

$2\ SO_2 + 2\ Ca(OH)_2 = 2\ CaSO_3.1/2H_2O + H_2O$

Certain additives such as formic acid and some dibasic organic acids, e.g. adipic acid, act as buffer agents in the slurry, and their use improves the conversion of the calcium compounds.

If gypsum is to be the end product, oxidation is completed by pumping air through the slurry in the Oxidiser:

$2\ CaSO_3.1/2H_2O + O_2 + 3\ H_2O = 2\ CaSO_4.2H_2O$

Where pure gypsum is required, any unreacted limestone or hydrated lime in the slurry is first sulphated by reaction with sulphuric acid:

$CaCO_3 + H_2O + H_2SO_4 = CaSO_4.2H_2O + CO_2$

$Ca(OH)_2 + H_2SO_4 = CaSO_4.2H_2O$

When magnesium hydroxide is used as the absorbent, the absorption, oxidation and neutralisation reactions are similar to those described above for calcium hydroxide, and the end-products are magnesium sulphite or magnesium sulphate.

CATEGORY S22

Process Code S22.1 – Lime Slurry Scrubbing/Spray Drying Process

Outline of Process [203]: A simplified block diagram of the process is presented in Figure 2.11. Gas from the Boiler, at a temperature of 120–160°C, enters the top of a Spray Dryer into which a slurry of lime is sprayed. The lime, which is hydrated and slurried in the Slaker, is pumped via a Feed Tank to the slurry atomiser nozzles. The gas passes down the Spray Dryer co-current with the slurry spray droplets; the sulphur oxides and halogen acids in the gas react with the lime, and water is evaporated from the slurry droplets to produce a dry product which also contains the particulates present in the ingoing gas. Coarse particles of product are collected at the base of the Spray Dryer, and fine particles are separated from the gas in a Baghouse or Electrostatic Precipitator. The gas leaves the system at 60–80°C, and is exhausted via a Reheater (if required) to the stack.

The dry end product contains calcium sulphite,

sulphate and halides, together with some calcium carbonate, unreacted lime and coal ash. It is usual to recycle part of the dry end product to the slurry preparation stage to increase the lime utilisation. The acid halides and SO_3 are absorbed with very high efficiency (e.g. 99% removal).

The percentage capture of SO_2 depends on spray droplet size, particle size of the hydrated lime, and the Ca/S molar ratio used; typically, captures of 75% are obtained at a Ca/S = 1, and 95% at Ca/S = 2.

Chemistry of Process [122]: The sulphur oxides, SO_2 and SO_3, and acid halides, HF, HCl and HBr (represented below as 'HHa') react in the Spray Dryer with the lime in the slurry as follows:

$Ca(OH)_2 + SO_2 = CaSO_3 + H_2O$
(forming calcium sulphite)

$Ca(OH)_2 + SO_3 = CaSO_4 + H_2O$
(forming calcium sulphate)

$Ca(OH)_2 + 2HHa = CaHa_2 + 2H_2O$
(forming calcium halides)

Reaction occurs in two phases: the first while the droplets are still wet; and the second after dryout, when porous particles are formed. Reaction is rapid during the wet first phase, the rate being controlled by the rate of dissolution of lime through a calcium sulphite layer. For this reason, the gas temperature in the Spray Dryer is maintained slightly above the saturation temperature (e.g. within 20°C) to delay the drying out of the droplets sufficiently for efficient absorption. In the dry second phase, reaction is controlled by pore diffusion in the particle. The gas residence time in the Spray Dryer is around 10 seconds, with most of the reaction occurring within about 2 seconds.

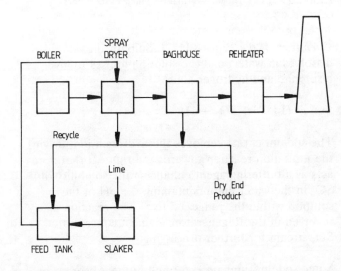

Figure 2.11 Block Diagram of Lime Slurry Scrubbing/Spray Drying FGD Process

CATEGORY S31

Process Code S31.1 – Wellman-Lord Process

Outline of Process [265]: A simplified block diagram of the process is presented in Figure 2.12. Gas flows, via a Prescrubber to remove particulates and acid halides, to the Absorber where it is scrubbed by a concentrated solution of sodium sulphite to form the bisulphite. The purified gas passes via entrainment separators to a steam-heated Reheater before being exhausted to stack. The sodium bisulphite solution is pumped to the Regenerator. Here it is converted to sodium sulphite, with evolution of sulphur dioxide, by thermal decomposition, using steam in a double-effect evaporator/crystalliser. Some sulphate formation occurs in the Absorber, and a side stream is treated by fractional crystallisation at 0°C to remove the sulphate and thiosulphate; the loss of sodium is made up by feeding fresh alkali. Caustic soda has to be used as the make-up alkali in the UK, as commercial soda ash cannot be used because of its chloride content.

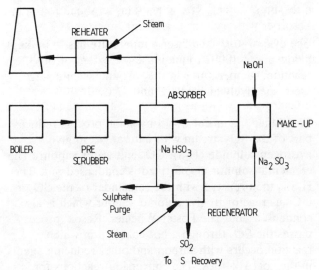

Figure 2.12 Block Diagram of Wellman-Lord FGD Process

The concentrated SO_2 stream leaving the Regenerator is cooled to condense water vapour, and is then compressed for further processing. It can be catalytically oxidised for the manufacture of sulphuric acid, but it is more usual to reduce it to elemental sulphur which is potentially marketable for a number of uses, but which can also be disposed of safely.

Chemistry of the Process [122, 137, 146]: Sodium sulphite solution reacts with sulphur dioxide in the Absorber to form the bisulphite:

$Na_2SO_3 + SO_2 + H_2O = 2 NaHSO_3$

In the Absorber, there is no phase separation, and the SO_2 partial pressure is first order with respect to

bisulphite ion concentration, [HSO_3^-]. Sulphur capture efficiencies of about 95% can be achieved. The reaction in the Regenerator is the reverse of that in the Absorber, but with the difference that sulphite crystallises out. The SO_2 partial pressure is then proportional to the product of the square of the sodium and bisulphite ion concentrations:
[Na^+]^2.[HSO_3^-]^2

Evaporation produces a small increase in concentration and therefore a large increase in SO_2 partial pressure, resulting in efficient stripping of SO_2 from the solution and conversion of the solution to sulphite.

Some sodium sulphate is formed in the Absorber by reaction of the sulphite with excess oxygen in the gas:

2 Na_2SO_3 + O_2 = 2 Na_2SO_4

Sodium sulphate is also formed, together with sodium thiosulphate, $Na_2S_2O_3$ by disproportionation in the Regenerator:

6 $NaHSO_3$ = 2 Na_2SO_4 + $Na_2S_2O_3$ + 2 SO_2 + 3 H_2O

The evaporator/crystalliser temperature has to be kept below about 100°C to minimise the extent of this reaction; temperatures in the first and second stage are respectively about 94°C and 77°C.

The Allied Chemical Co. conversion process reduces part of the SO_2 stream with natural gas to give hydrogen sulphide (H_2S), CO_2 and water vapour. The elemental sulphur also formed is condensed out. The H_2S is then reacted with the remainder of the SO_2 in a Claus reactor to give sulphur vapour which is also condensed out. The Foster Wheeler 'Resox' process passes the SO_2 through a hot coal bed in which reaction occurs with carbon and other reducing agents in the coal volatiles. The conversion reactions for the two processes can be represented as:

Allied process:
3 SO_2 + 2 CH_4 = 2 H_2S + 2 CO_2 + 2 H_2O + S

SO_2 + 2 H_2S = 3 S + 2 H_2O

Resox process: SO_2 + C = S + CO_2

Process Code S31.2 – Fläkt-Boliden Process

Outline of Process [137]: This process is one form of the Citrate process; the other form is described in Process Code S32.1. A simplified block diagram of the process is presented in Figure 2.13. The gas is first passed through a Prescrubber to remove particulates and acid halides, and is then passed into an Absorber where it is treated with sodium citrate solution. The solution, containing bisulphite ion, is pumped via a heat exchanger to the Regenerator, where SO_2 is removed by steam stripping. The regenerated solution is returned to the Absorber.

A side stream of the solution from the Absorber is treated by fractional crystallisation to remove the sulphate, and the loss of sodium is made up by feeding fresh alkali. Caustic soda has to be used as the make-up alkali in the UK, as commercial soda ash cannot be used because of its chloride content.

The concentrated SO_2 stream leaving the Regenerator is cooled to condense water vapour, and is then compressed for further processing. It can be catalytically oxidised for the manufacture of sulphuric acid, but it is more usual to reduce it in a sulphur recovery unit to elemental sulphur which is potentially marketable for a number of uses, but which can also be disposed of safely.

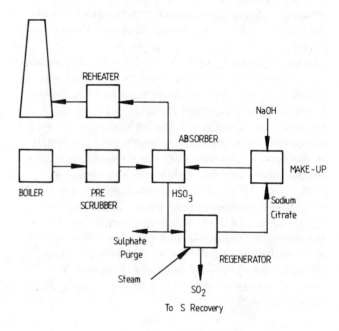

Figure 2.13 Block Diagram of Fläkt-Boliden FGD Process

Chemistry of the Process [122]: Sulphur dioxide dissolves in water, and the solution ionises to give bisulphite and hydrogen ions:

SO_2 + H_2O = HSO_3^- + H^+

The sodium citrate removes the hydrogen ion, driving the ionisation reaction towards the right. It therefore acts as a buffering agent enhancing the solubility of SO_2 in the water and maintaining the pH of the solution within the range 3.5 to 5. The reaction is reversed in the Regenerator, giving a concentrated SO_2 stream for further processing.

Some sodium sulphate is formed in the Absorber by reaction of the bisulphite with excess oxygen in the gas.

The Allied Chemical Co. conversion process reduces part of the SO_2 stream with natural gas to give hydrogen sulphide (H_2S), CO_2 and water vapour. The elemental sulphur also formed is condensed out. The H_2S is then reacted with the remainder of the SO_2 in a Claus reactor to give sulphur vapour which is also condensed out. The Foster Wheeler 'Resox' process passes the SO_2 through a hot coal bed in which reaction occurs with carbon and other reducing agents in the coal volatiles. The conversion reactions for the two processes can be represented as:

Allied process:
$$3\ SO_2 + 2\ CH_4 = 2\ H_2S + 2\ CO_2 + 2\ H_2O + S$$

$$SO_2 + 2\ H_2S = 3\ S + 2\ H_2O$$

Resox process: $SO_2 + C = S + CO_2$

Process Code S31.3 – Catalytic Inc./Institut Francais du Petrole (IFP) or Ammonia Scrubbing Process

Outline of Process [137]: A simplified block diagram of the process (which should not be confused with the throwaway ammonia process outlined in Process Code S11.3) is presented in Figure 2.14. After passing through a Prescrubber to remove particulates, acid halides and some SO_3, the gas from the Boiler is scrubbed with ammonium hydroxide solution in a multi-stage Absorber, and then with water in a separate Scrubber. The cleaned gas passes through the Reheater before it is exhausted to stack. The spent liquor, containing ammonium sulphite and bisulphite, is evaporated in the Regenerator to about 40% of its bulk at 150°C and is then thermally decomposed to produce SO_2, ammonia and water. Some ammonium sulphate is formed in the Absorber by absorption of SO_3 from the gas, and by oxidation of the ammonium sulphite with excess oxygen in the gas. To deal with this, a side stream of the concentrated slurry from the Regenerator is reduced with recycled molten sulphur in the Sulphate Removal system. This also produces SO_2, ammonia and water. The ammonia and water are returned to the Absorber, and the SO_2 is partially reduced with a reducing gas containing hydrogen and converted to elemental sulphur in a liquid-phase Claus Unit.

Chemistry of Process [137]: The absorption reactions occurring in the Absorber, which can remove more than 90% of the sulphur oxides in the gas, are:

Sulphite formation:
$$2\ NH_4OH + SO_2 = (NH_4)_2SO_3 + H_2O$$

Bisulphite formation:
$$(NH_4)_2SO_3 + SO_2 + H_2O = 2\ NH_4HSO_3$$

Sulphate formation:
$$(NH_4)_2SO_3 + SO_3 = (NH_4)_2SO_4 + SO_2$$

$$2\ (NH_4)_2SO_3 + O_2 = 2\ (NH_4)_2SO_4$$

In the Regenerator, thermal decomposition of the ammonium sulphite and bisulphite occurs at about 150°C, releasing ammonia, water and sulphur dioxide:

$$(NH_4)_2SO_3 = 2\ NH_3 + H_2O + SO_2$$

$$NH_4HSO_3 = NH_3 + H_2O + SO_2$$

The ammonium sulphate, however, does not decompose in the Regenerator, and a side stream of the concentrated slurry is reduced with molten recycled sulphur at 300–370°C in the Sulphate Removal system:

$$(NH_4)_2SO_4 = NH_4HSO_4 + NH_3$$

$$2\ NH_4HSO_4 + S = 3\ SO_2 + 2\ NH_3 + 2\ H_2O$$

$$(NH_4)_2SO_4 = SO_3 + 2\ NH_3 + H_2O$$

The ammonia and water from the Regenerator and Sulphate Removal system are returned, via the Ammonia Recovery system, to the Absorber, and the sulphur oxides are partially reduced in the SO_2 Reduction system with a reducing gas containing hydrogen to give a gas containing two volumes of H_2S per volume of unreduced SO_2. This reacts in a liquid phase Claus Unit to give elemental sulphur of more than 99% purity:

Reduction: $SO_2 + 3\ H_2 = H_2S + 2\ H_2O$

Sulphur Recovery: $2\ H_2S + SO_2 = 3\ S + 2\ H_2O$

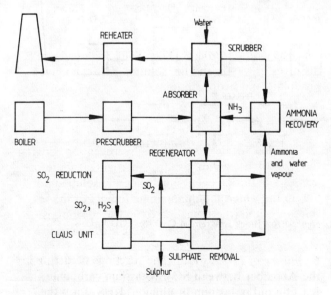

Figure 2.14 Block Diagram of IFP/Catalytic FGD Process

CATEGORY S32

Process Code S32.1–Citrate Process

Outline of Process [6, 122, 146]: Another form of this process, using thermal regeneration, is described in Process Code S31.2. A simplified block diagram of the process is presented in Figure 2.15. The gas is first passed through a Prescrubber to remove particulates and acid halides, and is then passed into an Absorber where it is treated with sodium citrate solution. The solution, containing bisulphite ions, is pumped to the Regenerator, where SO_2 is reacted with hydrogen sulphide (H_2S) in a liquid-phase Claus reaction yielding elemental sulphur. The regenerated solution is returned to the Absorber.

A side stream of the solution from the Absorber is treated by fractional crystallisation to remove the sulphate, and the loss of sodium is made up by feeding fresh alkali. Caustic soda has to be used as the make-up alkali in the UK, as commercial soda ash cannot be used because of its chloride content.

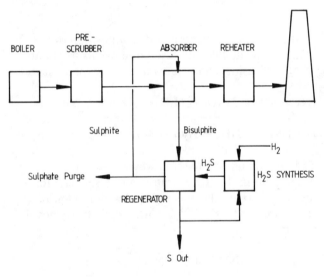

Figure 2.15 Block Diagram of Citrate FGD Process

Chemistry of the Process [122]: Sulphur dioxide disolves in water, and the solution ionises to give bisulphite and hydrogen ions:

$$SO_2 + H_2O = HSO_3^- + H^+$$

The sodium citrate removes the hydrogen ion, driving the ionisation reaction towards the right. It therefore acts as a buffering agent for enhancing the solubility of SO_2 in the water, maintaining the pH of the solution within the range 3.5 to 5. Sulphur capture efficiencies of 90% can be attained.

In the Regenerator, H_2S reacts with bisulphite and hydrogen ions to give elemental sulphur in a two-step process: a fast reaction giving thiosulphate ions; and a slow rate-controlling reaction giving sulphur:

Fast reaction:
$$2\,H_2S + 4\,HSO_3^- = 3\,S_2O_3^{--} + 2\,H^+ + 3\,H_2O$$

Slow reaction: $\quad 6\,H_2S + 3\,S_2O_3^{--} + 6\,H^+ = 3/2\,S_8 + 9\,H_2O$

Overall reaction: $2\,H_2S + HSO_3^- + H^+ = 3/8\,S_8 + 3\,H_2O$

The sulphur recovered has a purity of 99%. The H_2S is prepared by hydrogenation of part of the sulphur produced, using hydrogen derived from reforming natural gas.

Some sodium sulphate is formed in the Absorber by reaction of the bisulphite with excess oxygen in the gas.

Process Code S32.2–Conosox Process

Outline of Process [95]: A simplified block diagram of the process is presented in Figure 2.16. Gas from the Boiler is treated in the Absorber with a solution of potassium carbonate containing some potassium hydrosulphide, and then passes via the Reheater to stack. The solution absorbs SO_2 in the Absorber, and is then pumped to the Reducer where it is regenerated at about 230°C and 45 bar pressure. The reducing agent is carbon monoxide generated in the Partial Oxidiser from the reaction of oil with oxygen. The spent reducing gas from the Reducer, containing H_2S, passes to the Sulphur Recovery system, incorporating an H_2S absorber and stripper, together with a Claus unit to treat the concentrated H_2S stream. The regenerated absorbent solution from the Reducer contains some potassium sulphate which is separated from the solution in the Crystalliser before the solution is returned to the Absorber.

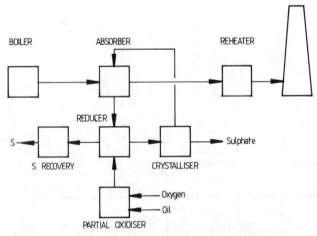

Figure 2.16 Block Diagram of Conosox FGD Process

Chemistry of Process [95]: The reactions occurring in the Absorber between SO_2, potassium carbonate, K_2CO_3, and potassium bisulphide, KHS, form the sulphite, bisulphite and thiosulphate:

Sulphite formation:
$$K_2CO_3 + SO_2 = K_2SO_3 + CO_2$$

Bisulphite formation:
$$K_2SO_3 + SO_2 + H_2O = 2\ KHSO_3$$

Thiosulphate formation:
$$2\ KHS + 4\ KHSO_3 = 3\ K_2S_2O_3 + 3\ H_2O$$

The thiosulphate is highly soluble, allowing the solution to contain absorbed sulphur in a highly concentrated form without risk of scale formation. The thiosulphate also reduces the equilibrium partial pressure of SO_2 in the gas, and inhibits the formation of potassium sulphate. Nevertheless, some sulphate is formed (equivalent to about 1% of the total sulphur captured) by oxidation of the sulphite, and this has to be separated from the regenerated solution in the Crystalliser:

$$2\ K_2SO_3 + O_2 = 2\ K_2SO_4$$

In the Reducer the carbonate and bisulphide are regenerated by reduction of thiosulphate with carbon monoxide at a temperature of about 230°C and pressure of 45 bar:

$$3\ K_2S_2O_3 + 12\ CO + 5\ H_2O$$
$$= 2\ K_2CO_3 + 2\ KHS + 4\ H_2S + 10\ CO_2$$

The CO is generated by partial oxidation of fuel oil in the Partial Oxidiser, where the reactions include the water gas shift reaction:

$$CO_2 + H_2 = CO + H_2O$$

The gas leaving the Reducer, containing H_2S and CO_2, passes to an absorption tower forming part of the Sulphur Recovery system. Here the H_2S and some of the CO_2 is absorbed in a mixed solution of polyethylene glycol dimethyl ethers. This solution passes to a stripper giving a concentrated H_2S-CO_2 gas (approximately equal volumes) which is treated in a Claus unit to produce elemental sulphur for marketing or safe disposal:

Oxidation: $\quad 2\ H_2S + 3\ O_2 = 2\ SO_2 + 2\ H_2O$

Claus Reaction: $2\ H_2S + SO_2 = 3\ S + 2\ H_2O$

Process Code S32.3 – Ispra Mark 13A Process

Outline of Process [272]: A simplified block diagram of the process is presented in Figure 2.17. Gas from the boiler is passed in turn through a Concentrator, a Reactor and a final Scrubber. In the Reactor and Scrubber, the gas is contacted with an aqueous solution of hydrobromic and sulphuric acids containing a small amount of bromine. The bromine reacts with the sulphur dioxide to form sulphuric acid. Part of the solution is pumped to the Concentrator, where the sensible heat of the gas is used to evaporate water and drive off bromine, leaving sulphuric acid. The remainder of the solution is pumped to an electrolyser where hydrobromic acid is converted to bromine and hydrogen. The cleaned gas flows through a Reheater before being exhausted to stack.

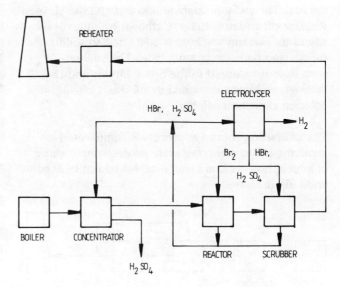

Figure 2.17 Block Diagram of Ispra Mark 13A FGD Process

Chemistry of Process [272]: The principal reaction occurring in the Reactor (temperature 60–70°C) and final Scrubber (temperature 40–50°C) is the oxidation of SO_2 to sulphuric acid (H_2SO_4) by bromine, which is reduced to hydrobromic acid (HBr):

$$SO_2 + 2\ H_2O + Br_2 = H_2SO_4 + 2\ HBr$$

The bromine is regenerated from the hydrobromic acid by electrolysis in the Electrolyser:

$$2\ HBr = Br_2 + H_2$$

Hydrogen is therefore one of the byproducts of the process. The other by-product, sulphuric acid at a concentration of 80% by weight, is obtained from the Concentrator which is operated at a temperature of 140°C. This process, which is developmental, is claimed to give over 90% sulphur capture.

Process Code S32.4 – Aqueous Sodium Carbonate Process

Outline of Process [6, 43, 137, 146]: A simplified block diagram of the process is presented in Figure 2.18. After clean-up in a Cyclone, the gas is treated with aqueous sodium carbonate in a Spray Dryer FGD unit. It then passes via an Electrostatic Precipitator to stack. The solids from the Spray Dryer and

Electrostatic Precipitator, containing sodium sulphite and sulphate, are converted to sodium sulphide by treatment in the Reactor with coal or coke in a molten salt bath at 900–1050°C. The melt is quenched and dissolved in water in the Quench vessel, and undissolved solids are separated from the solution. The gas from the Reactor contains carbon dioxide; it is cooled to remove volatilised chlorides, and these, together with carry-over solids, are separated from the gas. The sodium sulphide solution and the cleaned Reactor off-gas pass to the Carbonation tower, where the sodium sulphide is converted to sodium carbonate and hydrogen sulphide. The sodium carbonate is returned to the Spray Dryer, and the hydrogen sulphide is treated in a Claus Unit to produce elemental sulphur.

The overall regeneration process is complicated, involving about eighty separate process steps, many of which, however, are well-established in the paper and pulp industry.

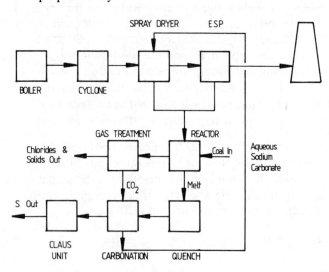

Figure 2.18 Block Diagram of Aqueous Sodium Carbonate FGD Process

Chemistry of the Process [122]: In the Spray Dryer, the SO_2 in the gas reacts with sodium carbonate, Na_2CO_3, and excess oxygen to form sodium sulphite and sulphate:

Sulphite formation: $SO_2 + Na_2CO_3 = Na_2SO_3 + CO_2$

Sulphate formation: $2 Na_2SO_3 + O_2 = 2 Na_2SO_4$

In the Reactor, the sodium sulphite and sulphate are both reduced at 900–1050°C to sodium sulphide (Na_2S) by carbon, supplied as coal or coke:

$2 Na_2SO_3 + 3 C = 2 Na_2S + 3 CO_2$

$Na_2SO_4 + 2 C = Na_2S + 2 CO_2$

Further carbon dioxide is generated in the Reactor by combustion of the fuel. The melt is quenched and dissolved in water in the Quench vessel, and undissolved solids are removed. The sodium sulphide solution is then passed to the Carbonation vessel. Here it is treated with CO_2-rich gases from the Reactor (after condensation and separation of volatilised chlorides from the gas, and removal of other solids carried over from the fuel) to form carbonate, bicarbonate and H_2S:

$Na_2S + CO_2 + H_2O = Na_2CO_3 + H_2S$

$Na_2CO_3 + CO_2 + H_2O = 2 NaHCO_3$

The bicarbonate solution is heated, releasing the carbon dioxide and restoring the sodium carbonate for return to the Spray Dryer. The gas from the Carbonation tower, containing H_2S, is sent to a Claus Unit, where part of the H_2S is oxidised to SO_2 which then reacts with the remainder of the H_2S:

$2 H_2S + 3 O_2 = 2 H_2O + 2 SO_2$

$2 H_2S + SO_2 = 3 S + 2 H_2O$

CATEGORY S41

Process Code S41.1–Magnesia Slurry Scrubbing Process

Outline of Process [6, 122, 146]: A simplified block diagram of the process is presented in Figure 2.19. Gas from the Boiler is passed through a Prescrubber to remove acid halides and particulates, and then to the Absorber, where sulphur oxides are absorbed by scrubbing with a hydrated magnesia slurry. The cleaned gas passes via a Reheater to the stack. Crystalline magnesium sulphite and sulphate are centrifuged out of the liquid and treated in the Dryer where they lose water of crystallisation. They are then calcined in the Regenerator in a reducing atmosphere at about 760–870°C to yield magnesia; addition of coke or carbon is required to reduce the sulphate. Sulphur dioxide evolved in the Regenerator is used to manufacture sulphuric acid or elemental sulphur for sale or disposal.

Chemistry of the Process [122]: Hydrated magnesia, $Mg(OH)_2$, reacts with sulphur dioxide in the Absorber to form magnesium sulphite trihydrate and hexahydrate. Some of the magnesium sulphite is converted to magnesium sulphate heptahydrate:

$Mg(OH)_2 + SO_2 + 2 H_2O = MgSO_3.3H_2O$

$Mg(OH)_2 + SO_2 + 5 H_2O = MgSO_3.6H_2O$

$2 MgSO_3 + O_2 + 14 H_2O = 2 MgSO_4.7H_2O$

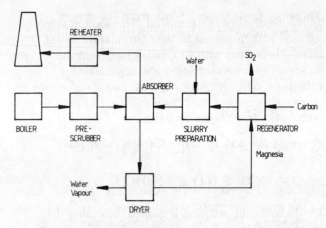

Figure 2.19 Block Diagram of Magnesia Slurry Scrubbing FGD Process

The sulphites and sulphate crystallise out, and are transferred to the Dryer where they lose their water of crystallisation. The anhydrous solids are then calcined in the Regenerator, to which carbon is added to reduce the sulphate. The regeneration reactions, which produce magnesia, MgO, and SO_2, occur at 760–870°C:

$$MgSO_3 = MgO + SO_2$$

$$2\,MgSO_4 + C = 2\,MgO + 2\,SO_2 + CO_2$$

The magnesia is then hydrated in the Slurry Preparation stage:

$$MgO + H_2O = Mg(OH)_2$$

CATEGORY S42

Process Code S42.1–Sulf-X Process

Outline of Process [175]: A simplified block diagram of the process is presented in Figure 2.20. Gas from the Boiler first passes through a Prescrubber to remove acid halides and particulates, and then into the Absorber where it is scrubbed with a slurry containing a mixture of iron sulphides in a sodium sulphide solution. The reactions with sulphur dioxide produce a range of insoluble iron/sulphur compounds, some soluble iron sulphate, and trace amounts of elemental sulphur. Some reduction of NO_x to elemental sulphur also occurs, but the liquid/gas flow ratio for efficient NO_x abatement is much higher than for SO_2 removal. The solid reaction products are separated from the solution in the Dewatering and Filtration system, dried in a steam-heated Dryer, and roasted at 650–750°C in an indirectly-fired Calciner to which coal or coke is fed as a reducing agent, regenerating the iron sulphides and

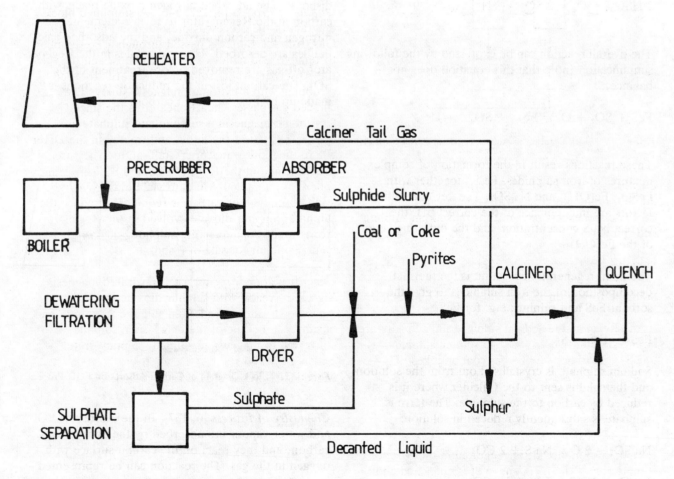

Figure 2.20 Block Diagram of Sulf-X FGD Process

releasing sulphur. Part of the solution from the Filtration stage is treated in a Sulphate Separation system to crystallise out sodium sulphate which is transferred to the calciner. The solids from the Calciner pass to the Quench, where they are mixed with liquid from the Filtration and Sulphate Separation stages to reform the sulphide slurry sent to the Absorber. Sulphur vapour in the tail gas from the Calciner is condensed for marketing or safe disposal, and the gas is returned to the Prescrubber inlet.

Chemistry of Process [95]: The sorbent slurry is a complex mixture of iron compounds in a sodium sulphide solution, with a pH controlled at 6.0–6.4 by the buffering action of $Fe(OH)_2$. The main absorbing compound is FeS. The reactions occurring in the Absorber include:

$$FeS + H_2O = Fe^{++} + HS^- + OH^-$$

$$SO_2 + H_2O = H^+ + HSO_3^-$$

$$2\ HS^- + 2\ HSO_3^- + O_2 = 2\ S_2O_3^{--} + H_2O$$

$$S_2O_3^{--} + H^+ = HSO_3^- + S$$

$$2\ HSO_3^- + O_2 = 2\ SO_4^{--} + H^+$$

The overall reaction can be expressed by the following simplification (note that this equation does not balance):

$$FeS + SO_2 + O_2 = Fe^{++} + SO_4^{--} + H^+$$

These reactions result in the formation of complex mixtures of iron sulphides, Fe_xS_y, together with $FeSO_4$, $Fe(OH)_2$ and Na_2SO_4. The key controlling factors are maintenance of the correct pH, the correct Na_2S concentration, and the buffering action of the $Fe(OH)_2$.

The main regeneration reaction is the thermal decomposition of the iron sulphides to give the sorbent FeS and sulphur; e.g. for FeS_2:

$$FeS_2 = FeS + S$$

Sodium sulphate is crystallised out from the solution, and the solid is sent to the Calciner where it is reduced by carbon to the sulphide. The ferrous sulphate is subsequently reduced in solution:

$$Na_2SO_4 + 2\ C = Na_2S + 2\ CO_2$$

$$FeSO_4 + Na_2S = FeS + Na_2SO_4$$

There are some losses of iron from the system, and these are made up by feeding pyrites, FeS_2, to the Calciner.

CATEGORY S51

Process Code S51.1 – Active Carbon Adsorption Process

Outline of Process [6, 137]: A simplified block diagram of the process is presented in Figure 2.21. Gas from the Boiler at 120–130°C flows horizontally through the Adsorber containing a moving bed of either activated carbon pellets sized 13 mm or of coke. The carbon adsorbs sulphur dioxide and water, which react with oxygen in the gas to form sulphuric acid. It also adsorbs nitrogen dioxide and acid halides. There is a temperature rise of 15–20°C across the Adsorber. The carbon bed acts as a panel bed filter and captures particulates from the gas. The cleaned gas then exhausts to stack. The carbon is screened to remove captured particulates, and then moves into the Regenerator, where it is heated in an inert atmosphere to 400–600°C by hot gas or by contact with hot sand. The sulphuric acid is decomposed, and the SO_3 released is reduced by carbon to SO_2 which can be used for manufacturing sulphuric acid for sale, or elemental sulphur for either sale or disposal. The adsorbed nitrogen dioxide reacts with carbon in the Regenerator to release elemental nitrogen and carbon dioxide, and the adsorbed acid halides are desorbed. The acid halides in the Regenerator off-gas are removed before treatment of the SO_2. The carbon lost in the reduction reactions is made-up by fresh activated carbon.

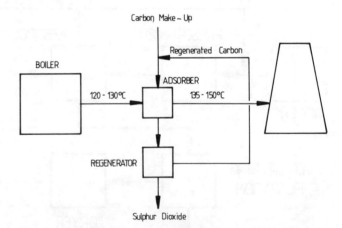

Figure 2.21 Block Diagram of Carbon Adsorption FGD Process

Chemistry of Process [6, 137]: In the Adsorber, SO_2 and water vapour are adsorbed by the activated carbon, and they react on the carbon surface with oxygen in the gas. The reaction can be represented by:

$2 SO_2 + O_2 + 2 H_2O = 2 H_2SO_4$

The sulphuric acid formed remains adsorbed in the pores of the carbon. The reaction is exothermic, and there is a 15–20°C rise in gas temperature. Nitrogen dioxide and acid halides are also adsorbed from the gas.

At the higher temperature (400–650°C) in the Regenerator, acid halides are desorbed. The adsorbed sulphuric acid decomposes to sulphur trioxide and water, and the trioxide is reduced by the carbon to form the dioxide:

$H_2SO_4 = SO_3 + H_2O$

$2 SO_3 + C = 2 SO_2 + CO_2$

The carbon also reduces adsorbed NO_2:

$2 NO_2 + 2 C = N_2 + 2 CO_2$

A carbon make-up is required to replace that consumed in the reduction reactions (0.09 kg/kg SO_2 removed, and 0.26 kg/kg NO_2 removed). The loss of carbon during reduction increases porosity, and hence the internal surface of the carbon remaining, which is therefore further activated. However, the increase in porosity weakens the carbon particles, so that there are carbon break-down losses which also have to be replaced, resulting in a carbon make-up rate that is typically 30–100% greater than the theoretical.

Active carbon can also catalyse the reduction of nitric oxide, and a modification of the process can be used for the simultaneous abatement of SO_2 and NO_x. This is described in Section 5.

CATEGORY S52

Process Code S52.1 – Wet Active Carbon Adsorption Process

Outline of Process [157]: A simplified block diagram of one form of the process, the Lurgi Sulfacid process, is presented in Figure 2.22. Gas from the Boiler flows through the Concentrator, where it is cooled by dilute (7%) sulphuric acid from the Adsorber. It then passes through the Adsorber containing active carbon, which adsorbs sulphur dioxide. Water is admitted to the Adsorber intermittently, without interrupting the gas flow, and this removes sulphuric acid formed in the pores of the carbon. The dilute acid formed is concentrated somewhat (to 15%) by evaporation in the Concentrator. In another form of the process, the Hitachi process (see block diagram presented in Figure 2.23), up to five Adsorbers are operated in parallel: four on stream at any time, and one being washed with dilute acid; a side stream of the acid is concentrated in a submerged combustion evaporation Concentrator.

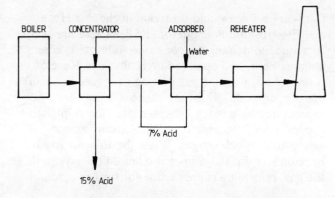

Figure 2.22 Block Diagram of Wet Active Carbon Adsorption (Lurgi Sulfacid) FGD Process

Chemistry of Process [6, 137]: In the Adsorber, SO_2 and water vapour are adsorbed by the activated carbon, and they react on the carbon surface with oxygen in the gas. The reaction can be represented by:

$2 SO_2 + O_2 + 2 H_2O = 2 H_2SO_4$

Washing with water or dilute acid removes the sulphuric acid from the carbon. The adsorption and stripping occur at relatively low temperatures, and the gas has to be reheated. There is not the loss of carbon associated with the dry active carbon adsorption process (see Process Code S51.1).

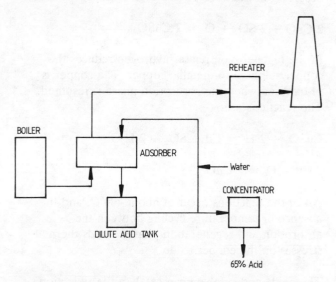

Figure 2.23 Block Diagram of Wet Active Carbon Adsorption (Hitachi) FGD Process

Process Code S52.2 – Copper Oxide Process

Outline of Process [6, 137]: A simplified block diagram of the process is presented in Figure 2.24. The process operates at a temperature of about 400°C, and the Reactors are therefore located after the Boiler

27

Convection Passes, and upstream of the Air Heater and Electrostatic Precipitator (ESP). There are two Reactors, containing copper oxide supported on an alumina base, and operated cyclically: one Reactor is on stream, with the gas flowing over (not through) the copper oxide, whilst the other is being regenerated by passing hydrogen over the sulphated copper oxide, reducing it to metallic copper and giving an SO_2-rich off-gas. When the Reactor is put back on stream, the copper is oxidised by oxygen in the gas, reforming copper oxide for further reaction.

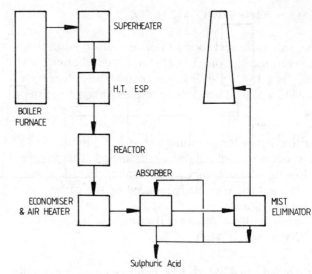

Figure 2.25   Block Diagram of Catalytic Oxidation FGD Process

Electrostatic Precipitator (H.T. ESP) and then through the Reactor containing a bed of vanadium pentoxide which catalyses the oxidation of SO_2 to SO_3. The gas leaving the Reactor is cooled to not less than about 230°C by passage through the Economiser and Air Heater, and is then scrubbed in the Absorber with dilute acid at about 110°C before being exhausted via a Mist Eliminator to the stack. The SO_3 forms sulphuric acid, and this is continuously withdrawn.

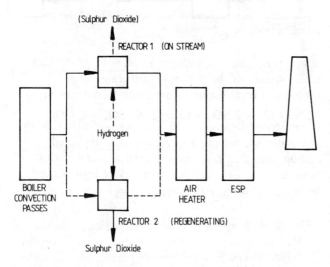

Figure 2.24   Block Diagram of Copper Oxide FGD Process

Chemistry of Process [6, 122, 137]: The reaction occurring in the absorption phase of the cycle is:

$2\ CuO + 2\ SO_2 + O_2 = 2\ CuSO_4$

In the regeneration phase, hydrogen reduces the copper sulphate to metallic copper. The copper is re-oxidised when the absorption phase is resumed. The reactions are:

$CuSO_4 + 2\ H_2 = Cu + SO_2 + 2\ H_2O$

$2\ Cu + O_2 = 2\ CuO$

All of the reactions occur at about 400°C, and the absence of temperature cycling between the absorption and regeneration phases avoids thermal stresses on the copper oxide.

The oxide and sulphate can catalyse the reduction of NO_x with ammonia, so that a modification of the process can be used for the simultaneous abatement of SO_2 and NO_x. This is described in Section 5.

Process Code S52.3 – Catalytic Oxidation Process

Outline of Process [157]: A simplified block diagram of the process is presented in Figure 2.25. Gas at about 500°C is passed through a High-Temperature

Chemistry of Process: The oxidation of SO_2 by oxygen in the flue gas in the Reactor is catalysed by vanadium pentoxide, as in the familiar contact process for the manufacture of the acid. Subsequent absorption in dilute acid forms sulphuric acid. The reactions are simply represented by:

Oxidation: $2\ SO_2 + O_2 = 2\ SO_3$

Acid formation: $SO_3 + H_2O = H_2SO_4$

CATEGORY S61

Process Code S61.1 – Hydrated Lime Injection Process

Outline of Process [137]: A simplified block diagram of the process is presented in Figure 2.26. Hydrated lime is injected into the gas and absorbs sulphur dioxide to form calcium sulphite, which is removed, together with unreacted lime and flyash, in the Baghouse or Electrostatic Precipitator. In the usual form of the process, the lime is injected into the gas upstream of the Economiser (Option A). A recent development [291] is to inject the lime downstream of the air heater (Option B). Option A is also used with magnesia or magnesium hydroxide to reduce SO_3 concentrations, and hence acid dewpoint temperatures.

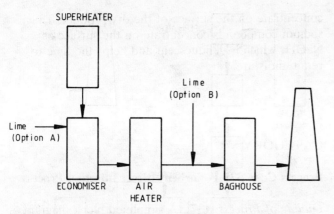

Figure 2.26 Block Diagram of Hydrated Lime Injection FGD Process

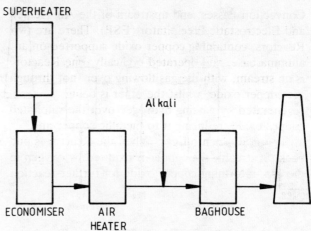

Figure 2.27 Block Diagram of Alkali Injection FGD Process

Chemistry of Process: The heterogeneous gas-solids reaction between hydrated lime and sulphur dioxide gives calcium sulphite, and is represented as:

$$Ca(OH)_2 + SO_2 = CaSO_3 + H_2O$$

A large excess of lime over the stoichiometric quantity is needed for efficient sulphur capture. The lime utilisation is improved by adoption of Option B–injection into the gas downstream of the Air Heater [291].

The improvement obtained with Option B is greatest when the gas temperature at the point of injection is only 75–80°C (i.e. only slightly above the water dewpoint) and when the hydrate contains 3–4% moisture; the moisture content is critical.

For Option B the lime should be 'dry hydrated', i.e. hydrated with the theoretical quantity of water to give a dry product. This has a different morphology from the lime prepared by 'wet slaking' with excess water: it contains more lattice defects which increase its reactivity. The actual form depends on the water temperature and on the additives in the water, which concentrate on the surface of the dry product. Thus, sodium compounds concentrate on the surface as NaOH, which is deliquescent and helps the lime to retain moisture.

Process Code S61.2–Alkali Dry Injection Process

Outline of Process [95]: A block diagram of the process is presented in Figure 2.27. Powdered alkali (sodium carbonate or bicarbonate) is injected into the gas downstream of the Air Heater. The temperature should be greater than 135°C when sodium bicarbonate is used. The reaction occurs in the duct and in the Baghouse used to remove the solids from the gas, before the gas is exhausted to stack.

Chemistry of Process [95]: The alkali reacts with sulphur dioxide and oxygen in the gas to form sodium sulphate:

$$2\ Na_2CO_3 + 2\ SO_2 + O_2 = 2\ Na_2SO_4 + 2\ CO_2$$

If sodium bicarbonate is used, it first decomposes to form sodium carbonate:

$$2\ NaHCO_3 = Na_2CO_3 + H_2O + CO_2$$

This decomposition is slow at temperatures below 135°C, giving less time for the subsequent absorption of SO_2. However, at higher temperatures, nahcolite (a naturally-occurring sodium bicarbonate found in the US) is a more effective injection reagent than sodium carbonate. This appears to be due to the porosity generated by the decomposition reaction. The effectiveness of the reagents is increased by reduction in particle size, e.g. to below 50 micron.

CATEGORY S62

Process Code S62.1–Lurgi Circulating Fluidised Bed (CFB) Lime Absorber Process

Outline of Process [293]: A simplified block diagram of the process is presented in Figure 2.28. Gas from the Boiler enters the base of the Circulating Fluidised Bed (CFB) Absorber via a venturi reactor, where it is intimately mixed with the fine-grained absorbent. The gas (at a velocity of 2–8 m/s) and solids pass up the Absorber and leave from the top to pass through a Cyclone and Electrostatic Precipitator (ESP), in which solids are removed from the gas. The gas is then exhausted via an induced draught fan to stack. The absorbent is hydrated lime which is fed pneumatically as a dry powder, and/or pumped as an aqueous slurry. The Absorber temperature is maintained at just above the water dewpoint temperature by controlling the proportions of slurry and dry powder. Solids separated by the Cyclone and ESP are recycled to the CFB Absorber, but a part of the separated solids is rejected for disposal.

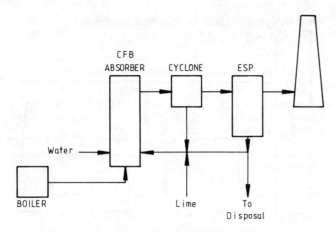

Figure 2.28 Block Diagram of Lurgi CFB Absorber FGD Process

concentrate on the surface of the dry product. Thus, sodium compounds concentrate on the surface as NaOH, which is deliquescent and helps the lime to retain moisture.

CATEGORY S71

Process Code S71.1 – Sorbent Direct Injection Process

Outline of Process [137]: A simplified block diagram of the process is presented in Figure 2.29. The finely-ground sorbent (usually limestone, but hydrated lime, soda ash or sodium bicarbonate can also be used) is injected into the Boiler Furnace with the fuel. The sorbent calcines, losing CO_2, and absorbs SO_2 to form sulphites and sulphates. When the process is applied to systems using modified burners for NO_x abatement (see Sections 3 and 5) the sorbent can also react with H_2S in the reducing regions of the flame to form the sulphide, some of which subsequently oxidises to the sulphates. The reacted and unreacted sorbent is removed, with flyash, in the Electrostatic Precipitator (ESP) or Baghouse.

Chemistry of Process: The heterogeneous gas-solids reaction between hydrated lime and sulphur dioxide gives calcium sulphite, and is represented as:

$$Ca(OH)_2 + SO_2 = CaSO_3 + H_2O$$

With a residence time of solids in circulation of up to 20 minutes, sulphur capture of over 95% can be achieved with an input calcium to sulphur molar ratio of about 1.7. The best performance is to be expected when the reaction temperature is only 75–80°C (i.e. only slightly above the water dewpoint) and when the hydrate contains 3–4% moisture; the moisture content is critical.

The lime should be 'dry hydrated', i.e. hydrated with the theoretical quantity of water to give a dry product. This has a different morphology from the lime prepared by 'wet slaking' with excess water: it contains more lattice defects which increase its reactivity. The actual form depends on the water temperature and on the additives in the water, which

Chemistry of Process [122]: Limestone is the sorbent usually used, and in the furnace it undergoes: calcination; sulphite formation; sulphate formation; and (with burners modified for NO_x abatement) sulphide formation:

Calcination: $\qquad CaCO_3 = CaO + CO_2$

Sulphite formation: $\quad CaCO_3 + SO_2 = CaSO_3 + CO_2$

$$CaO + SO_2 = CaSO_3$$

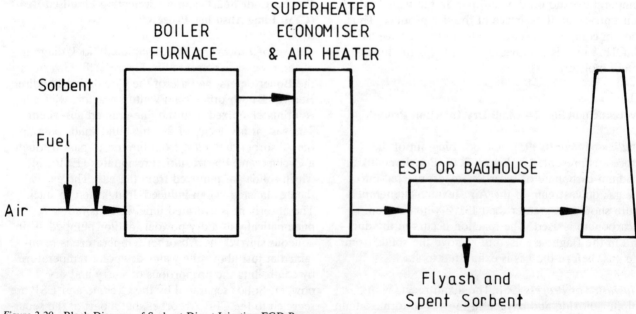

Figure 2.29 Block Diagram of Sorbent Direct Injection FGD Process

Sulphate formation:
$$2\ CaCO_3 + 2\ SO_2 + O_2 = 2\ CaSO_4 + 2\ CO_2$$
$$2\ CaO + 2\ SO_2 + O_2 = 2\ CaSO_4$$
$$2\ CaSO_3 + O_2 = 2\ CaSO_4$$
$$CaS + 2\ O_2 = CaSO_4$$

Sulphide formation:
$$CaCO_3 + H_2S = CaS + H_2O + CO_2$$
$$CaO + H_2S = CaS + H_2O$$
$$4\ CaSO_3 = 3\ CaSO_4 + CaS$$

Similar reactions occur with the other sorbents that can be used.

A large excess of sorbent is usually required because:

- The flame temperature is in the region of 1700–2000°C, i.e. well above the calcium sulphate decomposition temperature of 1200°C or calcium sulphite temperature of 450°C.

- Calcination at the flame temperature results in formation of an inactive ('dead-burned') lime resulting from melting of a calcium carbonate/calcium oxide eutectic. This gives a non-porous structure which inhibits diffusion of SO_2 into the solid particles.

In consequence the sulphur capture is less than 50% even with twice the stoichiometric ratio of limestone. However, with the burners modified for NO_x abatement, the peak flame temperatures are generally lower and may be located away from the main sulphur capture zone. Furthermore, the possibility exists of sulphur capture by sulphide formation. Such burners therefore allow the attainment of more efficient sulphur capture.

2.3 General Appraisal of Processes

This Section presents available information on the characteristics of the FGD process types: status, applicability, space and typical land area requirements, fresh water and water treatment requirements, reagent consumption, end-product or waste materials disposal requirements, typical power consumptions and reductions in combustion plant efficiency, operating experience and process developments.

Where available, published data are presented, but in some instances the information on the process characteristics listed above is also (or alternatively) derived from published estimates for specific cases (all at 100% full load), referred to as Cases 1 to 6.

The background data for these estimates are presented in Table 2.2.

Table 2.2 Background Data on Published Estimates

Case	1	2	3	4	5	6
Reference	95	95	6	277	280	*
MWe	2×500	2×500	800	3×660	500	2000
Coal: Sulphur wt.%	4.0	0.48	3.6	2.0	4.0	2.0
Chlorine wt.%	0.1	0.07	0.1	0.4	–	–

* Davy McKee promotional literature
– Indicates data unavailable

The system used to classify flue gas desulphurisation processes is illustrated in Figure 2.1; the codes of processes dealt with in this Section are listed in Table 2.1.

CATEGORY S11

The status of Category S11 processes is indicated in Table 2.3, which presents information on the size of plants built, year of completion, and percentage sulphur capture.

Process Code S11.1–Sea Water Scrubbing Process

Applicability: This process type is suitable for retrofit and new build applications. It has also been applied in the UK using water taken from and discharged to the River Thames [156]. The capital costs are comparable with those for process type S11.2, but its operating costs are considerably lower [IHI promotional literature].

Typical land area requirements: ***

Typical fresh water and water treatment requirements: The Fläkt-Hydro plants are required to treat flue gas from boilers burning 3% sulphur oil, and a liquid/gas ratio of about 8 litre/Nm^3 is used [651].

The water in the Electrolytic Zinc Co. of Australia Ltd. (Risdon, Tasmania) plants is used [667] to remove 99% of the input SO_2 (1.5 vol. %) from the gas with a liquid/gas ratio of 27 litre/Nm^3.

Consumption of reagents: The CEGB plants neutralised the waste water with chalk or with waste sludge from water softening plant [156].

The Fläkt-Hydro plants are located at the Porsgrunn Chemical Plant of Norsk Hydro, and the effluent is neutralised with wastes from a magnesium production plant, and neutralised with waste $Ca(OCl)_2$ [651].

The Risdon plants return the effluent to the sea at a pH of 2.5.

End-product or waste disposal requirements: The CEGB plants used river water and were required to discharge water of suitably low sulphur compounds concentration [156]; the concentration was below the solubility limits.

Typical pressure losses: At the Fläkt-Hydro plants: 25–35 mbar [651].

Typical power requirements: For the Fläkt-Hydro plants, 1.1–1.6% of the equivalent power generation.

Typical reductions in combustion plant efficiency: ***

Operating experience: The CEGB plants [156] and Fläkt-Hydro plants [651] have given some problems with corrosion, but the availability of the Risdon plant [667] is very high.

The process is similar to the 'Inert Gas System' installed by IHI on 15 ships, handling 11,000–36,000 m^3/h of gas (aggregate of 302,000 m^3/h) with 94–99% sulphur capture and 84–92% dust removal [IHI promotional literature].

Process developments: Little development is likely, owing to the limited applicability of the process.

Appraisal:

1.	Information available	1
2.	Process simplicity	3
3.	Operating experience	1
4.	Operating difficulty	1
5.	Loss of power	1*
6.	Reagent requirements	2
7.	Ease of end-product disposal	1
8.	Process applicability	1
	Total	11

*Assumed in absence of data

Process Code S11.2 – Alkali Scrubbing Process

Applicability: This process type is suitable for retrofit and new build applications. Depending upon operating conditions, sulphur capture can exceed 95%.

Table 2.3 Status of Processes Using Non-Regenerable Solution-Based Wet Reagents: Wet End-Products

Code No.	Vendor	No. of Plants	Plant size Range MWe	Total MWe	Dates	% S Capture	Ref.
S11.1	A/S Ardal & Sundall	–	–	–	–	80–90	651
	CEGB	4	120–228	754	1933–62	90–95	156, 650
	El'ytic Zinc Co.	2	–	194*	1948–51	99	651
	Flakt/Hydro	4	10–20	70	1975–	96–99	651
	IHI	4	272–370*	1272*	1972–75	–	**
S11.2	Andersen 2000	170	0.5–50	–	–	95–99	135
	B&W Co.	3	550	1650	–	86.5	135, 125
	FMC	5a	–	–	–	90–95	135
	IHI	65	6–480*	2285*	1968–75	–	**
	Thyssen/CEA	4	125–295	670	–	–	125
	UOP Air Correction Div	2	330–550	880	–	98	125, 655
S11.3	IHI	9	13–217*	1458*	1966–81	–	**
	Simon-Carves	1	17	17	1957	90	156
S11.4	Chiyoda 101	14b	14–1050*	4426*	1972–75	–	95,**
S11.5	Dowa Mining Co.	10	4–150*	761*	1972–80	90–98	95
S11.6	Airpol	70	–	–	–	95–99	135
	Andersen 2000	2	0.5–250	–	–	95–97	135
	FMC	12	265–420c	685c	1979–86	90–95	95c, 135
	Thyssen/CEA	2	85–1167*	1252*	1975–79	–	95,**
	Envirotech/GEESI	1	575	575	1979	90	95, 667
	Showa Denko	5	62–150	416	1973–74	–	95
	Kawasaki/Kureha	5	150–450	1950	1974–77	–	95

* Gas flow, thousand Nm^3/h ** Vendors' promotional literature
a Includes some calcium slurry scrubbing installations
b Total for 9 boilers; remaining 5 installations comprised other types of combustion plant
c For 2 utility installations
– Indicates information unavailable

Typical land area requirements: For a 550 MWe boiler, 0.56% sulphur coal, 98% sulphur capture [655]: 0.4 hectare for FGD plant; 61 hectare pond for waste disposal (estimated 10-year capacity).

Typical fresh water and water treatment requirements: Water requirements are 0.3–0.7 m^3/MWh [135]. For small units, e.g. 12 MWt a water consumption of 69 litres/GJt is estimated [160].

Consumption of reagents: Alkali: sulphur molar ratio of 1.5 [160]

End-product or waste disposal requirements: Some large units oxidise the end-product to form sodium sulphate for marketing; the quantity would be about equivalent to the sulphur captured. Others, including small units, e.g. 12 MWt, dispose of waste material as a sludge (containing sulphate, sulphite, bisulphite and particulates with about 85% water) having a volume of about 33 litres/GJt [160]. This would generally be difficult to dispose of in the UK in an environmentally acceptable manner.

Typical pressure losses: About 15 mbar [655].

Typical power requirements: For small units, e.g. 12 MWt, about 4 kW/MWt [160].

For a 550 MWe boiler, 0.56% sulphur coal, 98% sulphur capture [655]: about 7 kW/MWe.

Typical reductions in combustion plant efficiency: ***

Operating experience: Availability quite high. The main problems appear to be plugging of absorber spray nozzles and scale on absorber walls. Other problems (e.g. with dampers) have been solved [655].

Process developments: ***

Appraisal:

1.	Information available	1
2.	Process simplicity	2
3.	Operating experience	1
4.	Operating difficulty	1
5.	Loss of power	1*
6.	Reagent requirements	1
7.	Ease of end-product disposal	0
8.	Process applicability	2
	Total	9

*Assumed in absence of data

Process Code S11.3 – Ammonia Scrubbing Process

Applicability: This process type is suitable for retrofit and new build applications.

Typical land area requirements: ***

Typical fresh water and water treatment requirements: ***

Consumption of reagents: ***

End-product or waste disposal requirements: ***

Typical pressure losses: ***

Typical power requirements: ***

Typical reductions in combustion plant efficiency: ***

Operating experience: The CEGB Plant [156] operated successfully, but was shut down after about two years because of the difficulty of obtaining by-product ammonia, the impurity of the product and the objectional odour of the plume.

Process developments: The Walther process (Section 2.2 and 2.3) overcomes some of the objections to the simple ammonia scrubbing process.

Appraisal:

1.	Information available	0
2.	Process simplicity	2
3.	Operating experience	0
4.	Operating difficulty	1
5.	Loss of power	1*
6.	Reagent requirements	1
7.	Ease of end-product disposal	1
8.	Process applicability	2
	Total	9

*Assumed in absence of data

Process Code S11.4 – Sulphuric Acid Scrubbing Process

Applicability: This process type is suitable for retrofit and new build applications.

Typical land area requirements: ***

Typical fresh water and water treatment requirements: For a 500 MWe boiler, 1% sulphur oil [172], 0.08 m^3/MWh fresh water.

Consumption of reagents [172]: Limestone (to neutralise the sulphuric acid produced) about equivalent to the sulphur captured; catalyst (ferrous sulphate) about 0.13 kg/MWh.

End-product or waste disposal requirements: ***

Typical pressure losses: 55 mbar [172].

Typical power requirements [172]: For a 250 MWe boiler, 1% sulphur oil, 90% sulphur capture: 2.0% of the gross power produced.

Typical reductions in combustion plant efficiency: ***

Operating experience: A run of 50 days was reported in 1974 [172].

Process developments: ***

Appraisal:

1.	Information available	0
2.	Process simplicity	2
3.	Operating experience	1
4.	Operating difficulty	0*
5.	Loss of power	1*
6.	Reagent requirements	2
7.	Ease of end-product disposal	1
8.	Process applicability	2
	Total	8

*Assumed in absence of data

Process Code S11.5 – Dowa Process

Applicability: This process type is suitable for retrofit and new build applications.

Typical land area requirements: ***

Typical fresh water and water treatment requirements: See Table 2.4 for estimates.

Consumption of reagents: See Table 2.4 for estimates.

End-product or waste disposal requirements: See Table 2.4 for estimates.

Table 2.4 Published Estimated Requirements

Case No. (see Table 2.2)		1
Reference		95
MWe		2×500
Coal: Sulphur	wt. %	4.0
Chlorine	wt. %	0.1
Sulphur capture	%	90
Typical fresh water and water treatment requirements per MWh:		
Raw water		0.37 m³
Steam (28 bar, 300°C)		0.10 tonne
Consumption of reagents, kg/MWh:		
Limestone at a Ca/S molar ratio of 1.0		53.3
Alum make-up at 0.0034 mol/mol limestone		0.8
End-product or waste disposal requirements, m³/MWh:		
Waste gypsum sludge volume		0.09
Fly-ash volume		0.06
Typical pressure losses, mbar:		11
Typical power requirements, MWe:		14.8
Typical reductions in combustion plant efficiency, %:		
Loss in power sent out		4.6

Typical pressure losses: See Table 2.4 for estimates.

Typical power requirements: See Table 2.4 for estimates.

Typical reductions in combustion plant efficiency: See Table 2.4 for estimates.

Operating experience: Applied in nine Japanese installations (acid plants, smelters and one oil-fired boiler), and tested by EPRI at a US coal-fired boiler [95].

Process developments: ***

Appraisal:

1.	Information available	1
2.	Process simplicity	1
3.	Operating experience	1
4.	Operating difficulty	0*
5.	Loss of power	1
6.	Reagent requirements	1
7.	Ease of end-product disposal	1
8.	Process applicability	2
	Total	8

*Assumed in absence of data

Process Code S11.6 – Dual Alkali Process

Applicability: This process type is suitable for retrofit and new build applications.

Typical land area requirements [654]: For a 265 MWe boiler, 3.35% sulphur coal, 90% sulphur capture: 0.29 hectare for the FGD plant.

Typical fresh water and water treatment requirements: See Table 2.5 for estimates.

Consumption of reagents: See Table 2.5 for estimates.

End-product or waste disposal requirements: See Table 2.5 for estimates.

Typical pressure losses: See Table 2.5 for estimates.

Typical power requirements: See Table 2.5 for estimates.

Typical reductions in combustion plant efficiency: See Table 2.5 for estimates.

Operating experience: The system using lime as the regenerating reagent has been applied to about 50 industrial and utility installations in the US and Japan, with many others under construction or planned. The system using limestone has been applied to full-scale installations only in Japan [95].

A summary of problems reported by Battelle in their visit reports [653, 654, 664, 667, 683] is as follows (total number of plants visited, numbers experiencing problems in various plant areas, and those with problems remaining unsolved):

No. visited	5
Total equiv. MWe	1564
Total streams	14

Plant Area	Total	Unsolved	Solved
Prescrubber	1	0	1
Absorber	4	2	2
Demisters	2	1	1
Flues or stack	1	1	0
Dampers	3	3	0
Sorbent preparation	0	0	0
Waste/product system	2	1	1
Other	2	2	0
Totals	15	10	5

Process developments: The system using limestone as the regenerating reagent is now being offered (by FMC only) in the US [95], where there are four installations [125] in the range 265–390 MWe (total 1310 MWe) and three [664] each handling 160,000 Nm³/h.

Appraisal:

1.	Information available	2
2.	Process simplicity	1
3.	Operating experience	1
4.	Operating difficulty	0
5.	Loss of power	1
6.	Reagent requirements	1
7.	Ease of end-product disposal	1
8.	Process applicability	2
	Total	9

CATEGORY S12

The status of Category S12 processes is indicated in Table 2.6, which presents information on the size of plants built, year of completion, and percentage sulphur capture.

Process Code S12.1 – Alkali Scrubbing/Spray Drying Process

Applicability: This process type is suitable for retrofit or new build applications. Depending upon operating conditions, up to 95% sulphur capture can be achieved. However, the reagent is expensive unless naturally-occurring deposits (such as nahcolite in the US) are available; and the waste product is highly water-soluble, making environmentally acceptable disposal difficult.

Typical land area requirements: ***

Typical fresh water and water treatment requirements: ***

Consumption of reagents: Reagent is supplied at up to twice stoichiometric requirements, but the supply rate is usually lower. The Fläkt plant (20% of 550

Table 2.5 Published Estimated Requirements

Case No. (see Table 2.2)		1	1	3
Reference		95	95	6
MWe		2×500	2×500	800
Coal: Sulphur	wt. %	4.0	4.0	3.6
Chlorine	wt. %	0.1	0.1	0.1
Sulphur Capture	%	90	90	95
Regeneration by:		Limestone	Lime	Lime
Typical fresh water and water treatment requirements per MWh:				
Raw water	m³	0.27	0.28	–
Steam (29 bar, 300°C)	tonne	0.10	0.10	–
Consumption of reagents, kg/MWh:				
NaOH		1.7	1.0	1.0
Regenerant		55.9	31.1	28.5
Sludge fixative lime		4.2	4.5	–
End-product or waste disposal requirements:				
Sludge volume	m³/MWh	0.13	0.13	0.09
Fly-ash	wt. %	30	30	–
Moisture	wt. %	30	30	–
Typical pressure losses, mbar:		15	15	–
Typical power requirements, MWe:		14.3	12.9	–
Typical reductions in combustion plant efficiency, %:				
Loss in power sent out		4.5	4.9	–

– Indicates information unavailable

Table 2.6 Status of Processes Using Non-Regenerable Solution-Based Wet Reagents: Dry End-Products

Code No.	Vendor	No. of Plants	Plant size Range MWe	Total MWe	Dates	% S Capture	Ref.
S12.1	Flakt	2	3.5–110	114	1980–81	73–90	655, 671
	United McGill	11a	–	–	–	90	135
	Rockwell/Wheelabrator Frye	1	410	410	1981	70	223
S12.2	Walther	4	64–440	980	1984–88	Over 92	29, 659

a Includes some lime spray dryer plants
– Indicates information unavailable

MWe, 0.4% sulphur coal, 73% sulphur capture) operates with a Na/inlet S molar ratio of 1.2 or Ca/S molar ratio of 1.6 [671].

The Rockwell International/Wheelabrator-Frye plant (410 MWe, 0.87% sulphur coal) achieved 70% sulphur capture with a stoichiometric ratio of 1 mol/mol input S [684].

End-product or waste disposal requirements: ***

Typical pressure losses: 40 mbar [223]

Typical power requirements: For a 410 MWe boiler, 0.87% sulphur coal, about 70% sulphur capture: 0.5–0.6% of generation [684].

Typical reductions in combustion plant efficiency: ***

Operating experience: A summary of problems reported by Battelle in their visit reports [655, 666, 671, 684] is as follows (total number of plants visited, numbers experiencing problems in various plant areas, and those with problems remaining unsolved):

No. visited	2
Total equiv. MWe	520
Total streams	2

Plant area	Number of Problems		
	Total	Unsolved	Solved
Absorber	2	1	1
Flues or stack	0	0	0
Dampers	0	0	0
Sorbent preparation	0	0	0
Waste/product system	0	0	0
Other	1	1	0
Totals	3	2	1

Process developments: ***

Appraisal:

1.	Information available	1
2.	Process simplicity	3
3.	Operating experience	0
4.	Operating difficulty	0
5.	Loss of power	1*
6.	Reagent requirements	1
7.	Ease of end-product disposal	0
8.	Process applicability	2
	Total	8

*Assumed in absence of data

Process Code S12.2 – Walther Process

Applicability: This process type is suitable for retrofit and new build applications.

Typical land area requirements: For handling 50% of the flue gas from a 475 MWe boiler, 1.1% sulphur coal, over 92% sulphur capture, the FGD plant requires 0.2 hectare [659].

Typical fresh water and water treatment requirements: The water requirement is estimated at about 0.17 m^3/MWh [Walther promotional literature].

Consumption of reagents: The consumption of ammonia is about equivalent to the sulphur captured [Walther promotional literature].

End-product or waste disposal requirements: The end product, ammonium sulphate, is potentially marketable as a fertiliser.

Typical pressure losses: 20 mbar can be assumed [Walther promotional literature].

Typical power requirements: ***

Typical reductions in combustion plant efficiency: ***

Operating experience: No operating data are yet available.

Process developments: ***

Appraisal:

1.	Information available	1
2.	Process simplicity	2
3.	Operating experience	0
4.	Operating difficulty	0*
5.	Loss of power	1*
6.	Reagent requirements	1
7.	Ease of end-product disposal	1
8.	Process applicability	2
	Total	8

*Assumed in absence of data

CATEGORY S21

The status of Category S21 processes is indicated in Table 2.7, which presents information on the size of plants built, year of completion, and percentage sulphur capture. The list includes plants operating with limestone and lime slurries, with alkaline fly-ash added in some instances; plants operating with formic, adipic or succinic acid or DBA additives, or with magnesia-based, sodium thiosulphate or calcium chloride additives; and plants operating with and without forced oxidation of the calcium sulphite hemihydrate produced.

Process Code S21.1 – Limestone or Lime Slurry Scrubbing Processes

Applicability: This process type is suitable for retrofit and new build applications. Depending on the operating conditions, 90–95% sulphur capture is readily attained.

Typical land area requirements: Values reported are very variable, probably partly because they do not all encompass areas for raw material and product processing and handling (excluding areas for waste disposal). They are:

Table 2.7 Status of Processes Using Non-Regenerable Slurry-Based Wet Reagents: Wet End-Products

Code No.	Vendor	No. of Plants	Plant size Range MWe	Total MWe	Dates	% S Capture	Ref.
S21.1	AAF	7	65–495	1955	1975–81	–	125, 685
	Babcock-Hitachi	1	250	250	1976	93	678
	Babcock Power Ltd.	23	25–1628*	15,938*	1974–87	90–95	135,**
	B&W Co.	16	55–875	7320	1977–83	90–99	125, 654, 680
	Bahco	3	102a	102a	1969–	98	652
	Bechtel	2	778	1555	1983	–	125
	Bischoff	33	10–1500*	29,727*	1971–88	Over 95	135,**
	Chiyoda 121	7	23–200	700	1978–84	97–99	95, 680
	Combustion Engineering	31	72–890	12020b	1973–81b	75–90b	135
	Deutsche Babcock	31	150–470	8970	1985–88	90–95	29, 135
	Environeering/Riley Stoker	3	180–415	775	–	50–85	125, 662
	ESTS	2	309–600	909	1985–88	–	**
	Fläkt/Babcock	2	96–181	277	1987–88	–	29
	GEESI	24	175–1150	14305	1973–	70–85	125, 675
	IHI	33	10–2400*	13055*	1971–87	90–95	**, 678
	Kawasaki HI	2	156–250	406	1983–84	98	679
	Kobe Steel	1	720*	720.	–	90	684
	Lisop Oy	4	–	–	–	Over 80	135
	Mitsubishi	77	12–2450*	41,060*	1972–87	90–98	135,**
	Peabody P.S.	10	179–625	4525	–	–	125
	Pullman Kellogg	7	400–917	3995	1979–84	92–98	125, 661
	Research Cottrell	27	115–750	10,305	1973–88	95	135
	Saarberg-Holter	25	1–600	5760	1974–88	95	95, 135
		+ 7	4–131*	350	1972–82		
	Steinmuller	25	72–707c	6100c	1981–88c	85–95	135,**
	TVA	1	550	550	1977	80	125, 665
	Texcel	25	5 Min.	–	–	Over 95	135
	Thyssen/CEA	39	10–740	6430d	1975–88d	Over 95	135
	UOP Air Corr'n Div.	6	195–720	2990	–	80–85	125
	Waagner-Biro	3	160–330	690	1986	–	**
	Wheelabrator	1	750	750	–	–	125

* Gas flow, Thousand Nm³/h   **Vendor's promotional literature
a For 1 installation
b For 24 installations [125, 653, 669, 677]
c For 20 installations [29]
d For 24 plants in the range 77–500 MWe [29, 125, 653]
– Indicates information unavailable

Vendor	MWe	Hectare	Reference
B & W	55	0.22a	680
	364	0.21b	685
	618	4.0c	654
Environeering	2×180	0.40	662
GEESI	600	1.0	670
MHI/Thyssen	185	0.8	659
Peabody PS	400–447	0.46	685
UOP	532	0.19	656

Notes a: for FGD system; b: for absorbers and thickeners: c: for FGD system, sludge treatment and lime handling.

Typical fresh water and water treatment requirements: Water requirements are typically 0.1–0.4 m^3/MWh; the quantity appears to depend on the SO$_2$ concentration, though one vendor (Research-Cottrell) is reported to require only 0.03–0.05 m^3/MWh for inlet SO$_2$ concentrations as high as 10,000 mg/Nm^3 [135]. Cooling and reheating by gas-gas heat exchange reduces the requirement for water.

Waste water from systems with forced oxidation in Japan has 30–100 ppm chemical oxygen demand (COD), and contains 500–600 ppm fluorides, up to 500 ppm iron, 1000 ppm magnesium and 8000 ppm chlorides; it has to be treated with lime to precipitate the heavy metals and fluorides, followed by ion exchange to reduce the COD [184].

Further information in the form of published estimates is presented in Table 2.8 for limestone as absorbent, and Table 2.9 for lime as absorbent.

Consumption of reagents: The requirements for lime or limestone can be reduced if high-calcium fly-ash can be used [175], or by the use of fine grinding of the limestone, e.g. to about 95% finer than 874 micron [215]. Requirements for other reagents, e.g. organic acid or magnesia-based additives, depend on details of process variations, and are dealt with in Section 2.5.

End-product or waste disposal requirements: Estimates are presented in Table 2.8 for limestone as absorbent, and in Table 2.9 for lime as absorbent.

Typical pressure losses: Gas pressure drops across the FGD system are typically 15–20 mbar for units with spray towers; the Saarberg-Holter (25–34 mbar) and Chiyoda Thoroughbred 121 (40–50 mbar) pressure drops are higher, however. The higher pressure loss with the Saarberg-Holter process results from the

Table 2.8 Published Estimated Requirements Limestone Absorbent

Case No.		1a	1b	1c	3a	3b
Reference		95	95	95	6	6
MWe		2×500	2×500	2×500	800	800
Coal: Sulphur	wt. %	4.0	4.0	4.0	3.6	3.6
Chlorine	wt. %	0.1	0.1	0.1	0.1	0.1
Forced Oxidation		No	Yes	Yes*	No	Yes*
Sulphur capture	%	90	90	90	90	95
Molar ratio Ca/captured S		1.15	1.15	1.01	–	–
Typical fresh water and water treatment requirements per MWh:						
Cooling Tower blowdown	m^3	0.32	0.42	0.30	–	–
Steam	tonne	0.10	0.10	0.10	–	–
Typical sorbent requirements, kg/MWh:						
Limestone		60.8	60.8	53.7	44.9	45.0
Fixative lime		2.3	Nil	Nil	–	Nil
Formic acid		Nil	Nil	0.06	Nil	Nil
End-product or waste disposal requirements:						
Volume	m^3/MWh	0.13	0.15	0.17	0.11	0.13
Fly-ash	wt. %	30	30	30	–	–
Water	wt. %	30	20	20	–	–
Typical pressure losses, mbar:						
		15	15	40		
Typical power requirements, MWe:						
		20.6	20.5	18.8	–	–
Typical reductions in combustion plant efficiency, %:						
Loss of power sent out		5.2	4.9	5.1	–	–

* Chiyoda Thoroughbred 121 process
– Indicates information unavailable

use of venturi absorbers [192]. The use of venturi prescrubbers also increases the total pressure drop of plant with spray tower absorbers to about 50 mbar [175].

Further information in the form of published estimates is presented in Table 2.8 for limestone as absorbent, and Table 2.9 for lime as absorbent.

Typical power requirements: Estimates are presented in Table 2.8 for limestone as absorbent, and in Table 2.9 for lime as absorbent.

Typical reductions in combustion plant efficiency: Estimates are presented in Table 2.8 for limestone as absorbent, and in Table 2.9 for lime as absorbent.

Operating experience: This is the most extensively adopted FGD system, with over 400 installations worldwide indicated in Table 2.7 above. Early operating difficulties resulted from poor design and inadequate appreciation of details of the process chemistry; these have now been largely overcome, and operating reliability is high. At the TVA Paradise station, the major problems concerned: plugging of mist eliminators; collapse of eliminators due to temperature excursions; and heat exchanger failures due to poor design. Other problems included corrosion of outlet ducts and other components, and the use of unsuitable types of component. All of these problems were overcome by modifications to the plant [230].

Table 2.9 Published Estimated Requirements Lime Absorbent

Case No.		1d	1e	3c
Reference		95	95	6
MWe		2×500	2×500	800
Coal: Sulphur	wt.%	4.0	4.0	3.6
Chlorine	wt.%	0.1	0.1	0.1
Forced Oxidation		No	Yes*	No
Sulphur capture	%	90	90	95
Molar ratio Ca/captured S		1.10	1.02	–
Typical fresh water and water treatment requirements per MWh:				
Raw water	m³	0.12	0.10	–
Cooling tower blowdown	m³	0.18	0.16	–
Steam (29 bar, 300°C)	tonne	0.10	0.10	–
Typical sorbent requirements, kg/MWh:				
Lime		34.3	31.8	22.8
End-product or waste disposal requirements:				
Volume	m³/MWh	0.12	0.15	0.11
Fly-ash	wt.%	30	30	–
Water	wt.%	30	20	–
Typical pressure losses, mbar:		15	25	–
Typical power requirements, MWe:		17.4	17.0	–
Typical reductions in combustion plant efficiency, %:				
Loss of power sent out	%	4.9	4.9	–

* Saarberg-Holter process
– Indicates information unavailable

The older units in the US are operated without forced oxidation of the calcium sulphite, but most units in Japan and the FRG, and an increasing number of new units in the US, have forced oxidation in order to reduce end-product processing and disposal costs, and the environmental impact of waste disposal; for example, forced oxidation results in 40% reduction in area for disposal compared with ponding of natural-oxidation sludge [184]. However, retrofitting of forced oxidation to existing lime or limestone slurry FGD plant would incur high capital cost, as it involves changes in the process chemistry, water balance, pH control and product purity. Examples of FGD plant with forced oxidation are offered by Chiyoda, Combustion Engineering, GEESI, IHI, Pullman-Kellogg, Research Cottrell, Saarberg-Holter.

Testing of the Chiyoda Thoroughbred 121 system on a 23 MW pilot plant in the US demonstrated efficient and reliable operation when applied to a coal-fired boiler; four further units (including two units for 200 MWe utility boilers) are under construction [95]. Two 200 MWe units tested for over one year in Japan were found to give essentially 100% reliability [188].

Table 2.10 Summary of Problems Reported by Battelle

Absorbent	LST	LST	LST SH	L	L	L CT121
Forced Oxidation	No	Yes	Yes	No	Yes	Yes
No. of plants visited	12	2	1	15	7	1
Total equiv. MWe	7634	591	177	7687	6587	87
Total No. of streams	48	3	1	51	19	1
No. of problems in:						
Prescrubber	0	0	0	4	2	0
Unsolved	0	0	0	3	1	0
Absorber	3	0	0	9	3	1
Unsolved	2	0	0	8	3	0
Demister	2	0	0	8	3	0
Unsolved	1	0	0	6	3	0
Flue or stack	8	0	0	4	4	0
Unsolved	5	0	0	4	4	0
Damper	5	0	0	3	3	0
Unsolved	5	0	0	3	3	0
Sorbent prep.	2	1	1	3	1	0
Unsolved	2	1	1	2	1	0
Waste/prod. system	2	0	0	6	3	0
Unsolved	2	0	0	5	2	0
Other areas	3	1	1	9	4	0
Unsolved	3	1	1	8	3	0
Total of problems	25	2	2	46	23	1
Total unsolved	20	2	2	39	20	0

LST: Limestone
L: Lime
SH: Saarberg-Holter
CT121: Chiyoda Thoroughbred 121

A summary of problems reported by Battelle in their visit reports [650, 652–657, 659, 661–665, 668, 670, 674–682, 684, 685] is shown in Table 2.10 (total number of plants visited, numbers experiencing problems in various plant areas, and those with problems remaining unsolved).

Process developments: The use of additives, e.g. formic acid, adipic acid and 'dibasic acid, DBA' (a mixture of glutaric, adipic and succinic acids) allows improved performance and reduced scale formation with negligible increase in capital cost [95, 182] and without significant environmental impact [186]; the liquid/gas ratio needed in the absorber can be greatly reduced, e.g. by 50%, by the addition of 2 g/l of formic acid to the slurry [187]. Alternative additives are: sodium thiosulphate at a concentration of 100–700 ppm in the slurry which reduces scale-formation and produces a sulphite sludge containing 85% solids [184]; magnesia (Thiosorbic lime process [181]) added at the rate of 4–8% by weight of the solids. The magnesia forms soluble magnesium sulphite which is then the principal absorbing reagent, forming magnesium bisulphite and sulphate which react with lime to regenerate the magnesium sulphite.

Simultaneous absorption and forced oxidation, as in e.g. the Chiyoda Thoroughbred 121 process, improves sulphur capture by removing sulphite and bisulphite from the system. The development of spray towers in single units for 1000 MWe installations is held to be feasible [173]. The pressure drop associated with gas-gas heat exchangers (about 10 mbar), or the expenditure of fuel for reheating, can be avoided by exhausting the purified gas to atmosphere in the power station cooling tower with a suitable acid-resisting lining [187].

Appraisal:

1. Information available 2
2. Process simplicity 2
3. Operating experience 3
4. Operating difficulty 0
5. Loss of power 1
6. Reagent requirements 1
7. Ease of end-product disposal 1
8. Process applicability 2
   Total 12

CATEGORY S22

The status of Category S22 processes is indicated in Table 2.11, which presents information on the size of plants built, year of completion, and percentage sulphur capture.

Process Code S22.1–Lime Slurry Scrubbing/Spray Drying Process

Applicability: This process type is suitable for retrofit and new build applications. All of the utility installations in the US have been for combustion systems firing low-sulphur coal, for which this process type is more economical than wet slurry scrubbing systems [210], though the process is technically capable of application to high-sulphur coals [222] and has been applied to industrial boilers firing coal of up to 5% [198]. Depending upon operating conditions, any sulphur capture up to about 90% is readily attained. The process is capable of dealing with variable loads [197].

Typical land area requirements: Values depend upon the size of the plant as follows:

Table 2.11 Status of Processes Using Non-Regenerable Slurry-Based Wet Reagents: Dry End-Products

Code No.	Vendor	No. of Plants	Plant size Range MWe	Total MWe	Dates	% S Capture	Ref.
S22.1	B&W	2	447–575	1017	1982–85	85–87	95, 222
	Combustion Engineering	6	100–850	–	–	Over 90	135
	DB Gas Cleaning	10	–	–	–	95 Max	135
	Ecolaire	1	79*	79*	1981	65–90	682
	Fläkt (Drypac)	23	133–3,152a	12,188a	1982–88	70–90	29,** 125, 135, 655, 671, 681
	Fläkt/Niro/Joy	31	43–4300a	33,294a	1980–89	Over 95	135,**
	GEESI	1	44	44	1980	80	95
	Mikropul	2	3–100	–	–	90–95	135
	Research-Cottrell	9	10–60	–	–	90	135
	Rockwell/Wheelabrator	2	276–440	716	1981–85	70	95
		+2	50–113*	163*	1980–83	70	222, 654, 676
	Texcel	5	4 minimum	–	–	Over 95	135

* Steam production, tonne/h
** Vendor's promotional literature
– Indicates information unavailable
a Gas flow, thousands Nm³/h

Vendor	MWe	Steam tonne/h	Gas Nm³/h	Hectare	Reference
Ecolaire		79	93,000	0.06	682
Flakt		106	128,000	0.09	685
Joy/Niro	440		2,507,000	0.49	678
	316		1,539,000	0.37	684
		181	175,000	0.10	683
Research-Cottrell	60		243,000	0.08	684

Typical fresh water and water treatment requirements: Water requirements depend upon the gas temperature and other site-specific factors, and can be in the range of 0.02–0.2 m³/MWh [95, 135]. Water for slaking should be of higher quality than needed for limestone slurry processes, and dilution water for slurry preparation should be of low hardness to avoid blinding of the lime surface [222].

Further information is presented in Table 2.12 in the form of published estimates.

Consumption of reagents: The consumption of lime is between 1.3 and 1.85 times stoichiometric. However, in some instances the ash of the coal provides a significant amount of lime, and the fresh added lime requirement is thereby reduced. The lime is slaked on site; high-calcium, soft-burned, low-inerts pebble lime should be used; spontaneous slaking during storage should be avoided; and care is needed to produce a reactive, finely sized material.

Further information is presented in Table 2.12 in the form of published estimates.

End-product or waste disposal requirements: Published estimates are presented in Table 2.12.

Typical pressure losses: Observed pressure losses are in the range 18 mbar [678] to 37 mbar [654].

Further information is presented in Table 2.12 in the form of published estimates.

Typical power requirements: Measurements of power consumption for a number of installations in the range 60–440 MWe, firing low-sulphur coals (0.3–0.8% S) were equivalent to 0.5–0.9% of the gross power generated [678, 684].

Further information is presented in Table 2.12 in the form of published estimates.

Typical reductions in combustion plant efficiency: Published estimates are presented in Table 2.12.

Operating experience [222]: Early problems included: solids deposition from incomplete drying due to design faults, and to excessive water content of the slurry (ameliorated by recycle of the dry product, which reduces the water content of the slurry and increases the surface area); mechanical problems with rotary atomisers and plugging of nozzle atomisers; lime slaking and slurry piping problems. These problems can be overcome by better design and operating practices.

Corrosion due to condensation is not common but can occur at part-load conditions, and insulation of potentially cold spots has been found desirable; it is advantageous to reheat the gas to avoid such problems. Particulate removal (electrostatic precipitators, or the more commonly used fabric filters, both of which give enhanced sulphur capture performance [211, 212]) have not generally given problems.

A summary of problems reported by Battelle in their visit reports [652, 654, 663, 668, 670, 671, 676, 678, 680–685] is as follows (total number of plants visited, numbers experiencing problems in various plant areas, and those with problems remaining unsolved):

Table 2.12 Published Estimated Requirements

Case No. (Table 2.2)		2
Reference		95
MWe		2×500
Coal: Sulphur	wt. %	0.48
Chlorine	wt. %	0.07
Sulphur capture	%	70
Typical fresh water and water treatment requirements:		
Raw water	m³/MWh	0.19
Consumption of reagents:		
Pebble lime, Ca/S molar ratio 0.9	kg/MWh	4.9
End-product or waste disposal requirements:		
Waste volume (20% moisture, 63% fly-ash)	m³/MWh	0.04
Typical pressure loss, mbar:		30
Typical power requirements MW:		12.8
Typical reduction in combustion plant efficiency:		
Loss of power sent out	%	1.2

No. visited	14
Total equiv. MWe	1733
Total streams	22

Problem area	Number of Problems		
	Total	Unsolved	Solved
Absorber	9	8	1
Flue or stack	0	0	0
Dampers	0	0	0
Sorbent preparation	3	2	1
Waste/product system	7	6	1
Other areas	3	3	0
Totals	22	19	3

Process developments: A development is the 'dry scrubbing' process in which the exit temperature of gas from the scrubber is as close as possible to the adiabatic saturation temperature; the 'dry' particles of the end-product then retain sufficient surface moisture to allow rapid absorption of SO_2 to continue [210] as in the processes based on dry injection into the flue gas (Option B described in Section 2.2).

Other developments, still in the laboratory or pilot plant stage, include: 'in-duct' scrubbing (i.e. spraying slurry into the horizontal duct, obviating the need for a separate spray-dryer vessel) [190]; use of calcium chloride as an activation agent, particularly for limestone slurries [195]; and the use of magnesia-based slurries [196].

Appraisal:

1.	Information available	2
2.	Process simplicity	3
3.	Operating experience	2
4.	Operating difficulty	0
5.	Loss of power	2
6.	Reagent requirements	1
7.	Ease of end-product disposal	1
8.	Process applicability	2
	Total	13

CATEGORY S31

The status of Category S31 processes is indicated in Table 2.13, which presents information on the size of the plants built, year of completion, and percentage sulphur capture.

Process Code S31.1 – Wellman-Lord Process

Applicability: This process type is suitable for new build and retrofit applications. It has been installed on utility and industrial coal-fired and oil-fired boilers as well as on Claus plant and sulphuric acid plant. Depending upon operating conditions, it is capable of giving 90–99% sulphur capture. It is capable of dealing with a variable load and SO_2 input.

Land area requirements: Data reported by Battelle are presented in Table 2.14. Published estimates are presented in Table 2.15.

Typical fresh water and water treatment requirements: Published estimates are presented in Table 2.15.

Consumption of reagents: Data reported by Battelle are presented in Table 2.14. Published estimates are presented in Table 2.15.

End-product or waste disposal requirements: Published estimates are presented in Table 2.15.

Typical power requirements: Data reported by Battelle are presented in Table 2.14. Published estimates are presented in Table 2.15.

Typical reductions in combustion plant efficiency: Published estimates are presented in Table 2.15.

Operating experience: The operational reliability is quoted as being in the range 95–98%; one plant showed a reliability of 96% for the absorption area and 92% for the regeneration area. Prescrubbing is essential for coal-fired plants, to remove fly-ash, HCl, SO_3 and trace impurities [276].

For plants in which elemental sulphur is recovered, the sulphur condenser was a source of trouble: mainly sulphur deposition, leaks and valve failures. Another plant gave trouble from acidic conditions and unreliable steam supply [146].

A summary of problems reported by Battelle in their visit reports [658, 659] is given below (total number of plants visited, numbers experiencing problems in various plant areas, and those with problems remaining unsolved).

No. visited		2
Total equiv. MWe		1958
Total number of streams		7

Problem area	Number of Problems		
	Total	Unsolved	Solved
Prescrubber	1	1	0
Absorber	1	1	0

Table 2.13 Status of Processes Using Regenerable Solution-Based Wet Reagents: Thermal Regeneration

Code No.	Vendor	No. of Plants	Plant size Range MWe	Total MWe	Dates	% S Capture	Ref.
S31.1	Davy McKee	35	17–4640*	22276*	1970–88	90–99%	**, 266
S31.2	Fläkt R'ch/Boliden AB	3	0.1–5*	–	1976–77	–	95
S31.3	Air Products/Catalytic/IFP	2 (Pilot plant)	2–25	27	–	95%	6, 137

* Thousand m³/h
** Vendors' promotional literature
– Indicates information unavailable

Table 2.14 Requirements Reported by Battelle

Reference	658	658	659
MWe	711	1068	3×60
Fuel S content, %	0.8	0.8	5.8a
Sulphur capture, %	90	84	90
End-product	S (by Allied Process) or acid		Sulphuric Acid
Typical land area requirements, hectare:			
Absorbers	0.43	0.58	–
Regeneration	0.21*		–
Crystallisation	0.15*		–
Sulphur production	0.14*		–
Acid production (alternative	0.16*		–
Total area (FGD plant)	1.51–1.53*		2.4
Steam consumption:			
Tonne/tonne S removed	7.8*		–
Tonne/MWh (3.8 bar)	–		0.20–0.23
(42 bar from acid plant)	–		0.06
Liquid/gas ratio, litre/m³:			
Venturi pre-scrubber	0.9*		3.1
Absorber: liquid feed	0.1*		0.06
recirculation	0.24		0.24
Consumption of reagents:			
Soda ash, kg/kg S removed	0.19*		Nil
NaOH, kg/MWh	Nil*		about 4
Natural gas, m³/kg S removed	0.39*		Nil
Pressure drop, mbar	70–90	less than 70	75
Power consumption, MWe:	27	–	11

Note a: Mean of fuels supplied (petroleum coke, refinery fuel gas and oil)
* For (or shared by) both the 711 MWe and the 1068 MWe plant
– Indicates information unavailable

Demisters	2	0	2
Flues or stack	0	0	0
Dampers	0	0	0
Sorbent preparation	1	1	0
Product system	1	1	0
Other	2	2	0
Totals	8	6	2

Process developments: The formation of sulphate is now being reduced by 75% at some plants by modified operational techniques and the use of an anti-oxidant. This may allow the sulphate crystallisation part of the system to be dispensed with in future plants, with sulphate concentrations being controlled by a small purge [276].

Appraisal:

1.	Information available	2
2.	Process simplicity	1
3.	Operating experience	2
4.	Operating difficulty	0
5.	Loss of power	0
6.	Reagent requirements	2
7.	Ease of end-product disposal	2
8.	Process applicability	2
	Total	11

Process Code S31.2 – Fläkt-Boliden Process

Applicability: This process type is suitable for retrofit and new build applications.

Typical land area requirements: ***

Typical fresh water and water treatment requirements: Published estimates are presented in Table 2.16.

Consumption of reagents: Published estimates are presented in Table 2.16.

End-Product or waste disposal requirements: Published estimates are presented in Table 2.16.

Typical power requirements: Published estimates are presented in Table 2.16.

Typical reductions in combustion plant efficiency: Published estimates are presented in Table 2.16.

Operating experience: The process has been operated only on pilot plants and a few small commercial plants.

Process developments: ***

Table 2.15 Published Estimated Requirements

Case No. (Table 2.3)		1	3	4	6
Reference		95	6	265	**
MWe		2×500	800	3×660	2000
Coal: Sulphur	wt. %	4.0	3.6	2.0	2.0
Chlorine	wt. %	0.1	0.1	0.4	–
S Capture	%	90	95	90	91
S Recovery process		Claus	Resox	Claus	Claus
Typical land area requirements, hectare:					
Absorption plant		–	–	0.9	–
Chemical plant		–	–	1.1	–
Limestone stock (6 weeks supply)		–	–	0.6	–
Rail siding		–	–	2.3	–
Total (Approx.)		–	–	5	–
Typical fresh water and water treatment requirements per MWh:					
Process water	m^3	0.26	–	–	–
Cooling water	m^3	7.8	–	–	3.0
High pressure steam	tonne	0.10	–	)0.11–	)
Low pressure steam	tonne	0.20	–	)0.14	)0.08
Consumption of reagents per MWh (See Note a):					
NaOH (solid)	kg	1.1	2.0	Nil	Nil
(47% liquor)	kg	Nil	Nil	2.6	1.2
Natural gas	Nm^3	7.0	Nil	2.2	2.3
Coal or coke	kg	Nil	15.3	Nil	Nil
Limestone (for prescrubber water treatment)	kg	–	0.6	2.7	Nil
Lime (for ditto)	kg	–	Nil	Nil	1.25
End-product or waste disposal requirements per MWh:					
Sulphur	kg	15.5	12.8	5.5	5.8
Sodium sulphate:					
solution	m^3	0.002	Nil	Nil	Nil
solid	kg	Nil	3.4	2.2	1.0 (b)
Calcium chloride (25% soln.)	m^3	–	–	–	0.01
Slurry	m^3	0.06	0.004	0.06	–
Typical power requirements, MWe:		28.8	–	50	38
Typical reductions in combustion plant efficiency:					
Loss in power sent out	%	11	–	7.1	–

**Vendor's promotional literature
– Indicates information unavailable
Note a: Additive (NA$_4$ EDTA) is also required at a rate of 0.006 kg/MWh [135]
Note b: Composition [vendor's promotional literature]: Na$_2$SO$_4$ 55%, Na$_2$SO$_3$ 17%, Na$_2$S$_2$O$_5$ 26%, Na$_2$S$_2$O$_3$ 2%

Appraisal:

1.	Information available	1
2.	Process simplicity	1
3.	Operating experience	0
4.	Operating difficulty	0*
5.	Loss of power	0
6.	Reagent requirements	2
7.	Ease of end-product disposal	2
8.	Process applicability	2
	Total	8

*Assumed in absence of data

Process Code S31.3 – Catalytic Inc./IFP Process

Applicability: This process type is suitable for retrofit and new build applications. It is capable of giving about 95% sulphur capture [6].

Typical land area requirements: ***

Typical fresh water and water treatment requirements: ***

Consumption of reagents: Published estimates are presented in Table 2.17.

End-product or waste disposal requirements: Published estimates are presented in Table 2.17.

Typical power requirements: ***

Typical reductions in combustion plant efficiency: ***

Operating experience: Experience is limited to operation of two pilot plants [6].

Process developments: A modification of the process

can simultaneously give about 70% removal of NO_x [6, 137].

Appraisal:

1.	Information available	1
2.	Process simplicity	1
3.	Operating experience	0
4.	Operating difficulty	0*
5.	Loss of power	0*
6.	Reagent requirements	2
7.	Ease of end-product disposal	2
8.	Process applicability	2
	Total	8

*Assumed in absence of data

CATEGORY S32

The status of Category S32 processes is indicated in Table 2.18, which presents information on the size of plants built, year of completion, applicability (i.e. suitability for retrofit); and percentage sulphur capture.

Process Code S32.1–Citrate Process

Applicability: This process type is suitable for new build and retrofit applications. A sulphur capture of over 90% can be attained [6].

Typical land area requirements: ***

Typical fresh water and water treatment requirements: ***

Consumption of reagents: Data reported by Battelle [653] are presented in Table 2.19. Published estimated requirements are presented in Table 2.20.

End-product or wasste disposal requirements: Published estimated requirements are presented in Table 2.20.

Pressure losses: Data reported by Battelle [653] are presented in Table 2.19.

Power consumption: Data reported by Battelle [653] are presented in Table 2.19.

Typical reduction in combustion plant efficiency: ***

Table 2.16 Published Estimated Requirements

Case No. (Table 2.2)		1	5
Reference		95	268
MWe		2×500	500
Coal: Sulphur	wt. %	4.0	4.0
Chlorine	wt. %	0.1	–
Sulphur capture	%	90	90
Typical fresh water and water treatment requirements per MWh:			
Process water	m^3	0.25	0.25
Cooling water	m^3	12.9	10.0
L.P. steam (4.5 bar)	tonne	0.01	0.02
H.P. steam (25 bar)	tonne	0.23 –0.34	0.34
Consumption of reagents per MWh:			
NaOH	kg	0.27	1.2
Citric acid	kg	0.05	0.04
Natural gas	m^3	7.2	–
End-product or waste disposal requirements per MWh:			
Sulphur	kg	15.8	21.8
Sodium sulphate decahydrate	kg	–	3.4
Fly-ash + waste solids, 20% moisture	m^3	0.06	–
Typical power requirements, MWe:		19.9	9.1
Typical reductions in combustion plant efficiency:			
Loss in power sent out	%	9.1	–

– Indicates information unavailable

Table 2.17 Published Estimated Requirements

Case No. (Table 2.2)		3
Reference		6
MWe		800
Coal: Sulphur	wt. %	3.6
Chlorine	wt. %	0.1
Sulphur capture	%	95
Consumption of reagents, kg/MWh:		
Ammonia		0.19
Limestone		0.57
Coal and coke		22
End-product or waste disposal requirements per MWh:		
Elemental sulphur (Over 99.9% pure)	kg	13.8
Dry solids	kg	2.0
Slurry waste	m^3	0.003

Table 2.18 Status of Processes Using Regenerable Solution-Based Wet Reagents: Chemical Regeneration

Code No.	Vendor	No. of Plants	Plant size Range MWe	Total MWe	Dates	% S Capture	Ref.
S32.1	–	2	1–60	61	1979	90	6, 653
S32.2	Conoco Coal Dev. Co.	2 (Pilot plants)	0.35–7	7.35	–	–	95
S32.3	–	–	–	–	–	–	–
S32.4	Rockwell	1	100	100	1983	92	95, 269

– Indicates information unavailable

Operating experience: ***

Process developments: ***

Appraisal:

1.	Information available	1
2.	Process simplicity	1
3.	Operating experience	0
4.	Operating difficulty	0*
5.	Loss of power	0*
6.	Reagent requirements	2
7.	Ease of end-product disposal	2
8.	Process applicability	2
	Total	8

*Assumed in absence of data

Process Code S32.2 – Conosox Process

Applicability: This process type is suitable for new build and retrofit applications.

Typical land area requirements: ***

Typical fresh water and water treatment requirements: Published estimated requirements are presented in Table 2.21.

Consumption of reagents: Published estimated requirements are presented in Table 2.21.

End-product or waste disposal requirements: Published estimated requirements are presented in Table 2.21.

Typical power requirements: Published estimated requirements are presented in Table 2.21.

Typical reductions in combustion plant efficiency: Published estimated requirements are presented in Table 2.21. It can be noted that the reduction in power sent out is smaller than the power consumed by the process owing to the credit for steam generated at 5.8 bar.

Table 2.19 Data Reported by Battelle

Reference		653
MWe		60
Coal: Sulphur	wt. %	3.05
Gas: SO_2 content	ppmv	2000
Sulphur capture	%	90
Liquid/gas ratio, litre/m³:		
Prescrubber		3.8
Absorber		1.0
Consumption of reagents, kg/tonne recovered sulphur:		
Citric acid		14.4
Caustic soda		46.0
Pressure loss, mbar:		15
Power consumption MWe:		1.6

Table 2.20 Published Estimated Requirements

Case No. (Table 2.2)		3
Reference		6
MWe		800
Coal: Sulphur	wt. %	3.6
Chlorine	wt. %	0.1
Sulphur capture	%	95
Consumption of reagents, kg/MWh:		
Sodium citrate		0.19
Sodium hydroxide		0.6
Sodium thiosulphate		0.9
Limestone		0.57
Coal and coke		13
End-product or waste disposal requirements per MWh:		
Sulphur (99% pure)	kg	13.5
Sodium sulphate decahydrate	kg	4.7
Waste solids and slurry	m³	0.004

Table 2.21 Published Estimated Requirements

Case No. (Table 2.2)		1
Reference		95
MWe		2×500
Coal: Sulphur	wt. %	4.0
Chlorine	wt. %	0.1
Sulphur capture	%	90
Typical fresh water and water treatment requirements, m³/MWh:		
Raw water		0.03
Cooling water		5.7
Cooling tower blowdown		0.3
Consumption of reagents per MWh:		
Potassium hydroxide	kg	1.0
Liquid oxygen	kg	35
Fuel oil	m³	0.027
Methane	Nm³	1.33
End-product or waste disposal requirements per MWh:		
Sulphur	kg	17.1
Steam (5.8 bar)	tonne	0.034
Fly-ash to landfill (20% moisture)	m³	0.06
Typical power requirements, MWe:		9.2
Typical reductions in combustion plant efficiency:		
Loss in power sent out	%	0.03

Operating experience [95]: This process type has been operated on the pilot plant scale only.

Process developments [95]: Requires demonstration on the 100 MWe operating scale.

Appraisal:

1.	Information available	1
2.	Process simplicity	1
3.	Operating experience	0
4.	Operating difficulty	0*
5.	Loss of power	0
6.	Reagent requirements	1
7.	Ease of end-product disposal	2
8.	Process applicability	2
	Total	7

*Assumed in absence of data

Process Codes S32.3–Ispra Mark 13A Process

Applicability: ***

Typical land area requirements: ***

Typical fresh water and water treatment requirements: ***

Consumption of reagents: ***

End-product or waste disposal requirements: Published estimates are presented in Table 2.22.

Typical power requirements: Published estimates are presented in Table 2.22.

Table 2.22 Published Estimated Requirements

Reference		272
MWe		400
Coal: sulphur	wt. %	1.0
Sulphur capture	%	93
End-product or waste disposal requirements per MWh:		
Hydrogen	Nm^3	2.1
Sulphuric acid (80%)	kg	5.4
Typical power requirements, MWe:		5.2

Typical reductions in combustion plant efficiency: ***

Operating experience: The process has been operated only on the bench scale (about 10 m^3/h gas) at the Ispra (Italy) Joint Research Centre of the Commission of the European Communities. The Commission has invited proposals for construction and operation of a 20,000 m^3/h pilot plant.

Process developments [273]: To make the process more compatible with the gas temperature from power stations without sacrificing the ability to concentrate the sulphuric acid produced, the final stages of concentration could be effected by using a small side stream of the flue gas at 380°C. Part of the gas leaving the air heater would then pass to the first-stage concentrator, and the remainder would flow to the reactor via a recuperator which would be used for reheating the purified gas.

Appraisal:

1.	Information available	1
2.	Process simplicity	1
3.	Operating experience	0
4.	Operating difficulty	0*
5.	Loss of power	0*
6.	Reagent requirements	2
7.	Ease of end-product disposal	1
8.	Process applicability	2
	Total	7

*Assumed in absence of data

Process Code S32.4–Aqueous Sodium Carbonate Process

Applicability: This process type is suitable for new build and retrofit applications.

Typical land area requirements: ***

Typical fresh water and water treatment requirements: Published estimates are presented in Table 2.23.

Consumption of reagents: Published estimates are presented in Table 2.23.

End-product or waste disposal requirements: Published estimates are presented in Table 2.23.

Typical power requirements: Published estimates are presented in Table 2.23.

Typical reductions in combustion plant efficiency: Published estimates are presented in Table 2.23.

Table 2.23 Published Estimated Requirements

		1	3
Case (Table 2.2)			
Reference		95	6
MWe		2×500	800
Coal: Sulphur	wt. %	4.0	3.6
Chlorine	wt. %	0.1	0.1
Sulphur capture	%	90	95
Typical fresh water and water treatment requirements, m^3/MWh:			
Process water, filtered, softened		0.28	–
Cooling water, filtered		19	–
Consumption of reagents, kg/MWh:			
Sodium hydroxide		1.0	0.53
Coke		29	18.7
Diatomaceous earth filter precoat		0.25	–
End-product or waste disposal requirements per MWh:			
Sulphur	kg	16.4	15.5
Dry solids	kg	Nil	3.9
Wet waste	m^3	0.06	0.002
Typical power requirements, MWe:		37.6	–
Typical reductions in combustion plant efficiency:			
Loss in power sent out	%	3.6	–

– Indicates information unavailable

The 100 MW unit at the Huntley station used 4–5% of the station power [269].

Operating experience [269]: A 100 MW spray dryer with a 60 MW regeneration system, 2.5–3.5% sulphur coal, has been operated. The spray dryer gave deposition problems. Regenerator problems arose because it was rated for only 60% of the absorber capacity, and because some items (e.g. reducer feed pumps, equipment for handling molten sulphur) operated at the low end of their design range. The reducer lost its refractory lining.

Process developments [269]: Design changes to increase operating efficiency would have included

Table 2.24 Status of Processes Using Regenerable Slurry-Based Wet Reagents: Thermal Regeneration

Code No.	Vendor	No. of Plants	Plant size Range, MWe	Total MWe	Dates	% S Capture	Ref.
S41.1	United Engrs & Constructors	4	120–360	980	1974–82	90–98	95, 279
	Chemico	2	155–190	345	1972–73	–	95
	–	2	–	160	–	–	95

– Indicates information unavailable

installation of a mechanical particulates collector upstream of the spray dryer to collect 85% of the particulates (the remainder would be taken out with the filter cake in the regeneration unit), and of a fabric filter to collect the dry product from the spray dryer. However, research on the project terminated in 1984 because of lack of funding [681].

Appraisal:

1.	Information available	1
2.	Process simplicity	1
3.	Operating experience	0
4.	Operating difficulty	0*
5.	Loss of power	1
6.	Reagent requirements	1
7.	Ease of end-product disposal	2
8.	Process applicability	2
	Total	8

*Assumed in absence of data

CATEGORY S41

The status of Category S41 processes is indicated in Table 2.24, which presents information on the size of plants built, year of completion, and percentage sulphur capture.

Process Code S41.1–Magnesia Slurry Scrubbing Process

Applicability: This process type is suitable for new build and retrofit applications. Depending upon operating conditions, it is capable of attaining 98% or more sulphur capture [279].

Typical land area requirements: ***

Typical fresh water and water treatment requirements: Published estimates are presented in Table 2.25.

Consumption of reagents: Published estimates are presented in Table 2.25.

End-product or waste disposal requirements: Published estimates are presented in Table 2.25.

Typical pressure losses: Published estimates are presented in Table 2.25.

Typical power requirements: Published estimates are presented in Table 2.25.

Typical reductions in combustion plant efficiency: Published estimates are presented in Table 2.25.

Operating experience [95, 269, 278, 279]: The earliest plants, built by Chemico, gave generally satisfactory service, but are no longer operational as the boilers have been shut down.

The absorption systems of the United Engineers & Constructors (UE&C) plants operate with lower pH in the slurry than the Chemico plants (6.3 vs. 8.5); the process is tolerant of variations in the pH of the slurry. They had an average availability of over 99% under base load conditions. Early difficulties were mainly concerned with: slaking of the magnesia; erosion of centrifuge scrapers; and sulphite carry-over from the dryer cyclones. Modifications to correct these faults have been reported.

The UE&C plants use off-site regeneration systems, with oil- or gas-fired fluidised bed calciners (compared with rotary calciners in the Chemico plants). The regeneration system availability has risen from 40% to over 80%; initial problems

Table 2.25 Published Estimated Requirements

Case No. (Table 2.2)		1	3
Reference		95	6
MWe		2×500	800
Coal: Sulphur	wt. %	4.0	3.6
Chlorine	wt. %	0.1	0.1
S Capture	%	90	95
Typical fresh water and water treatment requirements, m^3/MWh:			
Raw water		0.36	–
Consumption of reagents, kg/MWh:			
Magnesia		3.2	0.47
Limestone		–	0.57
Fuel oil		0.45	–
Coal and coke		–	37.1
End-product or waste disposal requirements per MWh:			
Sulphuric acid (93%)	kg	48	Nil
Sulphur	kg	Nil	14.5
Waste material	m^3	0.06	0.006
Typical pressure losses, mbar:		20	–
Typical power requirements, MWe:		19.7	–
Typical reductions in combustion plant efficiency:			
Loss of power sent out	%	5.2	–

– Information unavailable

included: temperature control difficulties in the calciners leading to 'dead-burning' of the MgO; failure of the waste heat boiler to achieve its design performance; and mechanical problems with a booster fan.

The last two plants shown in Table 2.24 are operating in Japan: one on smelter gas, the other on oil-fired boiler and Claus gases; their availability is up to 95% in spite of some residual mechanical problems.

A summary of problems reported by Battelle [674] from a visit to a power station with UE&CFGD plants is as follows (problems in various plant areas, and remaining unsolved):

Total equiv MWe 700
Number of absorption streams 5

Plant area	Problems Unsolved	Solved
Prescrubber	No	Yes
Absorber	No	No
Demisters	Yes	No
Flues or stack	No	Yes
Dampers	No	No
Sorbent preparation	No	Yes
Product system	No	No
Other	No	Yes

Process developments: Design modifications to the regeneration system for the UE&C plants were expected [279] to increase availability to 85%.

Appraisal:

1.	Information available	2
2.	Process simplicity	1
3.	Operating experience	0
4.	Operating difficulty	1
5.	Loss of power	1
6.	Reagent requirements	2
7.	Ease of end-product disposal	2
8.	Process applicability	2
	Total	11

CATEGORY S42

The status of Category S42 processes is indicated in Table 2.26, which presents information on the size of plants built, year of completion, and percentage sulphur capture.

Process Code S42.1–Sulf-X Process

Applicability: This process type is suitable for new build and retrofit applications. It is capable of giving very high (99%) sulphur capture and some abatement of NO_x emissions, but the high liquid/gas ratio needed for high NO_x emission abatement would increase the absorber size and cost unduly.

Typical land area requirements: ***

Typical fresh water and water treatment requirements: Published estimates are presented in Table 2.27.

Consumption of reagents: Published estimates are presented in Table 2.27.

End-product or waste disposal requirements: Published estimates are presented in Table 2.27.

Typical pressure losses: Published estimates are presented in Table 2.27.

Typical power requirements: Published estimates are presented in Table 2.27.

Typical reductions in combustion plant efficiency: Published estimates are presented in Table 2.27.

Operating experience [277]: Experience limited to 1.5 MWt scale, with some stages of the regeneration system omitted. Initial difficulties, claimed to have been overcome, included mechanical transfer of the filter cake to and through the dryer, and oxidation of the wet cake in the calciner. Operation ceased in 1984, and for contractual reasons is not expected to be resumed [281].

Process developments: Forecast in 1983 [277]: tests of alternative, less expensive reducing agents for the calciner; long-term reliability verification; scale-up to the 10–60 MWe scale.

Appraisal:

1.	Information available	1
2.	Process simplicity	0
3.	Operating experience	0
4.	Operating difficulty	0*
5.	Loss of power	1
6.	Reagent requirements	2
7.	Ease of end-product disposal	2
8.	Process applicability	2
	Total	8

*Assumed in absence of data

CATEGORY S51

The status of Category S51 processes is indicated in Table 2.28, which presents information on the size of plants built, year of completion, and percentage sulphur capture.

Table Table 2.26 Status of Processes Using Regenerable Slurry-Based Wet Reagents: Chemical Regeneration

Code No.	Vendor	No. of Plants	Plant size Range MWe	Total MWe	Dates	% S Capture	Ref.
S42.1	P'burgh Env. Systems Inc. (PENSYS)	3	1–1.5	3.5	1977–82	99	95, 277

Table 2.27 Published Estimated Requirements

Case No. (Table 2.2)		1
Reference		95
MWe		2×500
Coal: Sulphur	wt. %	4.0
Chlorine	wt. %	0.1
S Capture	%	90
Typical fresh water and water treatment requirements per MWh:		
Raw water	m³	0.06
Cooling water blowdown	m³	0.27
High pressure steam	tonne	0.09
Low pressure steam	tonne	0.004
Consumption of reagents per MWh:		
Sodium sulphide	kg	0.52
Pyrite	kg	5.1
Coke	kg	9.6
Fuel oil	litre	1.4
End-product or waste disposal requirements per MWh:		
Sulphur	kg	16.2
Waste to landfill	m³	0.06
Fly-ash	wt. %	72
Scrubber blowdown solids	wt. %	8
Moisture	wt. %	20
Typical pressure losses, mbar:		40
Typical power requirements, MWe:		25.6
Typical reductions in combustion plant efficiency:		
Loss in power sent out	%	5.5

Process Code S51.1 – Active Carbon Adsorption Process

Applicability: This process type is suitable for new build and retrofit applications, and is capable of attaining greater than 98% sulphur capture; it also removes hydrogen chloride and fluoride. A modification of the process (see Section 5) provides simultaneous SO_2 and NO_x emission abatement.

Typical land area requirements: ***

Typical fresh water and water treatment requirements: ***

Consumption of reagents: Published estimates are presented in Table 2.29.

End-product or waste disposal requirements: Published estimates are presented in Table 2.29.

Table 2.29 Published Estimated Requirements

Case No. (Table 2.2)		3
Reference		6
MWe		800
Coal: Sulphur	wt. %	3.6
Chlorine	wt. %	0.1
Sulphur Capture	%	95
Sulphur recovery process		Resox
Consumption of reagents, kg/MWh:		
Coal or coke		41
Active carbon		3.2
End-product or waste disposal requirements, kg/MWh:		
Sulphur		14.5
Dry solids		5.6

The Reinluft process uses granular low-temperature coke (prepared by devolatilising peat, lignite or certain bituminous coals at temperatures below 700°C), sized 2.5–3.5 mm, in place of the specially prepared active carbon used by the Bergbau-Forschung process [157].

Typical power requirements: ***

Typical reductions in combustion plant efficiency: ***

Operating experience [157]: The combined Bergbau Forschung/Foster Wheeler process was tested in a 20 MWe unit at the Scholtz station (Gulf Power Co.). The major problems were: hot spots in the adsorber; poor reliability of the carbon/sand separator; plugging of the Resox system condenser with sulphur and carbon. The 35 MWe prototype at Lunen, which used a Claus sulphur recovery system, gave better performance in the adsorption and sulphur recovery systems.

In the Reinluft process, the coke make-up reaches its peak reactivity in 3–10 cycles. Some operating difficulties were found in the operation of the four plants: fires in the coke bed; corrosion of equipment

Table 2.28 Status of Processes Using Regenerable Dry Reagents: Thermal Regeneration

Code No.	Vendor	No. of Plants	Plant size Range*	Total *	Dates	% S Capture	Ref.
S51.1	BF/Uhde/Mitsui	5	1–1100	1505	1973–87	80–98	424, 435
	Sumitomo H.I.	1	300	300	–	90	424
	Chemiabau Reinluft	4	–	–	–	–	157

* Thousand Nm³/h
– Indicates information unavailable

by sulphuric acid. A new design has been developed to overcome these problems.

Process developments [157]: Westvaco are developing a fluidised bed adsorption and regeneration process in which adsorbed SO_2 is catalytically oxidised to sulphuric acid on the carbon, and regeneration of the carbon is by reaction with hydrogen. A 560 Nm^3/h pilot plant has operated for 350 hours with one short interruption from sulphur-plugging. Hydrogen consumption is 3–4 mol/mol SO_2.

Appraisal:

1.	Information available	1
2.	Process simplicity	2
3.	Operating experience	0
4.	Operating difficulty	0*
5.	Loss of power	2*
6.	Reagent requirements	2
7.	Ease of end-product disposal	2
8.	Process applicability	2
	Total	11

*Assumed in absence of information

CATEGORY S52

The status of Category S52 processes is indicated in Table 2.30, which presents information on the size of plants built, year of completion, and percentage sulphur capture.

Process Code S52.1 – Wet Active Carbon Adsorption Process

Applicability: This process type is suitable for new build and retrofit applications.

Typical land area requirements: ***

Typical fresh water and water treatment requirements: ***

Consumption of reagents [157]: The carbon consumption for the Hitachi process is about 2% p.a.

End-product or waste disposal requirements [157]: The Lurgi process produces 7% sulphuric acid which is concentrated to 15% in the gas cooler. The Hitachi process produces 20% acid which is concentrated to 65% using a submerged combustor.

Typical power requirements: ***

Typical reductions in combustion plant efficiency: ***

Operating experience [157]: The Hitachi process encountered minor problems initially, but its general performance is claimed to be very successful. In one plant the dilute (17%) acid was not concentrated, but used to produce gypsum from limestone.

Process developments [157]: A continuous process, with water washing of the carbon simultaneously with adsorption, is at the laboratory stage of development.

Appraisal:

1.	Information available	1
2.	Process simplicity	1
3.	Operating experience	0
4.	Operating difficulty	1
5.	Loss of power	2*
6.	Reagent requirements	2
7.	Ease of end-product disposal	1
8.	Process applicability	2
	Total	10

*Assumed in absence of data

Process Code S52.2 – Copper Oxide Process

Applicability: This process type is suitable only for new build applications.

Typical Land area requirements: ***

Typical fresh water and water treatment requirements: ***

Consumption of reagents: The hydrogen consumption [282] is 0.2 kg/kg sulphur captured. Additional information derived from published estimates is presented in Table 2.31.

Table 2.30 Status of Processes Using Regenerable Dry Reagents: Chemical Regeneration

Code No.	Vendor	No. of Plants	Plant size Range MWe	Total MWe	Dates	% S Capture	Ref.
S52.1	Lurgi	2	30–170*	200*	–	Over 90	157
	Hitachi	2	–	–	–	80	157
S52.2	Shell/UOP	1	0.6	0.6	1974	–	137
		+1	125*	125*	1983	90	282
S52.3	Monsanto	2	15–110	125	1960–72	Over 90	158, 137

* Thousand Nm^3/h
– Indicates information unavailable

End-product or waste disposal requirements: Published estimates are presented in Table 2.31.

Typical pressure losses: A plant handling 125,000 Nm³/h, 2500 ppmv SO_2, 90% sulphur capture [282] has a pressure drop of 20 mbar [282].

Table 2.31 Published Estimated Requirements

Case No. (Table 2.2)		3
Reference		6
MWe		800
Coal: Sulphur	wt. %	3.6
Chlorine	wt. %	0.1
Sulphur Capture	%	90
Sulphur recovery process		Claus
Consumption of reagents kg/MWh:		
Coal or coke		3
End-product or waste disposal requirements kg/MWh:		
Sulphur		13.7
Dry solids		5.0

Typical power requirements: ***

Typical reductions in combustion plant efficiency: ***

Operating experience [137]: The 0.6 MW unit was operated on a side stream from a 400 MWe coal-fired boiler at the Big Bend station, Florida, for two years. The copper oxide underwent 13,000 cycles without loss of desulphurisation reactivity.

The 125,000 Nm³/h plant mentioned above [282] has been in operation for over 2 years; initial problems in the recovered SO_2 concentration unit were resolved.

Process developments: Cost reductions for units larger than about 50 MWe (about 175,000 Nm³/h) would result from using 2–3 reactors on acceptance to 1 reactor on regeneration [282].

A fluidised bed version of the process is being developed [137].

A modification of the process can give simultaneous removal of SO_2 and NO_x (see Section 5). The process is no longer being licensed by Shell.

Appraisal:

1.	Information available	1
2.	Process simplicity	1
3.	Operating experience	0
4.	Operating difficulty	0*
5.	Loss of power	2*
6.	Reagent requirements	1
7.	Ease of end-product disposal	2
8.	Process applicability	2
	Total	8

*Assumed in absence of information

Process Code S52.3 – Catalytic Oxidation Process

Applicability: This process type is suitable for new build applications; with reheat of the gas upstream of the FGD system it is also suitable for retrofit applications. It is capable of attaining over 90% sulphur capture.

Typical land area requirements: ***

Typical fresh water and water treatment requirements: ***

Consumption of reagents: There is a catalyst loss of about 2.5% each time the catalyst is cleaned (about four times p.a.) [158].

End-product or waste disposal requirements: The 110 MW plant was expected [158] to produce 26 kg/MWh of 78% sulphuric acid.

Typical power requirements: ***

Typical reductions in combustion plant efficiency: ***

Operating experience: A plant of 110 MWt capacity was installed on a coal-fired boiler, with natural gas burners to achieve the gas temperature needed for the catalytic oxidation reaction. The plant was operated for 444 hours, but was then shut down to replace the natural gas burners with oil burners. Troubles encountered were: plugging of catalyst with soot from the oil burners (corrected by using an external burner giving better combustion); failure of the burner refractory lining; corrosion and plugging in the acid system. These problems stem mainly from attempting to operate the process as a retrofit system. The system has not been operated since 1975 [137, 157, 158].

Process developments: Development of this process is not being considered at present.

Appraisal:

1.	Information available	1
2.	Process simplicity	2
3.	Operating experience	0
4.	Operating difficulty	0
5.	Loss of power	1*
6.	Reagent requirements	2
7.	Ease of end-product disposal	1
8.	Process applicability	1
	Total	8

*Assumed in absence of information

Table 2.32 Status of Processes Using Non-Regenerable Dry Reagents Applied to Flue Gas: Dry Injection

Code No.	Vendor	No. of Plants	Plant size Range MWe	Total MWe	Dates	% S Capture	Ref.
S61.1	Conoco Coal Research	1	1	1	–	80	–
S61.2	–	–	–	–	–	–	–

– Indicates information unavailable

CATEGORY S61

The status of Category S61 processes is indicated in Table 2.32, which presents information on the size of plants built, year of completion, and percentage sulphur capture.

Process Code S61.1 – Hydrated Lime Injection Process

Applicability: ***

Typical land area requirements: ***

Typical fresh water and water treatment requirements: ***

Consumption of reagents: In field tests on a 1 MW scale by the Conoco Coal Research Division of Conoco Inc., hydrated lime, sized 80% below 43 micron, with a surface area of 23 m^2/g and a moisture content of 1.5%, was used at a calcium/sulphur molar ratio up to 2.7, giving sulphur capture of about 80%. The additive, sodium hydroxide, was supplied as an aqueous solution at NaOH/Ca(OH)$_2$ mass ratio of 0.1.

End-product or waste disposal requirements: ***

Typical power requirements: ***

Typical reductions in combustion plant efficiency: ***

Operating experience: ***

Process developments: ***

Appraisal:

1. Information available 0
2. Process simplicity 2
3. Operating experience 0
4. Operating difficulty 0*
5. Loss of power 2*
6. Reagent requirements 1
7. Ease of end-product disposal 1
8. Process applicability 2
   Total 8

*Assumed in absence of data

Process Code S61.2 – Alkali Injection Process

Applicability: ***

Typical land area requirements: ***

Typical fresh water and water treatment requirements: The requirement for water is negligible [95].

Consumption of reagents: For Nahcolite a feed rate equivalent to 1.2 times the stoichiometric rate is estimated to give about 75% sulphur capture [95].

End-product or waste disposal requirements: ***

Typical power requirements: For a 1000 MWe installation the power consumption is estimated to be about 7 MWe [95].

Typical reductions in combustion plant efficiency: ***

Operating experience: ***

Process developments: ***

Appraisal:

1. Information available 0
2. Process simplicity 2
3. Operating experience 0
4. Operating difficulty 0*
5. Loss of power 2*
6. Reagent requirements 1
7. Ease of end-product disposal 0
8. Process applicability 2
   Total 7

*Assumed in absence of information

Table 2.33 Status of Processes Using Non-Regenerable Dry Reagents Applied to Flue Gas: Dry Reactor

Code No.	Vendor	No. of Plants	Plant size Range*	Total *	Dates	% S Capture	Ref.
S62.1	Lurgi	–	100–300	3,000	–	94–98	**

* Thousand Nm^3/h ** Vendor's promotional literature
– Indicates information unavailable

CATEGORY S62

The status of Category S62 processes is indicated in Table 2.33, which presents information on the size of plants built, year of completion, and percentage sulphur capture.

Process Code S62.1 – Lurgi CFB Lime Absorber Process

Applicability: This process type is suitable for new build and retrofit applications.

Typical land area requirements: ***

Typical fresh water and water treatment requirements: ***

Consumption of reagents: Hydrated lime is typically used at a Ca/S molar ratio of 1.7 [Vendor's promotional literature].

End-product or waste disposal requirements: ***

Typical power requirements: A plant handling 250,000 Nm^3/h flue gas consumed 520 kW [Vendor's promotional literature].

Typical reductions in combustion plant efficiency: ***

Operating experience [Vendor's promotional literature]: A plant handling 250,000 Nm^3/h started operation in 1984. Problems encountered included: deposits at the base of the CFB at the ESP entry; coarse fly-ash entering the CFB; dust build-up in the second field of the ESP; wear on the suspension nozzle; and instability at part load. All of these were overcome by small design changes.

Process developments: ***

Appraisal:

1.	Information available	1
2.	Process simplicity	2
3.	Operating experience	0
4.	Operating difficulty	0
5.	Loss of power	2*
6.	Reagent requirements	1
7.	Ease of end-product disposal	1
8.	Process applicability	2
	Total	9

*Assumed in absence of information

CATEGORY S71

The status of Category S71 processes is indicated in Table 2.34, which presents information on the size of plants built, year of completion, and percentage sulphur capture.

Process Code S71.1 – Sorbent Direct Injection Process

Applicability: This process type is suitable for retrofit and new build applications.

Typical land area requirements: ***

Typical fresh water and water treatment requirements: ***

Consumption of reagents: The sorbent considered is usually limestone. A Ca/input S molar ratio of about 2 is needed to give 50–70% sulphur capture [296] though estimates [302] have been based on the assumption that this would give only 35% sulphur capture (about 65% if $Ca(OH)_2$ is used) and that the stone would be ground to 50% below 18 micron, 100% below 100 micron.

Further information is presented in Table 2.35 in the form of published estimates.

End-product or waste disposal requirements: Published estimates are presented in Table 2.35.

Typical power requirements: Published estimates are presented in Table 2.35.

Typical reductions in combustion plant efficiency: Published estimates are presented in Table 2.35.

Table 2.35 Published Estimated Requirements

Reference		302
MWe		4×500
Coal: Sulphur	Wt. %	2.0
Ash	Wt. %	20
Sulphur capture	%	35
Consumption of reagents, kg/MWh:		46
End-product or waste disposal requirements, kg/MWh: Additional		47
Typical power requirements, MWe:		11
Typical reductions in combustion plant efficiency: Reduction in power sent out	%	0.6

Table 2.34 Status of Processes Using Non-Regenerable Dry Reagents Applied in Furnace: Dry Injection

Code No.	Vendor	No. of Plants	Plant size Range MWe	Total MWe	Dates	% S Capture	Ref.
S71.1	Tampella	1	–	250	1986	50–70	290

– Indicates information unavailable

Operating experience: There is very little published information on operating experience. The process has not yet been applied on a commercial scale [302].

Process developments: The system incorporating Limestone Injection into Multistage Burners (LIMB) combines NO_x abatement with enhanced sulphur capture performance [296]. LIMB is described in Section 5.

Appraisal:

1.	Information available	0
2.	Process simplicity	2
3.	Operating experience	0
4.	Operating difficulty	0*
5.	Loss of power	2
6.	Reagent requirements	1
7.	Ease of end-product disposal	1
8.	Process applicability	2
	Total	8

*Assumed in absence of information

2.4 Processes for Detailed Study

The selection of processes for detailed study in this Volume has been based upon their suitability for application in the UK for the two datum combustion systems (Section 1.3) considered:

- Large (450 tonne steam/h) industrial boiler (Datum system 1).

- Small (25 tonne steam/h) factory boiler (Datum system 2).

In principle, all of the processes listed in Section 2.1 can be applied to all combustion plant, but the attraction of many processes diminishes with factors such as decrease in plant operating scale, and increases in FGD process complexity, reagent costs, and end-product disposal difficulty.

Appraisal of processes
To evaluate some of these factors, a rough appraisal of each process type has been made in Section 2.3 by assigning 'merit points' for a number of features; merit points have been awarded according to the scale:

0 Below average merit
1 Average merit
2 Above average merit
3 Outstandingly above average merit

The features to which these points have been assigned are described in Section 1.5; they are briefly:

(1) Information available
(2) Process simplicity
(3) Operating experience – extent and difficulties encountered
(4) Operating difficulty – availability, reliability
(5) Loss of power sent out – by installation of the FGD process
(6) Reagent requirements – quantities
(7) Ease of end-product disposal
(8) Process applicability – e.g. for retrofit

All of the processes listed in Section 2.1 and outlined in Section 2.2 are appraised in Section 2.3. The merit points assigned to the process types for each of the above features are summarised in Table 2.36. It should be noted that the number of points in the merit point system adopted in other Sections of the Manual are not strictly comparable with those considered here.

Table 2.36 Summary of FGD Process Appraisals

Code No.	Name	Merit Points for Feature No.								Total Points
		1	2	3	4	5	6	7	8	
S11.1	Sea water scrubbing	1	3	1	1	1	2	1	1	11
S11.2	Alkali scrubbing	1	2	1	1	1	1	0	2	9
S11.3	Ammonia scrubbing	0	2	0	1	1	1	1	2	8
S11.4	Sulphuric acid scrubbing	0	2	1	0	1	2	1	2	8
S11.5	Dowa	1	1	1	0	1	1	1	2	8
S11.6	Dual alkali	2	1	1	0	1	1	1	2	9
S12.1	Alkali spray drying	1	3	0	0	1	1	0	2	8
S12.2	Walther process	1	2	0	0	1	1	1	2	8
S21.1	Limestone slurry scrubbing	2	2	3	0	1	1	1	2	12
S22.1	Lime spray drying	2	3	2	0	2	1	1	2	13
S31.1	Wellman-Lord	2	1	2	0	0	2	2	2	11
S31.2	Flakt-Boliden	1	1	0	0	0	2	2	2	8
S31.3	Catalytic/IFP	1	1	0	0	0	2	2	2	8
S32.1	Citrate	1	1	0	0	0	2	2	2	8
S32.2	Conosox	1	1	0	0	0	1	2	2	7
S32.3	Ispra Mk. 13A	1	1	0	0	0	2	1	2	7
S32.4	Aqueous carbonate	1	1	0	0	1	1	2	2	8
S41.1	Magnesia scrubbing	2	1	0	1	1	2	2	2	11
S42.1	Sulf-X	1	0	0	0	1	2	2	2	8
S51.1	Active carbon	1	2	0	0	2	2	2	2	11
S52.1	Wet active carbon	1	1	0	1	2	2	1	2	10
S52.2	Copper oxide	1	1	0	0	2	1	2	1	8
S52.3	Catalytic oxidation	1	2	0	0	1	2	1	1	8
S61.1	Hydrated lime injection	0	2	0	0	2	1	1	2	8
S61.2	Alkali injection	0	2	0	0	2	1	0	2	7
S62.1	Lurgi circulating bed	1	2	0	0	2	1	1	2	9
S71.1	Sorbent injection	0	2	0	0	2	1	1	2	8

Features: 1. Information available
2. Process simplicity
3. Operating experience
4. Operating difficulty
5. Loss of power
6. Reagent requirements
7. Ease of end-product disposal
8. Process applicability

Processes suitable for the UK
The principal purpose of assigning merit points to each of the processes was to aid in the selection of processes that could be considered suitable for

application in the UK. It was arbitrarily assumed that suitable FGD processes would be those having more than 10 merit points.

Selection of processes for detailed study
All of the FGD process types are shown in Table 2.37 with an indication of which (if any) of the Datum combustion systems are considered, from the above criteria, to be suitable for application of the process in the UK.

Only one basic process type is considered to be suitable for the smallest operating scale dealt with in this Volume (Datum System 2) as well as for larger operating scales. This is the lime slurry scrubbing/spray dryer process (Process Code S22.1; merit rating 13 points). The relative simplicity of this process, the relative cheapness of the reagent (lime, manufactured from limestone which is abundant in the UK) and the production of an easily handled dry end-product, make it suitable for application to both Datum combustion systems. This process is evaluated in Section 2.5.

For large boilers, the simplest processes for consideration in the UK are:

– Sea water scrubbing process (Process Code S11.1; merit rating 11 points). This process is noteworthy for its extreme simplicity, but it is, of course, applicable only to coastal locations. The process is evaluated in Section 2.5.

– Limestone or lime slurry scrubbing processes (Process Code S21.1; merit rating 12 points). This uses cheap reagents and is a very well-established process, with more FGD plants of this basic type throughout the world than of all other processes put together. For UK application, only those processes embodying forced oxidation and producing gypsum (marketable, or easily disposed of according to circumstances) would be considered. Because of its importance, three examples of this basic process type are evaluated in Section 2.5, namely the IHI Gypsum process, the Chiyoda Thoroughbred 121 process and the Saarberg-Holter process.

Finally, the processes that are, in principle, the most complex are the regenerable reagent processes; their main attractions are: the reduced demand for reagent and the reduced production of wastes; and the potential for producing marketable products. Because of their major chemical processing content, they are of interest mainly for the largest operating scale, e.g. Datum System 1. Even for this operating scale, these processes might attract attention only if a reagent reprocessing industry were created in response to any future legislation imposing limits on sulphur oxides emissions in the UK. The processes to be considered are:

Table 2.37 Applications of FGD Processes Considered In Detail

Code No.	Name	Application to Datum System 1	Application to Datum System 2	Section
S11.1	Sea water scrubbing	Yes	–	2.5
S11.2	Alkali scrubbing	–	–	–
S11.3	Ammonia scrubbing	–	–	–
S11.4	Sulphuric acid scrubbing	–	–	–
S11.5	Dowa	–	–	–
S11.6	Dual alkali	–	–	–
S12.1	Alkali spray drying	–	–	–
S12.2	Walther process	–	–	–
S21.1	Limestone slurry scrubbing	Yes	–	2.5
S22.1	Lime spray drying	Yes	Yes	2.5
S31.1	Wellman-Lord	Yes*	–	2.5
S31.2	Flakt-Boliden	–	–	–
S31.3	Catalytic/IFP	–	–	–
S32.1	Citrate	–	–	–
S32.2	Conosox	–	–	–
S32.3	Ispra Mk. 13A	–	–	–
S32.4	Aqueous carbonate	–	–	–
S41.1	Magnesia scrubbing	Yes*	–	2.5
S42.1	Sulf-X	–	–	–
S51.1	Active carbon	Yes*	–	2.5
S52.1	Wet active carbon	–	–	–
S52.2	Copper oxide	–	–	–
S52.3	Catalytic oxidation	–	–	–
S61.1	Hydrated lime injection	–	–	–
S61.2	Alkali injection	–	–	–
S62.1	Lurgi circulating bed	–	–	–
S71.1	Sorbent injection	–	–	–

*In general, only if a centralised reagent reprocessing plant were available

– Wellman-Lord process (Process Code S31.1; 11 merit points)–a well-established process that has been evaluated by the CEGB [265]. This is evaluated in Section 2.5.

– Magnesia slurry scrubbing process (Process Code S41.1; 11 merit points) evaluated in Section 2.5. This is an example of a process that has been operated with spent reagent sent to an off-site reprocessing plant [279].

– Active carbon adsorption process (Process Code S51.1; 11 merit points) evaluated in Sections 2.5.

2.5 Evaluation of Selected FGD Processes

Evaluation of Sea Water Scrubbing FGD Process (Process Code S11.1)

See Section 2.2 for: outline of the basic process; its chemistry; block diagram.

See Section 2.3 for a list of manufacturers offering this type of equipment.

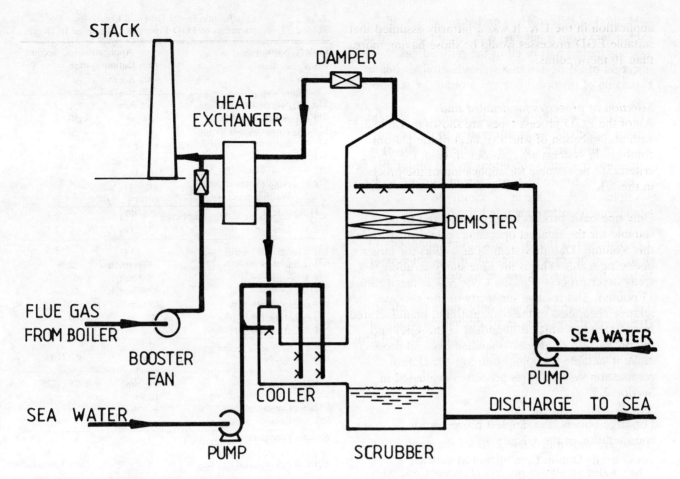

Figure 2.30 Sea Water Once-Through Process

See Section 2.3 for general appraisal of the basic process.

See Section 2.4 for the reason for choosing the Sea Water Scrubbing FGD Process.

This basic process type is considered (Section 2.4) to be suitable for application in the UK only to large boilers, and hence it is evaluated here only for Datum Combustion System 1.

Process Description

Figure 2.30 shows a simplified flow diagram for application of the process to an oil-fired water tube boiler.

Gas from the boiler, boosted by a Fan, is cooled in a Heat Exchanger, and is then scrubbed with sea water in the Absorber to remove sulphur oxides, acid halides and particulates. The purified gas then flows via a Demister to the Heat Exchanger, where it is reheated to above the dewpoint, and is then exhausted to the stack.

Status and Operating Experience

The status of the sea water scrubbing process is indicated in Section 2.3, where it is seen that more than fourteen plants have been built in the UK (using river water), Scandinavia, Japan and USA. The Japanese plants built by IHI are similar to their 'Inert gas systems' installed on 15 ships.

Information on operating experience quoted in Section 2.3 is limited. The CEGB plants [156] and Fläkt-Hydro plants [651] have given some trouble with corrosion, but the Risdon, Tasmania, plant has enjoyed a high availability [667].

Variations and Development Potential

The main potential variations in this process are:

– Design of the Absorber: a number of alternatives can be used: spray towers (the usual choice, giving the lowest gas pressure losses); venturi scrubbers with spray banks (higher efficiency but higher pressure loss); and turbulent contact absorbers (TCAs) with spray banks (high efficiency, but also with high pressure loss).

- Method of reheat: Heat exchange (in a heat exchanger) from input gas; combustion of liquid or gaseous fuel in part or all of the purified gas; injection of air heated in a steam-heated air heater; bypassing of uncooled flue gas around the scrubber.

- Treatment of effluent: Neutralisation of effluent with e.g. chalk, as an alternative to no treatment before discharging to sea. Where the effluent is treated, operating conditions should preferably be chosen to ensure that the calcium sulphite and sulphate formed are in sufficiently low concentration to remain in solution.

The process has the merit of extreme simplicity of design and operation, especially if it is unnecessary to treat the effluent water before discharge to sea.

The process can be applied only where the effluent (treated or untreated) can be discharged to sea; its application is therefore likely to be limited, and little development can be expected.

Process Requirements

These are shown in Table 2.38 for the application considered: Datum Combustion System 1.

It is assumed that:

- The SO_2 content of the gas is to be reduced to 650 mg/Nm³ (dry) equivalent to 87.0% sulphur capture.

- The NO_x and HCl contents of the gas are unaffected.

- The particulates content of the gas is reduced to 15 mg/Nm³ (dry).

- The effluent is not treated with chalk before discharge.

By-products and Effluents

The effluents from the process consist of sulphate and sulphite ions in solution, together with particulates.

The composition of the effluent water is summarised in Table 2.39.

Efficiency and Emission Factors

The efficiency and emission factors for the process are summarised in Table 2.40 for Datum Combustion System 1.

Table 2.38 Process Requirements Oil-Fired Combustion System 1

Inlet Gas at full load		
Volume flow	'000 Nm³/h	456
Dry gas	'000 Nm³/h	408
Water vap.	'000 Nm³/h	48
Actual volume flow	'000 m³/h	707
Temperature	°C	150
Particulates content	mg/Nm³ (dry)	39
SO_2 content	mg/Nm³ (dry)	5010
NO_x content	mg/Nm³ (dry)	1025
HCl content	mg/Nm³ (dry)	3
Exit Gas at full load		
Volume flow	'000 Nm³/h	422
Dry gas	'000 Nm³/h	407
Water vap.	'000 Nm³/h	15
Actual volume flow	'000 m³/h	577
Temperature	°C	100
Particulates content	mg/Nm³ (dry)	15 (17)
SO_2 content	mg/Nm³ (dry)	650 (1086)
NO_x content	mg/Nm³ (dry)	1025 (1025)
HCl content	mg/Nm³ (dry)	3 (3)
Reaction temperature	°C	27
Particulates removal		Simultaneous
Requirements at full load		
Sea water	tonne/h	2102
Electric Power	MWe	0.6
Manpower	men/shift	***
Average load factor	%	***

Figures in parentheses are annual average emissions for 90% FGD plant availability

Table 2.39 Estimated Properties of Waste Water Oil-Fired Combustion System 1

Water flow	tonne/h	2130
Composition of wet product:		
Ash	ppm	5
Added S as SO_4	ppm	1390

Total from four 500 MWe units

In calculating the efficiency and emission factors, it is assumed that at full load:

- The performance without the incorporation of the FGD plant would be as shown in Table 1.4.

- The FGD unit consumes 0.6 MWe electric power, regarded as equivalent to the combustion of a further 0.2 tonne/h of oil at a power station, assuming an overall power generation efficiency of 33%. This additional oil is arbitrarily assumed to be included with the oil burned in the boiler for calculating the efficiency factor.

For illustration purposes, the annual average emissions and emission factors for FGD plant availabilities of 100% and 90% are given in Tables 2.38 and 2.40. Further details are given in Section 1.6.

Little design variation is likely, and this factor is therefore ignored.

Application of the process is limited to locations close to the sea, where there is no environmental impediment to the discharge of the dilute effluent to sea.

Effect of load variations: ***

Costs:

Cost basis: ***

Capital costs: See Section 2.6–for retrofit and for new build applications.

Annual running costs and cost factors: See Section 2.6–for retrofit and for new build applications.

Effects of design variations and costs: No design variation need be considered (see above).

Effects of annual load patterns on annual running costs: ***

Table 2.40 Efficiency and Emission Factors–Oil-Fired Combustion System 1

Oil heat input (gross)	MWt	464
Oil fired	tonne/h	39.4
FGD plant power consumed	MWe	0.6
Equivalent oil input	tonne/h	0.2*
Useful energy from system	GJt/h	1468
Total equiv. oil input	tonne/h	39.6(a)
Efficiency factor	GJt/tonne	37.1(a)
Emissions		
Sulphur in SO_2	kg/h	133 (222)
Nitrogen in NO_x	kg/h	127 (127)
Chlorine in HCl	kg/h	1.2 (1.2)
Particulates	kg/h	6.1 (7.1)
Emission factors (per tonne oil)		(a)
Sulphur	kg/tonne	3.36 (5.61)
Nitrogen	kg/tonne	3.21 (3.21)
Chlorine	kg/tonne	0.03 (0.03)
Particulates	kg/tonne	0.15 (0.18)

*Calculated assuming overall power generation efficiency = 0.33
(a) Based on oil fired to boiler plus oil equivalent to electric power consumed
Figures in parentheses are annual average emissions for 90% FGD plant availability

Process Advantages and Drawbacks

Reasonably well-established process, with a pedigree dating from 1933; simple chemistry; simple equipment; land not required for stocking reagent (unless neutralisation with chalk is required) or for holding or disposing of product; high sulphur capture efficiency (over 95% capture can be attained); potential for environmentally acceptable effluent disposal, with adjustment of pH if required.

Disadvantages are that the design is limited to applications at certain coastal locations; careful design and control needed to avoid scaling problems.

Evaluation of IHI Limestone (or Gypsum) FGD Process
(Process Code S21.1–Limestone or Lime Slurry Scrubbing Processes)

See Section 2.2 for: outline of the basic process; its chemistry; block diagram.

See Section 2.3 for a list of manufacturers offering this type of equipment.

See Section 2.3 for general appraisal of the basic process.

See Section 2.4 for the reason for choosing the IHI Limestone/Lime Slurry Scrubbing process.

This basic process type is considered (Section 2.4) to be suitable for application in the UK only to large boilers, and hence it is evaluated here only for Datum Combustion System 1. The process presented here, which is of the type producing gypsum as an end product, is offered by Ishikawajima-Harima Heavy Industries (IHI) Company Limited (see Appendix 2 for details of this manufacturer).

See also below for evaluations of the Chiyoda Thoroughbred 121 Limestone-gypsum process and the Saarberg-Hölter-Lurgi lime-gypsum process for two other versions of this basic process type.

Process Description

Figure 2.31 shows a simplified flow diagram for application of the process to an oil-fired water tube boiler.

The system comprises two main sections: the sulphur oxides removal section; and the gypsum production section.

It is assumed that impure gypsum will be produced as a waste product, so that complete conversion of the limestone to gypsum will not be required.

For sulphur oxides removal, gas from the boiler, boosted by a Fan, is cooled in the Heat Exchanger and by injection of water into the gas, and is then

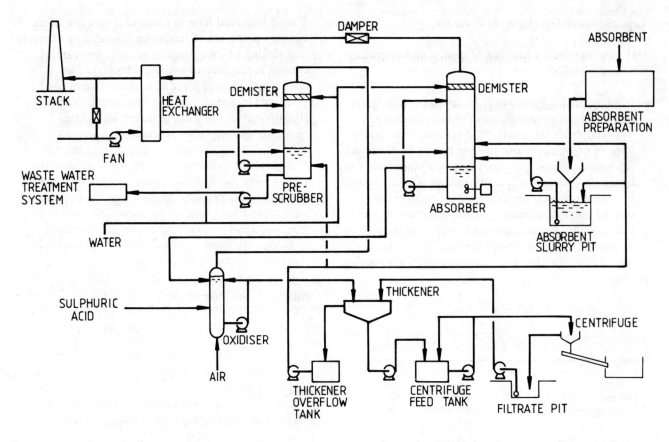

Figure 2.31 I.H.I. Limestone-Gypsum FGD Process

scrubbed with limestone slurry circulated through the Absorber to remove sulphur oxide. The purified gas then flows via a Demister to the Heat Exchanger, where it is reheated by exchange with incoming gas, before flowing via a damper to the Stack.

Limestone slurry is prepared from raw limestone, which is ground and mixed with water in the Limestone Slurry Preparation System. The slurry is circulated through the Absorber; a proportion of the circulating slurry is bled off to the Gypsum Production Section.

In the Gypsum Production Section the slurry is circulated through the Oxidiser, in which calcium sulphite hemihydrate reacts with the oxygen in air blown through the slurry. Sulphuric acid is added to adjust the pH of the slurry to the optimum for oxidation. Air leaving the Oxidiser joins the gas entering the Absorber. Part of the slurry circulating through the Oxidiser is bled off to a Thickener, where the gypsum is concentrated.

The overflow from the Thickener flows via the Thickener Overflow Tank, from which it is pumped to the Absorber, and the Limestone Slurry Preparation System. The thickened underflow is pumped via a pump tank to the Centrifuges, from which gypsum, containing less than 10% water, is removed. The centrifuge filtrate is pumped back to the Thickener.

About 8% of the nitrogen oxides content of the gas is removed by the Absorber. A modification of the process (Section 5) allows for simultaneous removal of the sulphur and nitrogen oxides.

Status and Operating Experience

The status of the IHI process is indicated in Section 2.3, where it is seen that since 1971, thirty-three plants handling between 10,000 and 2,400,000 Nm³/h of gas (total 13,055,000 Nm³/h) have been built or are under construction. Of these, nine plants have been erected for utility coal-fired and oil-fired boilers for generating capacities between 26 MWe and 700 MWe, and a further four (156 to 500 MWe coal-fired utility units) are under construction.

Information on operating experience is limited. Operating experience for a plant for a 700 MWe coal-fired boiler was found [678] to be generally satisfactory. However, it has been reported [189] that for the same plant, scaling occurred in the demister, but the scale formation was corrected by changing the spray pattern on the demister. Small extents of scaling occurred on other surfaces in contact with slurry – this is a hazard in the presence of calcium sulphate, which is slightly soluble and crystallises out.

Variations and Development Potential

IHI have adopted a number of design and operating variations, including:

- Cooling of the gas entirely by injection of a water spray instead of using a gas-gas heat exchanger. This simplifies the plant, as it removes one major item of equipment, but the gas has to be reheated (to restore plume buoyancy) e.g. by combustion of liquid or gaseous fuel, or by mixing with air heated in a steam-heated air heater. This reduces the overall thermal efficiency of the combustion plant, but gives greater design simplification and flexibility as regards stack gas temperature. This variation would therefore be worth consideration for smaller plants.

- Use of hydrated lime in place of limestone. This entails calcination of limestone (usually by others) and slaking of lime (usually on site, but careful control is required to avoid problems). The advantages include a reduced requirement for fine grinding (hydrated lime is a fine powder) and a slightly higher solubility in water and higher reactivity (giving faster reaction).

- Design of the Absorber: a number of alternatives can be used: spray towers (the usual choice, giving the lowest gas pressure losses); venturi scrubbers with spray banks (higher efficiency but higher pressure loss); and turbulent contact absorbers (TCAs) with spray banks (maximum efficiency, but also with high pressure loss).

A number of design and operating variations also exist between manufacturers. The most important of these are:

- Elimination of the oxidation stage. This eliminates a major plant item, but gives rise to the need for disposing of calcium sulphite hemihydrate (see Section 2.2) to a settling lagoon. This material is difficult to concentrate and dewater, so that its disposal gives major problems.

- Combining absorption and oxidation in one vessel, eliminating a process stage, as in the Chiyoda Thoroughbred 121 process (see below), the Saarberg-Hölter-Lurgi process (see below) and the Babcock-Hitachi process. This also reduces the requirements for lime or limestone.

- Use of additives such as formic acid and adipic acid, as in the Saarberg-Hölter-Lurgi process (see below). These act as buffering agents which result in more rapid reaction between sulphur oxides and limestone, but they increase the cost for the reagent slurry.

In essence, this is a well-established and proven design.

The basic simplicity of the process allows for scaling-up of the plant over very wide ranges, and

Table 2.41 Process Requirements–Oil-Fired Combustion System 1

Inlet Gas at full load		
Volume flow	'000 Nm^3/h	456
Dry gas	'000 Nm^3/h	408
Water vapour	'000 Nm^3/h	48
Actual volume flow	'000 m^3/h	707
Temperature	°C	150
Particulates content	mg/Nm^3	39
SO$_2$ content	mg/Nm^3	5010
NO$_x$ content	mg/Nm^3	1025
HCl content	mg/Nm^3	3
Exit Gas at full load		
Volume flow	'000 Nm^3/h	482
Dry gas	'000 Nm^3/h	407
Water vapour	'000 Nm^3/h	75
Actual volume flow	'000 m^3/h	641
Temperature	°C	90
Particulates content	mg/Nm^3 (dry)	15 (17)
SO$_2$ content	mg/Nm^3 (dry)	650 (1086)
NO$_x$ content	mg/Nm^3 (dry)	945 (953)
HCl content	mg/Nm^3 (dry)	3 (3)
Reaction temperature	°C	55
Particulates removal		Simultaneous
Reagent		Limestone
Particle size		90% below 75 micron
Concentration in slurry	wt. %	14
Ca/S Molar Ratio		1.05
Requirements at full load		
Limestone (97% pure)	tonne/h	3.46
Water	GJ/h	22.1
Electric Power	MWe	2.6
Manpower	men/shift	1*
Average load factor	%	***

Figures in parentheses are annual average emissions for 90% FGD plant availability

Table 2.42 Estimated Properties of Gypsum Produced–Oil-Fired Combustion System 1

Rate of production:		
Wet	tonne/h	6.08
Dry	tonne/h	5.48
Composition of wet product:		
Moisture	wt. %	10.0
Particulates	wt. %	0.3
Calcium carbonate	wt. %	9.4
Other impurities	wt. %	1.7
Gypsum	wt. %	78.6
Purity of dry product	wt. %	87.3

adoption of modifications to suit the particular needs of the combustion plant. For example, IHI [173] have described conceptual designs for plant to handle the gas from a 1000 MWe utility boiler in a single Absorber spray tower. The sulphur capture efficiency of the Absorber can be controlled by varying the liquid flow/gas flow ratio (L/G ratio) in the Absorber, and by controlling the pH of the slurry; maximum efficiency of capture is at a pH of about 5.3.

Process Requirements for Each Application Considered

These are shown in Table 2.41 for the two applications considered: Datum Combustion Systems 1.

It is assumed that:

- The SO_2 content of the gas is to be reduced to 650 mg/Nm³ (dry)

- The NO_x content of the gas is reduced by 8%.

- The HCl content of the gas is unaffected.

- The particulates content of the gas is reduced to 15 mg/Nm³ (dry).

- The end product is gypsum which will be disposed of to a land fill. High purity is not required, and only minor quantities of sulphuric acid will be required to adjust the pH for oxidation.

By-products and Effluents

The effluent from the process is gypsum, which has to be disposed of, e.g. to a landfill, unless:

- It is produced in sufficient purity–achieved without difficulty by addition of sulphuric acid to convert unreacted limestone to gypsum; however, this measure is probably worthwhile only on the largest scale.

- There is a market for it e.g. for manufacturing plasterboard–this will apply in only a small number of instances.

Some care has to be taken in the selection of a suitable site for disposal, as calcium sulphate is slightly soluble in water, and the sulphate anion is harmful to concrete.

The rate of production and impurities content of gypsum produced are summarised in Table 2.42 for the Datum Combustion System considered.

Efficiency and Emission Factors

The efficiency and emission factors for the process are summarised in Table 2.43 for the application considered: Datum Combustion System 1.

In calculating the efficiency and emission factors, it is assumed that at full load:

- The performance without the incorporation of the FGD plant would be as shown in Table 1.4.

- The FGD unit consumes 2.6 MWe electrical power, regarded as equivalent to the combustion of a 0.7 tonne/h of oil at a power station, assuming an overall power efficiency of 33%. This additional oil is arbitrarily assumed to be included with the oil burned in the boiler for calculating the efficiency and emission factors.

For illustration purposes, the annual average emissions and emission factors for FGD plant availabilities of 100% and 90% are given in Tables 2.41 and 2.43. Further details are given in Section 1.6.

Effect of Load Variations: ***

Effect of Design Variations: ***

Limitations: ***

Table 2.43 Efficiency and Emission Factors

Oil heat input (gross)	MWt	464
Oil fired	tonne/h	39.4
FGD plant power consumed	MWe	2.6
Equivalent oil input	tonne/h	0.7*
Useful energy from system	GJt/h	1468
Total equiv. oil input	tonne/h	40.1(a)
Efficiency factor	GJt/tonne	36.6(a)
Emissions		
Sulphur in SO_2	kg/h	133 (222)
Nitrogen in NO_x	kg/h	117 (118)
Chlorine in HCl	kg/h	1.2 (1.2)
Particulates	kg/h	6.1 (7.1)
Emission factors (per tonne oil)		(a)
Sulphur	kg/tonne	3.32 (5.54)
Nitrogen	kg/tonne	2.92 (2.94)
Chlorine	kg/tonne	0.03 (0.03)
Particulates	kg/tonne	0.15 (0.18)

*Calculated assuming overall power generation efficiency = 0.33
(a) Based on oil fired to boiler plus oil equivalent to electric power consumed
Figures in parentheses are annual average emissions for 90% FGD plant availability

Costs

Cost basis: See Section 2.6

Capital costs: See Section 2.6–for retrofit and for new build applications.

Annual running costs and cost factors: See Section 2.6–for retrofit and for new build applications.

Effects of Design Variations on Costs: ***

Effects of Annual Load Patterns on Annual Running Costs: ***

Process Advantages and Drawbacks

Well-established process; simple chemistry; high sulphur capture efficiency (over 95% capture can be attained); low reagent/sulphur stoichiometric ratio; waste product potentially marketable, and reasonably benign if disposal is required.

The disadvantages are that a large land area is required (reduced if gypsum is produced); fine grinding needed (for limestone but not for lime); careful design and control needed to avoid scaling problems.

Evaluation of Chiyoda Thoroughbred 121 FGD Process
(Process Code S21.1–Limestone or Lime Slurry Scrubbing Processes)

See Section 2.2 for: outline of the basic process; its chemistry; block diagram.

See Section 2.2 for a list of manufacturers offering this type of equipment.

See Section 2.3 for general appraisal of the basic process.

See Section 2.4 for the reason for choosing the Limestone/Lime Slurry Scrubbing Process.

This basic process type is considered (Section 2.4) to be suitable for application in the UK only to large boilers, and hence it is evaluated here only for Datum Combustion System 1. The process presented here, which is of the type producing gypsum as an end product, is offered by Chiyoda Chemical Engineering and Construction Company Ltd, Yokohama, Japan (see Appendix 2 for details of this manufacturer).

See also the IHI limestone-gypsum process (above) and the Saarberg-Hölter-Lurgi lime-gypsum process (below) for two other versions of this basic process type.

Process Description

Figure 2.32 shows a simplified flow diagram for application of the process to an oil-fired water tube boiler.

The system comprises two main sections: the sulphur oxides removal section; and the gypsum dewatering section.

It is assumed that the gypsum will be produced as a waste product; however, as the process has a

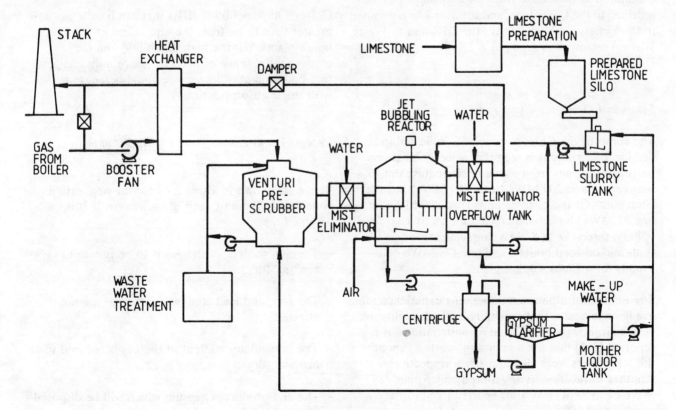

Figure 2.32 Chiyoda Thoroughbred 121 Limestone-Gypsum FGD Process

limestone conversion to gypsum approaching 99%, the gypsum produced will be of high purity.

For sulphur oxides removal, gas from the boiler, boosted by a Fan, is cooled in the Heat Exchanger and by injection of water, and then enters a Jet Bubbling Reactor (JBR) in which the following processes occur simultaneously: absorption of sulphur oxides, by reaction with a limestone slurry, to form calcium sulphite and sulphate; oxidation of calcium sulphite to sulphate by reaction with oxygen in air pumped into the JBR; and precipitation of the oxidation product (gypsum). In the JBR, the gas is bubbled through the slurry, via a large number of pipes dipping into the slurry. Most of the limestone dissolves in the slurry due to the existence of a low pH (in the range 3.5–5.0). The oxidising air is admitted via sparge pipes at the base of the JBR. The purified gas leaving the JBR, together with nitrogen and unreacted oxygen from the oxidising air, flows via a Mist Eliminator and Damper to the Heat Exchanger, where it is reheated by exchange with incoming gas, before flowing to the Stack.

Limestone slurry is prepared from raw limestone, which is ground and mixed with water in the Limestone Slurry Preparation System. The slurry is pumped to the JBR. Gypsum slurry is pumped to the Gypsum Dewatering Section.

In the Gypsum Dewatering Section the slurry is pumped with a Gypsum Slurry Pump to a Basket Centrifuge which yields a product containing about 10% water, and a filtrate which flows into a Gypsum Clarifier. The underflow from the Gypsum Clarifier is returned to the Centrifuge, and the overflow is pumped to the Limestone Slurry Tank, the JBR and to Waste Water Treatment as required.

Status and Operating Experience

The status of the Chiyoda process is indicated in Section 2.3, where it is seen that since 1978, seven units handling gas from plants of size equivalent to between 23 and 200 MWe (total 700 MWe) have been built. Of these, three units (two 200 MWe and one 23 MWe) have been erected for utility coal-fired boilers, three (52 to 87 MWe equivalent) for coal- and oil-fired boilers, and one (75 MWe equivalent) for a Claus sulphur plant.

The limited information on operating experience available indicates that reliability and availability are very high. For the two 200 MWe installations, it is reported [263] that for a period of nearly 11 months the reliabilities were 99.9 and 100% respectively. Another installation at an oil-fired boiler plant (87 MWe equivalent) was also reported [685] to have a reliability of 100%.

Variations and Development Potential

These can include:

– Cooling of the gas entirely by injection of a water spray instead of using a gas-gas heat exchanger. This simplifies the plant as it removes one major item of equipment, but the purified gas has to be reheated to 90–150°C, e.g. by burning liquid or gaseous fuel in it, or by injecting air from a steam-heated air heater, in order to restore plume buoyancy. This reduces the overall thermal efficiency of the combustion plant, but gives greater design simplification and flexibility as regards stack gas temperature. This variation would therefore be worth consideration for smaller plants.

– Reheating, as described above, in conjunction with use of a heat exchanger in locations where greater reheat is needed for plume buoyancy.

Chiyoda claim that the adoption of the JBR qualifies this as a second-generation limestone-gypsum FGD process, in that it eliminates the need for slurry recycle pumps, slurry spray nozzles, separate reaction vessels for absorption and oxidation reactions, thickeners and other pre-watering equipment, and ancillary equipment such as pumps, agitators, rakes and process instrumentation. The low pH allows the limestone to dissolve, resulting in more rapid reaction with sulphur oxides and freedom from scaling and plugging in the mist eliminator. Further, rapid dissolution of limestone allows a coarser material (e.g. 90% below 75 micron) to be used.

Chiyoda have not built JBRs that can handle gas flows greater than those from the equivalent of 200 MWe boiler plants, but the high reliability, and the consequent absence of need for spare units, suggest that boiler plant of larger size could be equipped with single absorption trains.

Process Requirements for Each Application Considered

These are shown in Table 2.44 for the application considered: Datum Combustion System 1. It is assumed that:

– The SO_2 content of the gas is to be reduced to 650 mg/Nm^3 (dry).

– The NO_x and HCl contents of the gas are not affected.

– The particulates content of the gas is reduced to 15 mg/Nm^3 (dry).

– The end product is gypsum which will be disposed of to a landfill.

- There is one JBR system

- The molar ratio of (fresh sorbent)/(captured sulphur) is 1.01 mol Ca per mol S absorbed, giving a limestone utilisation of 99%.

By-products and Effluents

The effluents from the process consist of gypsum and waste water requiring treatment. The gypsum has to be disposed of, e.g. to a landfill, unless:

- It is produced in sufficient purity.

Table 2.44 Process Requirements – Oil-Fired Combustion System 1

Inlet Gas at full load		
Volume flow	'000 Nm³/h	456
Dry gas	'000 Nm³/h	408
Water vapour	'000 Nm³/h	48
Actual volume flow	'000 m³/h	707
Temperature	°C	150
Particulates content	mg/Nm³ (dry)	39
SO₂ content	mg/Nm³ (dry)	5010
NOₓ content	mg/Nm³ (dry)	1025
HCl content	mg/Nm³ (dry)	3
Exit Gas at full load		
Volume flow	'000 Nm³/h	471
Dry gas	'000 Nm³/h	407
Water vapour	'000 Nm³/h	64
Actual volume flow	'000 m³/h	625
Temperature	°C	90
Particulates content	mg/Nm³ (dry)	15 (17)
SO₂ content	mg/Nm³ (dry)	650 (1086)
NOₓ content	mg/Nm³ (dry)	1025 (1025)
HCl content	mg/Nm³ (dry)	3 (3)
Reaction temperature	°C	53
Particulates removal		Simultaneous
Reagent		Limestone
Purity	Wt. %	97
Particle size		90% below 150 micron
Concentration in slurry	Wt. %	21
Ca/(Captured S) Molar Ratio		1.01
Requirements at full load		
Limestone (97% pure)	tonne/h	2.9
Water	tonne/h	35
Electric Power	MWe	1.9
Manpower	men/shift	***
Average load factor	%	***

Figures in parentheses are annual average emissions for 90% FGD plant availability

- There is a market for it e.g. for manufacturing plasterboard – this will apply in only a small number of instances.

It is assumed that the gypsum produced is disposed of to a landfill, and some care has to be taken in the selection of a suitable site for disposal, as calcium sulphate is slightly soluble in water, and the sulphate anion is harmful to concrete.

The rate of production and impurities contents of gypsum produced are summarised in Table 2.45 for the Datum Combustion System considered.

The requirements for water treatment are dependent on the composition of the waste water leaving the plant, and on the local waste water disposal consent requirements. The main constituents of waste water to a water treatment plant will be particulates, together with calcium, sulphite, sulphate and bicarbonate ions.

Table 2.45 Estimated Properties of Gypsum Produced – Oil-Fired Combustion System 1

Rate of production:		
Wet	tonne/h	5.4
Dry	tonne/h	4.8
Composition of wet product:		
Moisture	wt. %	10.0
Inerts	wt. %	1.6
Particulates	wt. %	0.2
Calcium carbonate	wt. %	0.5
Gypsum	wt. %	87.7
Purity of dry product	wt. %	97.5

Efficiency and Emission Factors

The efficiency and emission factors for the process are summarised in Table 2.46 for Datum Combustion System 1.

In calculating the efficiency factors, it is assumed that at full load:

- The performance without the incorporation of the FGD plant would be as shown in Table 1.4.

- The power consumption of 1.9 MWe by the FGD plant is equivalent to the combustion of a further 0.5 tonne/h of oil at a power station, assuming an overall power generation efficiency of 33%. This additional oil is arbitrarily assumed to be included with the oil burned in the boiler for calculating the efficiency and emission factors.

For illustration purposes, the annual average emissions and emission factors for FGD plant availabilities of 100% and 90% are given in Tables 2.44 and 2.46. Further details are given in Section 1.6.

Table 2.46 Efficiency and Emission Factors – Oil-Fired Combustion System 1

Oil heat input (gross)	MWt	464
Oil fired	tonne/h	39.4
FGD plant power consumed	MWe	1.9
Equivalent oil input	tonne/h	0.5*
Useful energy from system	GJt/h	1468
Total equiv. oil input	tonne/h	39.9(a)
Efficiency factor	GJt/tonne	36.8(a)
Emissions		
Sulphur in SO_2	kg/h	133 (222)
Nitrogen in NO_x	kg/h	127 (127)
Chlorine in HCl (assumed)	kg/h	1.2 (1.2)
Particulates	kg/h	6.1 (7.1)
Emission factors (per tonne coal)		(a)
Sulphur	kg/tonne	3.33 (5.56)
Nitrogen	kg/tonne	3.18 (3.18)
Chlorine (assumed)	kg/tonne	0.03 (0.03)
Particulates	kg/tonne	0.15 (0.18)

*Calculated assuming overall power generation efficiency = 0.33
(a) Based on oil fired to boiler plus oil equivalent to electric power consumed
Figures in parentheses are annual average emissions for 90% FGD plant availability

Effect of load variations: ***

Effect of design variations: ***

Limitations: ***

Costs

Cost basis: See Section 2.6.

Capital costs: For retrofit and for new build applications – see Section 2.6.

Annual running costs and cost factors: See Section 2.6.

Effects of Design Variations on Costs: ***

Effects of Annual Load Patterns on Annual Running Costs: ***

Process Advantages and Drawbacks

Process well-established in Japan, and successfully tested in USA on 23 MWe scale; very high reliability; simple chemistry; high sulphur capture efficiency (over 95% capture can be attained); very low reagent/Sulphur stoichiometric ratio; land area required is smaller than for many other limestone-gypsum processes; waste product potentially marketable, and reasonably benign if disposal is required.

The disadvantages are that fine grinding is needed (but coarser than for other limestone-gypsum processes). No large scale experience – may be scale up problems unless multiple parallel units are employed.

Evaluation of Saarberg-Hölter FGD Process (Process Code S21.1 – Limestone or Lime Slurry Scrubbing Processes)

See Section 2.2 for: outline of the basic process; its chemistry; block diagram.

See Section 2.3 for list of manufacturers offering this type of equipment.

See Section 2.3 for general appraisal of the basic process.

See Section 2.4 for the reason for choosing the Limestone/Lime Slurry Scrubbing Process.

This basic process type is considered (Section 2.4) to be suitable for application in the UK only to large boilers, and hence it is evaluated here only for Datum Combustion System 1. The process presented here is of the type using lime or limestone slurry with an organic acid (formic acid) additive and producing gypsum as an end product. The process is offered in Europe by Saarberg-Hölter-Lurgi GmbH, and in the United States by the Davy McKee Corporation (see Appendix 2 for details of these manufacturers).

See also the IHI limestone-gypsum process (above) and the Chiyoda Thoroughbred 121 limestone-gypsum process (below) for two other versions of the same basic process.

Process Description

Figure 2.33 shows a simplified flow diagram for application of the process to an oil-fired water tube boiler.

The system comprises two main sections: the sulphur oxides removal section, and the gypsum dewatering section.

It is assumed that the gypsum will be produced as a waste product; however, as the process has a lime conversion to gypsum of about 98% the gypsum produced will be of high purity.

For sulphur oxides removal gas from the boiler, boosted by a Fan, is cooled in the Heat Exchanger and by injection of water before entering a 'Rotopart Scrubber'. This is divided into two sections, each provided with sprays through which the absorbent is circulated. In the first stage, the incoming gas flows downwards, co-current with the absorbent spray; in the second section, the gas flow is upwards counter-current to the spray. The following processes

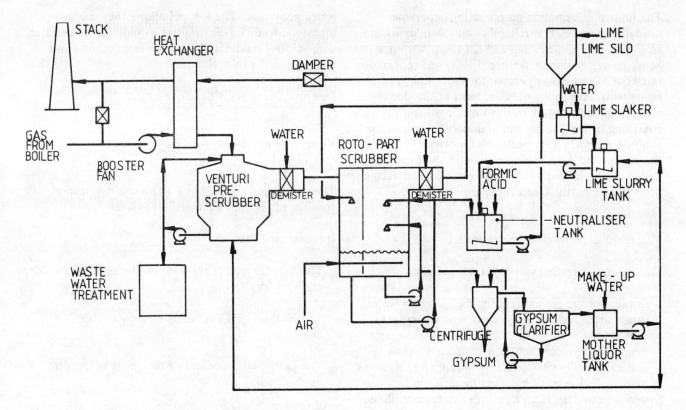

Figure 2.33 Saarberg-Hölter-Lurgi Lime-Gypsum FGD Process

occur simultaneously in the Scrubber: absorption of sulphur oxides, by reaction with the lime or limestone slurry, to form calcium sulphite, bisulphite and sulphate; oxidation of calcium sulphite and bisulphite to sulphate by reaction with oxygen in air pumped into the Rotopart Scrubber; and precipitation of the oxidation product (gypsum).

Most of the lime or limestone dissolves in the slurry due to the existence of a low pH (in the range 3.5–5.0) resulting from the formation of soluble calcium bisulphite. A liquid/gas flow ratio of about 3 litre/Nm^3 is required with lime as reagent, or 9–10 litre/Nm^3 with limestone (compared with up to about 18 litre/Nm^3 with limestone slurry scrubbers without integrated oxidation). The oxidising air is admitted via sparge pipes at the base of the Rotopart Scrubber. The purified gas leaving the Rotopart Scrubber, together with nitrogen and unreacted oxygen from the oxidising air, flows via a Demister and Damper to the Heat Exchanger, where it is reheated by exchange with incoming gas, before flowing to the Stack.

Lime slurry is prepared from raw lime (assumed to be calcined off-site), which is stored in the Lime Silo, slaked in the Lime Slaker with water, and slurried in the Lime Slurry Tank with recycled mother liquor from the Gypsum Separation System. The slurry is pumped to the Neutraliser Tank, where it is mixed with a bleed from the first section of the Rotopart Scrubber. The formic acid additive is also added in the Neutraliser Tank, and the absorbent is pumped from the Neutraliser Tank to the first section of the Rotopart Scrubber. Part of the absorbent in the base of the first section of the Scrubber is pumped to sprays in the second section. The absorbent in both sections is continuously treated with oxidising air admitted through sparge pipes, and this precipitates gypsum as a slurry. A bleed of slurry circulated in the second section is pumped to the Gypsum Dewatering Section.

In the Gypsum Dewatering Section the slurry is pumped with a Gypsum Slurry Pump to a Basket Centrifuge which yields a product containing about 10% water, and a filtrate which flows into a Gypsum Clarifier. The underflow from the Gypsum Clarifier is returned to the Centrifuge, and the overflow is pumped to the Mother Liquor Tank, and thence to the Lime Slurry Tank and the Prescrubber as required.

Status and Operating Experience

The status of the Saarberg-Hölter process is indicated in Section 2.3, where it is seen that since 1974, twenty-five units handling gas from plants of size equivalent to between 1 and 600 MWe (total 5760 MWe) have been built or are under construction; and a further seven units have been built handling 4000–131,000 Nm^3/h gas (total 350,000 Nm^3/h) from incinerators, Claus plants and boilers. One unit installed in the 230 MWe station of Saarbergwerke at Völklingen in FRG has been erected within the cooling tower, into which the exhaust gases are discharged, thus dispensing with the need for reheating and for a stack.

The limited information on operating experience available indicates that reliability and availability are very high. A plant treating 25% of the gas from a 700 MWe power station at Weiher, FRG, was reported in 1983 [192] to have operated for more than 12,500 hours with a reliability of 98%; most of the downtime was in the early history of this unit, involving balancing of the flue gas fan and modification of the solution pumps. Other problems reported [652] for the same plant were caking of lime in the Lime Silo, and difficulty in finding the right cloth for the vacuum filter (replacing the Centrifuge: see below).

Variations and Development Potential

Design and operating variations can include:

- Cooling of the gas entirely by injection of a water spray instead of using a gas-gas heat exchanger. This simplifies the plant, as it removes one major item of equipment, but the purified gas has to be reheated to 90–150°C, e.g. by burning liquid or gaseous fuel in it, or by injecting air from a steam-heated air heater, in order to restore plume buoyancy. This reduces the overall thermal efficiency of the combustion plant, but gives greater design simplification and flexibility as regards stack gas temperature. This variation would therefore be worth consideration for smaller plants.

- Reheating, as described above, in conjunction with use of a heat exchanger in locations where greater reheat is needed for plume buoyancy.

- Use of limestone as the absorbent in place of lime. This reduces reagent costs, but increases the liquid/gas flow ratio needed (to about 9 litre/Nm³ gas) and also increases the energy requirement for grinding the limestone.

- When lime is used as the absorbent, the sprays in the second section of the Rotopart Scrubber can be dispensed with. In this case, the base of the Rotopart Scrubber becomes little more than a narrow trough, and oxidation takes place in a separate oxidiser.

- Replacement of the Centrifuge for dewatering the gypsum by a thickener and rotary vacuum filter. The thickener also replaces the Neutraliser Tank, and the formic acid and lime slurry are added to the slurry flowing to the thickener.

The adoption of the Rotopart Scrubber, with simultaneous absorption of SO_2 and oxidation to gypsum, can eliminate the need for: slurry circulation pumps, or alternatively for separate reaction vessels for absorption and oxidation reactions; thickeners and other pre-dewatering equipment; and ancillary equipment such as agitators, rakes and some process instrumentation. The low pH allows the lime or limestone to dissolve, resulting in more rapid reaction with sulphur oxides and freedom from scaling and plugging in the mist eliminator.

Potential for process development: ***

Process Requirements for Each Application Considered

These are shown in Table 2.47 for the application considered: Datum Combustion System 1.

It is assumed that:

- The SO_2 content of the gas is to be reduced to 650 mg/Nm³ (dry).

- The NO_x and HCl contents of the gas are unaffected.

- The particulates content of the gas is reduced to 15 mg/Nm³ (dry).

- The end product is gypsum which will be disposed of to a landfill, and high purity is not required.

- There is one Rotopart Scrubber system.

- The absorbent is hydrated lime, prepared by slaking quicklime on site.

- The molar ratio of (fresh sorbent)/(captured sulphur) used is 1.02 mol Ca per mol S absorbed, giving a lime or limestone utilisation of 98%.

By-products and Effluents

The effluents from the process consist of gypsum and waste water requiring treatment. The gypsum has to be disposed of, e.g. to a landfill, unless:

- It is produced in sufficient purity.

- There is a market for it e.g. for manufacturing plasterboard – this will apply in only a small number of instances.

It is assumed that the gypsum produced is disposed of to a landfill, and some care has to be taken in the selection of a suitable site for disposal, as calcium sulphate is slightly soluble in water, and the sulphate anion is harmful to concrete.

The rate of production and impurities content of the gypsum produced are summarised in Table 2.48 for the Datum Combustion System considered.

Water treatment requirements are dependent on the

Table 2.47 Process Requirements – Oil-Fired Combustion System 1

Inlet Gas at full load		
Volume flow	'000 Nm³/h	456
Dry gas	'000 Nm³/h	408
Water vapour	'000 Nm³/h	48
Actual volume flow	'000 m³/h	707
Temperature	°C	150
Particulates content	mg/Nm³ (dry)	39
SO₂ content	mg/Nm³ (dry)	5010
NOₓ content	mg/Nm³ (dry)	1025
HCl content	mg/Nm³ (dry)	3
Exit Gas at full load		
Volume flow	'000 Nm³/h	462
Dry gas	'000 Nm³/h	407
Water vapour	'000 Nm³/h	55
Actual volume flow	'000 m³/h	613
Temperature	°C	90
Particulates content	mg/Nm³ (dry)	15 (17)
SO₂ content	mg/Nm³ (dry)	650 (1068)
NOₓ content	mg/Nm³ (dry)	1025 (1025)
HCl content	mg/Nm³ (dry)	3 (3)
Reaction temperature	°C	49
Particulates removal		Simultaneous
Reagent		Lime
Purity	Wt. %	94.8
Particle size		90% below 150 micron
Concentration in slurry	Wt. %	15
Ca/(Captured S) Molar Ratio		1.02
Additive		Formic acid
Concentration in slurry	Wt. %	0.08
Liquid/gas ratio	litre/m³	5.2
Requirements at full load		
Lime (94.8% pure)	tonne/h	1.67
Formic acid (87% soln.)	kg/h	7
Antifoulant	kg/h	10
Process air (0.5 bar)	tonne/h	2.7
Water	tonne/h	8
Electric power	MWe	1.4
Manpower	men/shift	***
Average load factor	%	***

Figures in parentheses are annual averages for 90% FGD plant availability

Table 2.48 Oil-Fired Combustion System 1

Rate of production:		
Wet	tonne/h	5.5
Dry	tonne/h	4.9
Composition of wet product:		
Moisture	wt. %	10.0
Calcium hydroxide	wt. %	0.7
Calcium formate	wt. %	0.2
Other impurities	wt. %	1.7
Gypsum	wt. %	87.4
Purity of dry product	wt. %	97.2

composition of the waste water leaving the plant, and on the local waste water disposal consent requirements. The main constituents of waste water to a water treatment plant will be particulates, together with calcium, sulphite, sulphate and bicarbonate ions.

Efficiency and Emission Factors

The efficiency and emission factors for the process are summarised in Table 2.49 for the application considered: Datum Combustion System 1.

In calculating the efficiency factors, it is assumed that at full load:

– The performance without the incorporation of the FGD plant would be as shown in Table 1.4.

– The power consumption of 1.4 MWe by the FGD plant is equivalent to the combustion of a further 0.4 tonne/h of oil at a power station, assuming an overall power generation efficiency of 33%. This additional oil is arbitrarily assumed to be included with the oil burned in the boiler for calculating the efficiency factor.

Table 2.49 Efficiency and Emission Factors – Oil-Fired Combustion System 1

Oil heat input (gross)	MWt	464
Oil fired	tonne/h	39.4
FGD plant power consumed	MWe	1.4
Equivalent Oil input	tonne/h	0.4*
Useful energy from system	GJt/h	1468
Total equiv. oil input	tonne/h	39.8(a)
Efficiency factor	GJt/tonne	36.9(a)
Emissions		
Sulphur in SO₂	kg/h	133 (222)
Nitrogen in NOₓ	kg/h	127 (127)
Chlorine in HCl	kg/h	1.2 (1.2)
Particulates	kg/h	6.1 (7.1)
Emission factors (per tonne coal)		(a)
Sulphur	kg/tonne	3.34 (5.58)
Nitrogen	kg/tonne	3.19 (3.19)
Chlorine	kg/tonne	0.03 (0.03)
Particulates	kg/tonne	0.15 (0.18)

*Calculated assuming overall power generation efficiency = 0.33
(a) Based on oil fired to boiler plus oil equivalent to electric power consumed
Figures in parentheses are annual average emissions for 90% FGD plant availability

For illustration purposes, the annual average emissions and emission factors for FGD plant availabilities of 100% and 90% are given in Tables 2.47 and 2.49. Further details are given in Section 1.6.

Effect of load variations: ***

Effect of design variations: ***

Limitations: ***

Costs

Cost basis: See Section 2.6.

Capital costs: For retrofit and for new build applications – see Section 2.6.

Annual running costs and cost factors: See Section 2.6.

Effects of design variations on costs: ***

Effects of annual load patterns on annual running costs: ***

Process Advantages and Drawbacks

Process well-established in FRG; high reliability; land area required is smaller than for many other lime-gypsum processes (and has been demonstrated to be virtually zero when the FGD plant has been built inside a cooling tower); fine grinding not needed if lime is used as absorbent (calcium hydroxide is a fine powder); absorbent dissolves under operating conditions; simple chemistry; high sulphur capture efficiency (over 95% capture can be attained); very low reagent/sulphur stoichiometric ratio (98% utilisation of lime or limestone); product is easily dewatered; product potentially marketable, and reasonably benign if disposal is required.

The disadvantages are the tendency for aeration system to become plugged by gypsum build-up; low pH necessitates use of corrosion-resistant materials.

By-products and Effluents from Limestone Slurry Scrubbing

Liquid Effluents: The quantity and composition of the liquid effluents depend on whether the scrubbing liquid circuit is operated on a closed loop or an open loop basis. Closed loop operation means that all liquid effluents are recycled after treatment to remove salts and suspended particulate matter that would otherwise interfere with the functioning of the scrubbing process and the correct operation of the mist eliminators.

Open loop operation means that liquid is discharged continuously in substantial quantities from the scrubbing plant, to be replaced by fresh water from a river, piped mains supply or a waste water discharge from some other process (e.g. boiler blowdown).

In closed loop operation, liquid effluent discharge rate is very low, or zero, the process being an overall consumer of water. This is because the treated flue gases have a higher dew-point than the gases entering the FGD system, and because some moisture is removed with the solid product or waste residue. Maintenance of a correct water balance in closed loop operation is often very difficult.

In open loop operation (except for sea-water or estuary-water scrubbing) a high proportion of the scrubbing liquid is likewise recycled after treatment, because this reduces both the water input to the process and the amount of contaminated effluent that has to be discharged. Because of the difficulty of maintaining a water balance in closed loop operation, open loop is preferable, except where water is in severely limited supply. Many inland US plants employ closed loop operation.

The principal source of contaminants in the effluent from the SO_2 absorption circuit are impurities present in the limestone or lime used as sorbent. Apart from silica and alumina compounds, magnesium and iron are the most common impurities. All are present in very variable amounts and, in some applications of the slurry scrubbing process, magnesium is deliberately introduced to improve absorption of SO_2 and to reduce scaling.

Part of the alumina and most of the iron and magnesium dissolve in the scrubbing liquor. The silica and undissolved aluminosilicates are carried through to form harmless impurities in the gypsum product. There is a risk of iron being precipitated as ferric hydroxide in the oxidiser, resulting in undesirable discoloration of the product, but magnesium and aluminium sulphates remain in solution although the aluminium sulphate may be partly hydrolysed to aluminium hydroxide which finishes up as an impurity in the gypsum.

The removal of magnesium from the product is achieved by thorough washing of the gypsum after separation from the supernatant liquor.

The principal substances present in the effluent derived from the scrubbing circuit are therefore calcium sulphate at saturation SO_2 level (about 0.2% by weight) and iron, magnesium and possibly aluminium sulphates, together with trace amounts of other elements derived from the sorbent.

The quantity of liquid effluent arising from the SO_2 absorption stage varies according to the rate of build-up of impurities in the scrubbing circuit and hence with the limestone composition. Typically, the rate is of the order of 60 tonne/h, but it may not be necessary to discharge the whole of this – it may be environmentally or economically attractive to recycle part of the effluent (or whole of it) following treatment to remove soluble salts and suspended particulate matter.

The processes used for water treatment consist of several steps. In the first step, sulphite and sulphates are precipitated as their calcium salts by the addition of lime. In the second step, the pH is raised to 9–10 by the addition of caustic soda, resulting in the precipitation of metal oxides and hydroxides. Many of these come down in the form of flocculant precipitates and their removal is often assisted by adding ferric chloride or a polyelectrolyte coagulant. Settling of the precipitate is effected in a thickener, the underflow from which forms a sludge that may be further dewatered in a filter press for ease of handling and disposal.

The overflow from the thickener is treated for separation of oil, if necessary, and further purified by ion exchange using weak anion exchange resins to remove thiosulphate and other reducing anions. The pH of the effluent is then adjusted before discharge.

For discharge to surface watercourses there are usually restrictions on concentrations of suspended and dissolved solids, chemical oxygen demand and pH, as well as on concentrations of specific toxic elements.

If it is proposed to recycle the effluent, a lesser degree of purification would probably be required.

Solid Residues and Products: These include gypsum and waste materials for disposal, mainly residues from water treatment.

Kyte and Cooper [144] have discussed the requirements for gypsum quality to satisfy existing market applications, and have estimated the rate of production from UK power stations equipped with FGD plants in relation to the size of existing UK market for gypsum. Table 2.50 taken from [144] shows some typical guideline specifications used in the UK, USA, Germany and Japan. In Table 2.51, the quality requirements for gypsum in cement and in wallboard manufacture are compared with the quality of gypsum produced in Japanese power stations [677]. The FGD products are seen to be well within the requirements for both applications, apart from that for purity (gypsum content) in the wallboard application, where some samples fell outside the limit.

A comparison of the quality of gypsum from Scholven F power station with that of natural gypsum is given in Table 2.52. In contrast to some of the Japanese samples, purity here is high–far superior to that of the natural gypsum with which it is compared (which would not meet the Japanese requirements for either cement or wallboard manufacture).

For wallboard manufacture, large crystal size is wanted, and this can be achieved more easily by carrying out the forced oxidation process in a separate vessel. Hence, if it is intended to market the gypsum for wallboard manufacture, processes such as the IHI process, which use a separate vessel for oxidation, are superior to those which carry out oxidation in the absorber.

In the production of plaster of paris, which is the basic material for making wallboard and other plaster products such as stucco, the gypsum ($CaSO_4.2H_2O$) is carefully calcined to form the alpha- and beta-crystalline forms of calcium sulphate hemihydrate ($CaSO_4.1/2H_2O$). For some uses the alpha form is preferred and for others the beta form. Proportions of each are controlled by using appropriate conditions for calcination. The plaster sets rapidly on mixing with water by rehydrating to the dihydrate: setting may be delayed by incorporating small amounts of an additive, e.g. borax or certain organic materials.

If calcination is carried beyond the hemihydrate stage, anhydrous $CaSO_4$ is produced and, unless 'hard burnt', this also rehydrates on mixing with water. Keene's cement and a number of other proprietary plasters are of this type, containing various additives to control setting rate and the properties to the hardened material.

The application of gypsum in the cement industry is mainly as a setting retardant and the quality of gypsum is less critical than for wallboard manufacture. The rate of addition represents only a small percentage of the cement, so the market for this use is not large.

The rate of gypsum production from a 2000 MWe power station (burning 2% sulphur coal) has been

Table 2.50 Guidelines Issued by Users for Gypsum Purity [144]

	National Gypsum Co.	Georgia Pacific Co.	US Gypsum Co.	Westroc	German/ Japanese	UK
Gypsum content min %	94	90	–	95	95	95
Calcium sulphite, %	0.5	–	–	–	0.25	0.25
Sodium ion max ppm	500	200	75	80	600	500
Chloride max ppm	800	200	75	80	100	100
Magnesium max ppm	500	–	50	50	1000	1000
Free water, max %	15	10	12	10	10	8
pH	6–8	3–9	6.5–8	6–9	5–9	5–8
Particle size, microns	–	–	20–40	–	–	16–63

Table 2.51 By-Product Gypsum Quality in relation to requirements

Property	General Requirements		Quality ex FGD	
	For Cement	For wallboard	Takasaqo Takehara	Matsushima
Surface H_2O %	10–12 max	10–12 max	6–8	6
Purity %	90 min	95 min	90–97	99
Size, microns	50 min	50 min	50–150	50
Cryst. shape	–	–	pillar and plate shaped	
pH	–	5–7	6–8	6.8
Composition %:				
$CaCO_3$	2 max	0.5–2 max	0.5–1	0.1
CaO (Combined)	–	–	32.4	32.3
SO_3	–	44 min	45.3	46.2
SO_2	–	–	0.05	0.07
MgO	–	0.08 max	0.03	0.01g(−Mg)
Na_2O	–	0.04 max	0.03(−Na)	0.003(Na)
Carbon	–	–	0.1 max	0
Wet tensile strength Kg/cm^2	–	8 min	12–13	13
Adhesion to paper	–	acceptable	acceptable	acceptable

estimated to be 15% of the existing UK market for gypsum [144]. Hence, if a large number of stations were to be equipped with gypsum-producing FGD plant it would probably be necessary to dispose of some of the gypsum produced. The disposal of unwanted, sub-standard gypsum is referred to as 'stacking', and for disposal in this way the quality of the product would not be of such importance. There would not, for example, be a need to convert excess limestone to the sulphate, but pH control in the oxidiser may still be necessary in order to achieve a satisfactory degree of oxidation.

It is possible that other outlets may be found for gypsum, if large amounts become available from flue gas desulphurisation. Research has shown that it may be used to stabilise other wastes which have to be tipped, and for this purpose a small amount of $Ca(OH)_2$ may also have to be added [328, 330]. In one patented technique [325] for the use of FGD gypsum as a stablising agent for the treatment of wastes, such as radioactive or toxic materials, the waste is mixed with FGD gypsum and a smaller amount of water-dispensable melamine formaldehyde resin.

It is however, unlikely that the waste stabilisation application will command anything like the same market value as its use as plaster.

It is reasonable to enquire whether, in the event of the product having to be dumped, the oxidation stage could not be omitted. The majority of lime/limestone slurry scrubbing processes installed in the USA are, in fact, of this type and do not use forced oxidation. However, many of the more recently installed plants do employ forced oxidation, and some of the older ones have had it installed, even though it is not proposed to market the gypsum produced.

Table 2.52 Gypsum Composition

Component %	From Scholven FPS	Natural Gypsum
$CaSO_4.2H_2O$	98–99	78–85
$CaSO3.1/2H_2O$	0–0.7	0
Na_2O	0.01	0.03
MgO	0.1	0.5–3.0
Fe	0.05	–
Silicate	–	10–20
Moisture	7–10	1

The reasons why oxidation is being adopted in the USA is because experience has shown that the calcium sulphite/sulphate sludge produced in the absence of forced oxidation is very troublesome to handle and to dewater, and many plant failures have occurred. On top of this, the sludges tend to be thixotropic, making them less than ideal for landfill. In order to create stable tips, the dewatered sludges have to be carefully blended with the correct proportion of lime. Following tipping, care has to be taken to avoid contamination of water courses or subterranean water reserves with leachates from the tipped material. Monitoring of ground water in the vicinity of the tipping area on a routine basis is insisted on by some authorities.

Water treatment sludges will be produced in relatively small quantities, the amounts depending largely on the purity of the sorbent used. Tipping, after fixation with a suitable stabilising agent (possibly gypsum plus lime) should result in no serious environmental risks.

Evaluation of Fläkt-Niro FGD Process (Process Code S22.1 – Lime Slurry Spray Dryer Processes)

See Section 2.2 for: outline of the basic process; its chemistry; block diagram.

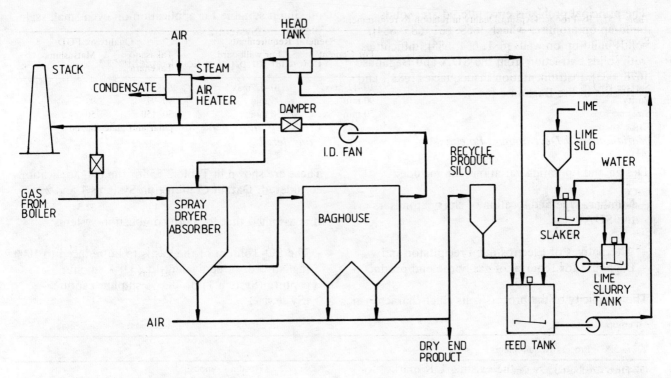

Figure 2.34 Lime Spray Drying FGD Process

See Section 2.3 for a list of manufacturers offering this type of equipment.

See Section 2.3 for general appraisal of the basic process.

See Section 2.4 for the reason for choosing the Fläkt-Niro process to illustrate the Code S22.1 basic process.

This basic process type is considered (Section 2.4) to be suitable for application in the UK to large boilers and to small factory boilers, and hence it is evaluated here for Datum Combustion Systems 1 and 2. The process presented here is of the type using lime slurry in a Spray Dryer Absorber, producing a dry end product. The process is offered in Europe by Fläkt Industri AB, (Växjö, Sweden) in collaboration with A/S Niro Atomiser, (Copenhagen, Denmark), and in the USA by Joy-Niro. See Appendix 2 for details of these manufacturers.

Process Description

Figure 2.34 shows a simplified flow diagram for application of the process to an oil-fired boiler. It is assumed that the solid end-product will be disposed of to a land fill.

For dust and sulphur oxides removal, gas from the boiler enters a Spray Dryer Absorber (SDA) fed with a spray of calcium hydroxide slurry. Operating conditions are chosen to give a gas temperature leaving the SDA that is only 10–20°C above the saturation temperature, so as to delay the dry-out of the slurry; this is because absorption of sulphur oxides is more rapid when water is present, and they are therefore removed more efficiently when dry-out is delayed. The residence time of the slurry droplets in the SDA is about 10 s. The gas then passes to an electrostatic precipitator (ESP) or, more usually a Baghouse, to remove the gas-borne reaction products which have not been collected in the hopper base of the SDA. A significant proportion (10–20%) of the total sulphur oxides capture occurs in the ESP or Baghouse. The collection hoppers of the SDA and of the ESP or Baghouse are electrically heated to prevent solids build-up resulting from condensation. The cleaned gas then passes, via a Fan, to the Stack. If necessary, it is first reheated to 90–150°C by combustion of a gaseous or liquid fuel, or by mixing with air heated by steam.

Part of the collected solids is returned via a Recycle Product Silo to the Feed Tank, where it is mixed with fresh lime slurry. The fresh lime is slaked with water in the Slaker, and the calcium hydroxide is slurried with water in the Lime Slurry Tank before being pumped to the Feed Tank. The dry waste end product contains captured particulates, calcium sulphite and sulphate, and unreacted calcium hydroxide.

Status and Operating Experience

The status of the Fläkt-Niro process is indicated in Section 2.3, where it is seen that Fläkt, Niro, and Joy have installed, or are constructing, over 50 units.

Availability of operating experience is generally high.

The principal problems encountered are: solids build-up on atomiser wheels [668, 681, 683, 684]; solids build-up on walls [681, 684, 685]; difficulties with solids extraction from the SDA and Baghouse [678, 683]; instrumentation inadequacies [683]; and corrosion in the Baghouse [683].

Variations and Development Potential

Design and operating variations can include:

– Number, type and location of slurry atomisers in the SDA.

– Alternatives of Electrostatic Precipitator and Baghouse for removal of gas-borne end-product.

The simplicity of the process is its chief characteristic, making it suitable for application on even small-scale plant (e.g. System 2).

Potential for process development: ***

Process Requirements for Each Application Considered

These are shown in Table 2.53 for the two applications considered: Datum Combustion Systems 1 and 2.

It is assumed that for both combustion systems:

– The SO_2 content of the gas is to be reduced to 1000 mg/Nm^3 (dry), equivalent to 80.0% sulphur capture (System 1) or 78.6% sulphur capture (System 2).

Table 2.53 Process Requirements

Oil-Fired Combustion System		1	2
Inlet Gas at full load			
Volume flow	'000 Nm^3/h	456	23.2
Dry gas	'000 Nm^3/h	408	20.9
Water vap.	'000 Nm^3/h	48	2.30
Actual volume flow	'000 m^3/h	707	42.8
Temperature	°C	150	230
Particulates content	mg/Nm^3 (dry)	39	36
SO_2 content	mg/Nm^3 (dry)	5010	4675
NO_x content	mg/Nm^3 (dry)	1025	615
HCl content	mg/Nm^3 (dry)	3	3
Exit Gas at full load			
Volume flow	'000 Nm^3/h	479	25.5
Dry gas	'000 Nm^3/h	407	20.9
Water vap.	'000 Nm^3/h	72	4.60
Actual volume flow	'000 m^3/h	603	32.7
Temperature	°C	75	79
Particulates content	mg/Nm^3 (dry)	15 (17)	15 (17)
SO_2 content	mg/Nm^3 (dry)	1000 (1401)	1000 (1368)
NO_x content	mg/Nm^3 (dry)	1025 (1025)	615 (615)
HCl content	mg/Nm^3 (dry)	3 (3)	3 (3)
Reaction temperature	°C	53	59
Particulates removal		Simultaneous	Simultaneous
Reagent		Lime	Slaked Lime
Purity	Wt. %	94.8	96.0
Concentration in slurry	Wt. %	12*	6
Ca/(Captured S) Molar Ratio		1.3	1.3
Requirements at full load			
Quicklime (94.8%)	tonne/h	2.0	–
Slaked Lime (96%)	tonne/h	–	0.12
Water (slaking)	tonne/h	–	0.03
Water (slurrying)	tonne/h	19	1.9
Electric power	MWe	3.1	0.16
Manpower	men/shift	***	***
Average load factor	%	***	***

*As Ca$(OH)_2$
Figures in parentheses are annual average emissions for 90% FGD plant availability

- The NO_x and HCl contents of the gas are unaffected.

- The gas leaving the Baghouse has a particulates content of 15 mg/Nm^3 (dry).

- The same molar ratio of (fresh sorbent)/(captured sulphur) is used (1.3 mol Ca per mol S absorbed), giving a lime utilisation of about 75%.

By-products and Effluents

The effluent from the process, which consists of calcium sulphate, sulphite and chloride, together with inerts present in the lime as impurities, and particulates, has to be disposed of, e.g. to a landfill.

The rates of production and composition of the material produced are summarised in Table 2.54 for the two Datum Combustion Systems considered.

Table 2.54 Estimated Properties of Waste Solids Produced

Oil-Fired Combustion System		1	2
Rate of production:	tonne/h	4.3	0.20
Composition of product:			
Moisture	wt. %	1.8	1.8
Particulates	wt. %	0.2	0.2
Calcium hydroxide	wt. %	13.1	13.1
Calcium sulphate	wt. % (a)	26.4	26.4
Calcium sulphite	wt. % (b)	56.1	56.1
Other impurities	wt. %	2.4	2.4

(a) As gypsum   (b) As hemihydrate

Efficiency and Emission Factors

The efficiency and emission factors for the process are summarised in Table 2.55 for Datum Combustion Systems 1 and 2.

In calculating the efficiency factors, it is assumed that at full load:

- The performance without the incorporation of the FGD plant would be as shown in Table 1.4.

- For Datum System 1, the FGD unit consumes 3.1 MWe of electric power, equivalent to the combustion of a further 0.8 tonne/h of oil at a power station, assuming an overall power generation efficiency of 33%. This additional oil is arbitrarily assumed to be included with the oil burned in the boiler for calculating the efficiency and emission factors.

- For Datum System 2, the FGD unit consumes 0.16 MWe of electric power, equivalent to the combustion of a further 0.04 tonne/h of oil at a power station, assuming an overall power generation efficiency of 33%. This additional oil is arbitrarily assumed to be included with the oil burned in the boiler for calculating the efficiency and emission factors.

In both Cases the gas leaving the SDA is at a temperature sufficiently high to make reheating unnecessary.

Effect of plant availability: For illustration purposes,

Table 2.55 Efficiency and Emission Factors

Oil-Fired Combustion System		1	2
Oil heat input (gross)	MWt	464	22.2
Oil fired	tonne/h	39.4	1.88
FGD plant power consumed	MWe	3.1	0.16
Equivalent oil input	tonne/h	0.8*	0.04*
Useful energy from system	GJt/h	1468	67.8
Total equiv. oil input	tonne/h	40.2(a)	1.92(a)
Efficiency factor	GJt/tonne	36.5(a)	35.3(a)
Emissions			
Sulphur in SO_2	kg/h	204 (286)	10.5 (14.3)
Nitrogen in NO_x	kg/h	127 (127)	3.92 (3.92)
Chlorine in HCl	kg/h	1.2 (1.2)	0.06 (0.06)
Particulates	kg/h	6.1 (7.1)	0.31 (0.36)
Emission factors (per tonne oil)			(a)
Sulphur	kg/tonne	5.07	5.47
		(7.11)	(7.45)
Nitrogen	kg/tonne	3.16	2.04
		(3.16)	(2.04)
Chlorine	kg/tonne	0.03	0.03
		(0.03)	(0.03)
Particulates	kg/tonne	0.15	0.16
		(0.18)	(0.19)

*Calculated assuming overall power generation efficiency = 0.33
(a) Based on oil fired to boiler plus oil equivalent to electric power consumed
Figures in parentheses are annual average emissions for 90% FGD plant availability

the annual average emissions and emission factors for FGD plant availabilities of 100% and 90% are given in Tables 2.53 and 2.55. Further details are given in Section 1.6.

Effect of load variations: ***

Effect of design variations: ***

Limitations: ***

Costs

Cost basis: See Section 2.6.

Capital costs: See Section 2.6–for retrofit and for new build applications; Datum Combustion Systems 1 and 2 only.

Annual running costs and cost factors: See Section 2.6–for retrofit and for new build applications.

Effects of design variations on costs: ***

Effects of annual load patterns on annual running costs: ***

Process Advantages and Drawbacks

Basic process well-established in USA and Europe; simple process allowing application to small scale without high labour demands; fairly high reliability; land area required is small; fine grinding not needed as lime is used as absorbent (calcium hydroxide is a fine powder); simple chemistry; fairly high sulphur capture efficiency (over 90% capture) can be attained; waste product can be easily and safely disposed of.

The disadvantages are the tendency for deposits to form on SDA walls; absorbent more costly than limestone, and higher stoichiometric ratios needed.

Effluents and Residues from Lime Slurry Spray-Dry

Liquid Effluents: The outstanding feature of spray-dry FGD processes is that, when operated correctly, they do not produce any liquid effluent directly consequent upon the desulphurisation process. A boiler plant using the process will generate only those liquid effluents normally associated with oil-fired boiler plants, i.e. boiler water blowdown, site run-off from rainfall and hose-down operations, sewage, etc.

There remains a possibility that, in cases of solids build-up in the dryer, the use of high-pressure water jets to clear the obstruction could produce quantities of strongly alkaline water containing suspended solids and calcium salts in solution. The disposal of these could present problems of differing magnitude depending on the volume of water that it was necessary to use to clear the blockage and on the situation of the plant. Although a more disagreeable operation and one introducing a risk of damage to the plant, removal of deposits in the dry state might be preferable if disposal of contaminated water is likely to be difficult. It is present practice on some facilities to recycle the contaminated water to the lime slaker.

Solid Residues: As discharged from the baghouse or electrostatic precipitator, the solid residues from a lime slurry spray-dry FGD process are in the form of a dry, free-flowing powder [337] composed of small, irregular shaped clusters of reaction products [338]. The composition of the residues varies, depending on the quantity of lime necessary to achieve the required SO_2 emission level, and the amounts of sulphur dioxide and water reacting with lime.

Compounds present, according to Donnelly [310] are: calcium sulphite hemihydrate, anhydrous calcium sulphate, calcium sulphate hemihydrate and dihydrate, calcium hydroxide and calcium carbonate.

A considerable effort has been applied, mainly by Niro Atomizer, in the study of environmental effects of the dumping of the residues, and into possible uses for them. In the absence of opportunities for marketing the material, the probable method of disposal is as landfill. Research has shown [310] that the moistened residues, when properly tipped and compacted, undergo fixation reactions which yield a filled material of high strength and low permeability to water. It is recommended, as an extra precaution, that the run-off should be collected and treated as necessary before it is released to natural drainage. It can also be recycled to the lime slaker line. Moistening of the residues can be carried out in various types of mixer, such as a pug mill or double-bladed type mixer, and sufficient water should be added to lubricate the grains of dry material to facilitate compaction. Care must be taken not to use too much water as this would produce a sticky material that is difficult to handle.

The properties of the compacted mass that determine its suitability for landfill are its density, its compressive strength, its permeability, and the composition of any leachate generated. The density of the compacted mass is an indicator of the effectiveness of compaction and the optimum amount of water addition is that which gives the maximum density upon compaction, which generally corresponds to the maximum compressive strength. The optimum water addition is selected on the basis of laboratory tests for a given waste product.

An unconfined compressive strength of 172 kN/m^2 (25

p.s.i.) is considered a minimum for supporting the machinery used in tipping operations, and values of this property for a wide range of spray-dry residues have been shown to lie between 44 and 6900 kN/m^2 after 28 days curing at 23°C [310].

Tests showed very low permeability coefficients (of the order of 10^{-6} cm/s) after 14 days of curing. It was noted also that the actual migration of water through a properly constructed landfill is substantially lower than indicated by laboratory determinations of permeability coefficients. Water penetrates the landfill only when there is standing water on it.

Leachates showed initially high pH, up to about 12 but this gradually falls, becoming eventually acid with pH in the range 5 to 6.

Possible uses for lime spray-dry residues have been identified. They are: acid mine drainage control and co-disposal as a fixation agent for hazardous wastes [310]; pelletising and curing of the moistened residues to form a synthetic aggregate for use in civil engineering and construction projects [308, 310]; use, with cement additions, in the production of bricks and building blocks grouting materials, 'stabilisate' [e.g. sub-base material for roads, etc.) or use in concrete [337]; the production of blocks for the construction of artificial reefs for coastal protection (using similar processes to those used for making building blocks) [309]; incorporation with Portland cement to produce mortars [338]; and use in cement manufacture as a setting retarder [320].

Of particular concern, where the material is likely to come into contact with steel as, for example, in reinforced concrete, is the possible effect of sulphate ions on the protective passive layer of steel which is normally formed in contact with cement. Tests on several products from spray-dry FGD showed that there was generally no sulphate attack. Only one of the products tested showed an increase in corrosion, and that only in half of the six tests carried out with it [338].

A broad view of the conclusions of all the test work indicates that, on the basis of field exposure of laboratory-produced artefacts and synthetic aggregates, satisfactory quality products can be made from spray-dry residues by using mixes of the right proportions. The amount of cement that has to be added to the mix ranges from 0–5% for 'stabilisate' and grouting material; 15 to 20% for building blocks; and in the order of 70% as a cement constituent of a concrete mix (i.e. excluding fines and aggregate).

With residues from oil combustion, the main problems arise from the variable nature of the product due to variations in lime slurry feed rate to meet emission limits, and other process variables. This variation is likely to make marketing more difficult and will tend to depress market values for the material below those commanded by prime quality materials normally purchased.

Evaluation of Wellman-Lord FGD Process (Process Code S31.1)

See Section 2.2 for: outline of the basic process; its chemistry; block diagram.

See Section 2.3 for a list of manufacturers offering this type of equipment.

See Section 2.3 for general appraisal of the basic process.

See Section 2.4 for the reason for choosing the Wellman-Lord (Code S31.1) basic process.

This basic process type is considered (Section 2.4) to be suitable for application in the UK only to large boilers having access to reagent reprocessing facilities, and hence it is evaluated here only for Datum Combustion System 1. The process is offered by the Davy-McKee Corporation. See Appendix 2 for details of this manufacturer.

Process Description

Figures 2.35 and 2.36 show simplified flow diagrams for application of the process to an oil-fired boiler. It is assumed that the sulphur dioxide produced in the process is treated to recover elemental sulphur.

For Sulphur oxides removal, gas from the boiler, boosted by a fan, is scrubbed with water in a Venturi Prescrubber and its Mist Eliminator, to cool the gas and to remove particulates. The gas enters the Absorber, which is of the valve-tray design, where it is contacted counter-currently with sodium sulphite solution, which absorbs sulphur oxides from the gas to form sodium bisulphite, together with a small proportion of sodium sulphate. The gas passes through a mist eliminator at the top of the Absorber, and is then heated by mixing with air heated by steam in the Air Heater, before being exhausted to the Stack.

The prescrubber water is circulated to a Waste Water Treatment System.

The absorbent solution from the base of the Absorber is pumped to the Absorber Product Tank; a side stream is pumped to the Sulphate Purge Tank. The further treatment of these two streams is described below.

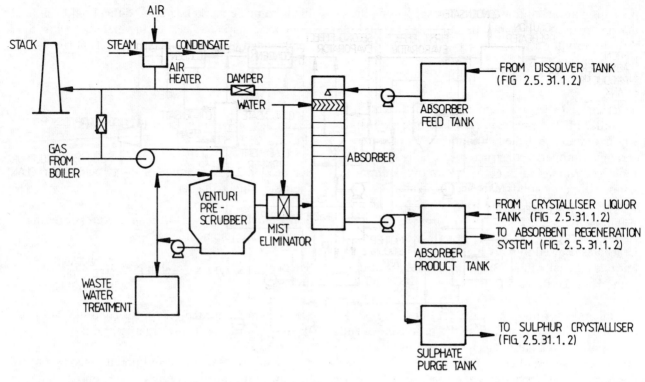

Figure 2.35 Wellman-Lord FGD Process: SO$_2$ and Dust Removal System

The absorbent regeneration system is shown in the simplified flow diagram, Figure 2.36. The solution from the Absorber Product Tank is pumped, via a Solution Preheater, to the First Effect Evaporator of a Double Effect Evaporator system. In the First Effect, the solution is heated with steam, with the condensate from the steam being used to preheat the ingoing solution in the Solution Preheater. Part of the solution circulating through the First Effect Evaporator is bled to the Second Effect Evaporator, where it is heated at a lower pressure with vapour from the First Effect Evaporator and from the Sulphate Crystalliser (see below). Condensate and vapour from the Second Effect Evaporator pass through a water-cooled Condenser. From here, condensate is treated with steam in a Condensate Stripper to strip out dissolved sulphur dioxide. The gas, with water vapour, passes through a second Condenser; condensate from this returns to the Condensate Stripper, and the stripped SO$_2$ is pumped by the SO$_2$ Compressor to the SO$_2$ Recovery Plant.

The sulphate purge treatment is also shown in Figure 2.36. The solution from the Sulphate Purge Tank is pumped via a Sulphate Purge Preheater to the Crystalliser, where it is concentrated by evaporation using heat supplied by steam; the steam condensate passes to the Sulphate Purge Preheater to preheat the ingoing sulphate purge. The vapour from the Sulphate Crystalliser passes to the steam jacket of the Second Effect Evaporator (see above). Concentrated slurry from the Crystalliser passes to a Centrifuge, in which sodium sulphate crystals are separated from the liquor. The liquor passes via the Crystalliser Liquor Tank to the Absorber Product Tank for regeneration of its bisulphite content.

As seen in Figure 2.36, part of the solution leaving the First Effect Evaporator is pumped to the Mother Liquor Tank. From here it passes in two streams: one to the Crystalliser Liquor Tank (and hence back to the Absorber Product Tank); and the other to the Dissolving Tank. The solution from the Second Effect Evaporator, containing regenerated sodium sulphite, passes directly to the Dissolving Tank, which also receives the stripped condensate from the Condensate Stripper. The losses from the system of sodium sulphate are made up by feeding sodium hydroxide to the solution in the Dissolving Tank; the sodium hydroxide forms sodium sulphite by reaction with SO$_2$ in the Absorber. The solution is pumped from the Dissolving Tank to the Absorber Feed Tank (Figure 2.35) and the start of a new cycle.

Status and Operating Experience

The status of the Wellman-Lord process is indicated in Section 2.3, where it is seen that thirty-five installations have been or are being erected, in the size range 17,000–4,636,000 m^3/h (total 22,276,000 m^3/h). These plants (twelve in USA, eighteen in Japan, four in the FRG and one in Austria) have been installed on: power station and other boilers burning coal, coke, lignite and oil; Claus plants; and sulphuric acid plants.

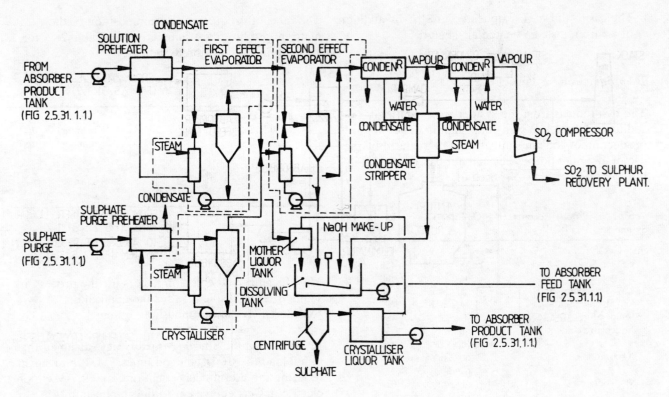

Figure 2.36 Absorbent Regeneration System of Wellman-Lord FGD Process

Operating experience shows (see Section 2.3) that availability and reliability are generally high. The principal problems that have been encountered appear not to have been process-related, but to have been caused by damage to the FGD plant resulting from difficulties in operation of the boiler plant, inexperienced FGD plant operation, and unsuitable materials of construction [658, 659]. Reliability has been quoted [276] as greater than 95%, and often over 98%.

Variations and Development Potential

Design and operating variations can include:

– Reheating of the effluent gas by heat exchange with ingoing unpurified gas.

– Adoption of mechanical vapour recompression in the evaporator system.

– Refrigerated vacuum crystallisation of the sulphate purge (this was employed on the earlier plants, but is less easy to operate than the steam-heated evaporation crystallisation) [276].

– Use of an oxidation inhibitor to reduce the extent of oxidation of absorbent to sulphate by about 75% [276].

– Alternatives for sulphur recovery: as elemental sulphur by a variety of processes (Alliance process, Claus process, Foster-Wheeler 'Resox' process); as sulphuric acid (by conventional catalytic oxidation of SO_2); or as liquid SO_2.

Although complex, the process can achieve high sulphur capture without excessive reagent make-up requirements or production of waste products. The sulphur produced can be marketed as sulphuric acid, liquid SO_2 or as elemental sulphur; or it can be disposed of safely as elemental sulphur. High operating flexibility is obtainable by installing large surge capacity for regenerated and unregenerated reagent.

Potential for process development: ***

Process Requirements for Each Application Considered

These are shown in Table 2.56 for the application considered: Datum Combustion System 1.

It is assumed that:

– The SO_2 content of the gas is to be reduced to 650 mg/Nm³ (dry) at the inlet to the reheater, equivalent to 87.0% capture of SO_2.

– The NO_x and HCl contents of the gas are unaffected.

– The particulates content of the gas is reduced to 15 mg/Nm³ (dry) at the inlet to the reheater.

79

– The gas to the stack is reheated to 81°C by addition of air from a steam-heated air-heater.

By-products and Effluents

The by-products from the system are crystallised sodium sulphate and elemental sulphur. The production of sodium sulphate can be reduced by the use of an anti-oxidant. The sulphur can be marketed, but can also be safely disposed of.

Table 2.56 Process Requirements–Oil-Fired Combustion System 1

Inlet Gas at full load		
Volume flow	'000 Nm³/h	456
Dry gas	'000 Nm³/h	408
Water vapour	'000 Nm³/h	48
Actual volume flow	'000 m³/h	707
Temperature	°C	150
Particulates content	mg/Nm³ (dry)	39
SO₂ content	mg/Nm³ (dry)	5010
NOₓ content	mg/Nm³ (dry)	1025
HCl content	mg/Nm³ (dry)	3
Exit Gas to stack at full load		
Volume flow	'000 Nm³/h	462*
Dry gas	'000 Nm³/h	407*
Water vap.	'000 Nm³/h	55
Actual volume flow	'000 m³/h	650**
Temperature	°C	81**
Particulates content	mg/Nm³ (dry)	15* (17*)
SO₂ content	mg/Nm³ (dry)	650* (1086*)
NOₓ content	mg/Nm³ (dry)	1025* (1025*)
HCl content	mg/Nm³ (dry)	3* (3*)
Reaction temperature	°C	53
Particulates removal		Separate
Reagent		NaOH
Liquid/gas flow ratio		
Prescrubber	litre/m³	3
Absorber (total)	litre/m³	0.35
Requirements at full load		
Caustic soda (47% liquor)	kg/h	120
Anti-oxidant	kg/h	0.4
Cooling water	tonne/h	870
Process water	tonne/h	29
HP steam	GJt/h	44
LP steam	GJt/h	35
Condensate return	GJt/h	−5
Natural gas	Nm³/h	360
Electric Power	MWe	2.7
Manpower	men/shift	***
Average load factor	%	***

* At inlet to reheater
**At inlet to stack, downstream of hot air reheater
Figures in parentheses are annual average emissions for 90% FGD plant availability

The rates of production and composition of the material produced are summarised in Table 2.57 for the Datum Combustion System 1.

Table 2.57 Estimated Rates of Output Solids Production–Oil-Fired Combustion System 1

Rate of production:		
Waste sulphate/sulphite	tonne/h	0.11
Sulphur (99%)	tonne/h	0.87
Composition of waste product:		
Na₂SO₃	wt. %	47
Na₂SO₄	wt. %	53

Efficiency and Emission Factors

The efficiency and emission factors for the process are summarised in Table 2.58 for the application considered: Datum Combustion System 1.

In calculating the efficiency factors, it is assumed that at full load the FGD unit consumes 79 GJt/h of steam, reducing the useful energy sent out; and 2.7 MWe of electric power, equivalent to the combustion of a further 0.7 tonne/h of oil at a power station, assuming an overall power generation efficiency of 33%. This additional oil is arbitrarily assumed to be included with the oil burned in the boiler for calculating the efficiency factor.

Table 2.58 Efficiency and Emission Factors–Oil-Fired Combustion System 1

Oil heat input (gross)	MWt	464
Oil fired	tonne/h	39.4
FGD plant power consumed	MWe	2.7
Equivalent oil input	tonne/h	0.7*
Steam for FGD plant	GJt/h	79
Condensate heat recovery	GJt/h	−5
Useful energy from system	GJt/h	1394
Total equiv. oil input	tonne/h	40.1(a)
Efficiency factor	GJt/tonne	34.8(a)
Emissions		
Sulphur in SO₂	kg/h	133 (222)
Nitrogen in NOₓ	kg/h	127 (127)
Chlorine in HCl	kg/h	1.2 (1.2)
Particulates	kg/h	6.1 (7.1)
Emission factors (per tonne oil)		(a)
Sulphur	kg/tonne	3.32 (5.54)
Nitrogen	kg/tonne	3.17 (3.17)
Chlorine	kg/tonne	0.03 (0.03)
Particulates	kg/tonne	0.15 (0.18)

*Calculated assuming overall power generation efficiency = 0.33
(a) Based on oil fired to boiler plus oil equivalent to electric power consumed
Figures in parentheses are annual average emissions for 90% FGD plant availability

Effect of plant availability: For illustration purposes, the annual average emission factors for FGD plant availabilities of 100% and 90% are given in Tables 2.56 and 2.58. Further details are given in Section 1.6.

Effect of load variations: ***

Effect of design variations: ***

Limitations: ***

Costs

Cost basis: See Section 2.6.

Capital costs: See Section 2.6-for retrofit and for new build applications.

Annual running costs and cost factors: See Section 2.6-for retrofit and for new build applications.

Effects of design variations on costs: ***

Effects of annual load patterns on annual running costs: ***

Process Advantages and Drawbacks

Process well-established in USA, Japan and Europe; fairly high reliability; simple chemistry; high sulphur capture efficiency (over 90% capture) can be attained; operating flexibility; no requirement for large quantities of reagent; no large quantities of waste products; by-product potentially marketable or can be easily and safely disposed of.

The disadvantages are complexity of process; need for experienced labour.

Evaluation of Magnesia Slurry Scrubbing FGD Process
(Process Code S41.1)

See Section 2.2 for: outline of the basic process; its chemistry; block diagram.

See Section 2.3 for a list of manufacturers offering this type of equipment.

See Section 2.3 for general appraisal of the basic process.

See Section 2.4 for the reason for choosing the Magnesia Slurry Scrubbing (Code S41.1) basic process.

This basic process type is considered (Section 2.4) to be suitable for application in the UK only to large boilers having access to reagent reprocessing facilities, and hence it is evaluated here only for Datum Combustion System 1. The process is offered by United Engineers and Constructors. See Appendix 2 for details of this manufacturer.

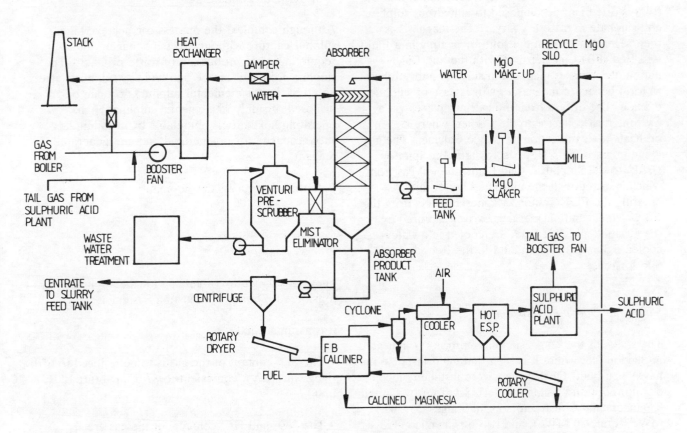

Figure 2.37 Magnesia Slurry Scrubbing FGD Process

Process Description

Figure 2.37 shows a simplified flow diagram for application of the process to an oil-fired boiler. It is assumed that the sulphur dioxide produced in the process is treated to recover sulphuric acid.

For sulphur oxides removal, gas from the boiler, boosted by a Fan, is cooled in a gas-gas Heat Exchanger and by injection of water. The gas enters the Absorber, which is of the spray grid design, where it is contacted counter-currently with magnesium hydroxide slurry, which absorbs sulphur oxides from the gas to form magnesium sulphite trihydrate $MgSO_3.3H_2O$, together with some magnesium sulphate heptahydrate $MgSO_4.7H_2O$. The gas passes through a mist eliminator at the top of the Absorber, and is then reheated by heat exchange with the ingoing gas in the Heat Exchanger. The absorbent slurry from the base of the Absorber flows to the Absorber Product Tank.

Regenerated magnesia pellets from the Recycle MgO Silo are ground in a Mill, and together with make-up magnesia, slaked with water in the MgO Slaker. The resultant magnesium hydroxide is slurried with water in the Feed Tank.

For absorbent regeneration and sulphur recovery the slurry from the Absorber Product Tank is pumped to a Centrifuge, where the crystallised reaction products are separated from liquid. The liquid centrate is returned to the slurry Feed Tank, and the solid is dried and decomposed to anhydrous sulphite and sulphate in a Rotary Dryer. The magnesia is then regenerated from the solid by heating in a Fluidised Bed (FB) Calciner fired with fuel oil. Calcination decomposes magnesium sulphite and sulphate to form pellets of magnesia, with release of sulphur dioxide. The off gas from the calciner passes via a Cyclone dust collector to the Cooler, where it preheats the combustion air to the Calciner. The gas is then passed via a Hot Electrostatic Precipitator (ESP) to the Sulphuric Acid Plant, the tail gas from which is mixed with the untreated boiler flue gas entering the FGD system. Magnesia pellets from the FB Calciner, and magnesia carryover captured by the Cyclone and Hot ESP, are cooled in a water-cooled Rotary Cooler and fed to the Recycle MgO Silo for reuse.

Status and Operating Experience

The status of the Magnesia Slurry process is indicated in Section 2.3, where it is seen that eight installations have been built. Of these, six were installed on coal- or oil-fired utility boilers in the USA: four by United Engineers and Constructors in the range 120–360 MWe (total 980 MWe), and two by Chemico (155 MWe and 190 MWe, totalling 345 MWe). The remaining two units are installed in Japan: one treating smelter gas, the other treating gases from an oil-fired boiler and Claus furnaces. The two Chemico plants, and the boilers to which they were connected, have been dismantled.

Operating experience shows (see Section 2.3) that availability and reliability are generally high. The principal problems that have been encountered appear to have arisen in the magnesia calcination system: temperature control of the calciner leading to dead-burning of the magnesia, and poor performance of a waste heat boiler. There have also been problems with a booster fan.

Variations and Development Potential

Design and operating variations can include:

– Regeneration by calcination in a rotary calciner (the technique adopted in the early Chemico plants).

– Use of coke to reduce sulphate in the Calciner; this allows a lower calcination temperature to be adopted.

– Alternatives for sulphur recovery: as elemental sulphur by a variety of processes (Alliance process, Claus process, Foster-Wheeler 'Resox' process); as sulphuric acid (by conventional catalytic oxidation of SO_2); or as liquid SO_2.

Although complex, the process can achieve high sulphur capture without excessive reagent make-up requirements or production of waste products. The sulphur produced can be marketed as sulphuric acid, liquid SO_2 or as elemental sulphur; or it can be disposed of safely as elemental sulphur. High operating flexibility is obtainable by installing large surge capacity for regenerated and unregenerated reagent.

Potential for process development: ***

Process Requirements

These are shown in Table 2.59 for the application considered: Datum Combustion System 1.

It is assumed that:

– The SO_2 content of the gas is to be reduced to 650 mg/Nm³ (dry), equivalent to 87.0% capture of SO_2.

– The NO_x and HCl contents of the gas are unaffected.

Table 2.59 Process Requirements – Oil-Fired Combustion System 1

Inlet Gas at full load		
Volume flow	'000 Nm³/h	456
Dry gas	'000 Nm³/h	408
Water vapour	'000 Nm³/h	48
Actual volume flow	'000 m³/h	707
Temperature	°C	150
Particulates content	mg/Nm³ (dry)	39
SO_2 content	mg/Nm³ (dry)	5010
NO_x content	mg/Nm³ (dry)	1025
HCl content	mg/Nm³ (dry)	3
Exit Gas at full load		
Volume flow	'000 Nm³/h	474
Dry gas	'000 Nm³/h	407
Water vapour	'000 Nm³/h	67
Actual volume flow	'000 m³/h	628
Temperature	°C	90
Particulates content	mg/Nm³ (dry)	15 (17)
SO_2 content	mg/Nm³ (dry)	650 (1086)
NO_x content	mg/Nm³ (dry)	1025 (1025)
HCl content	mg/m³ (dry)	3 (3)
Reaction temperature	°C	53
Particulates removal		Simultaneous
Reagent		$Mg(OH)_2$
Concentration in slurry	wt. %	20
Requirements at full load		
Magnesia (97% pure)	tonne/h	167
Process air	tonne/h	2.6
Process water	tonne/h	6
No. 6 fuel oil	tonne/h	3.5
Electric Power	MWe	1.3
Manpower	men/shift	***
Average load factor	%	***

Figures in parentheses are annual average emissions for 90% FGD plant availability

- The particulates content of the gas is reduced to 15 mg/Nm³ (dry).

- The gas to the stack is reheated to 90°C in the Heat Exchanger.

By-products and Effluents

The by-product from the system is sulphuric acid. There is some loss of magnesium sulphite and sulphate to waste, requiring a make-up of fresh magnesia.

The rates of production and composition of the sulphuric acid and waste material produced are summarised in Table 2.60 for the Datum Combustion System considered.

Table 2.60 Estimated Properties of Outputs – Oil-Fired Combustion System 1

Rate of production:		
Wet waste product	tonne/h	0.79
Dry waste product	tonne/h	0.73
Sulphuric acid (97%)	tonne/h	0.90
Composition of wet waste:		
Moisture	wt. %	7
$Mg_2SO_3.3H_2O$	wt. %	67
$Mg_2SO_4.7H_2O$	wt. %	25
Inerts	wt. %	0.6

Efficiency and Emission Factors

The efficiency and emission factors for the process are summarised in Table 2.61 for the application considered: Datum Combustion System 1.

In calculating the efficiency and emission factors, it is assumed that at full load the FGD unit consumes: 3.5 tonne/h fuel oil in the calciner; and 1.3 MWe of electric power, equivalent to the combustion of a further 0.3 tonne/h of oil at a power station, assuming an overall power generation efficiency of 33%. This additional oil is arbitrarily assumed to increase the effective oil consumption (but not the emissions).

Effect of plant availability: For illustration purposes, the annual average emissions and emission factors for FGD plant availabilities of 100% and 90% are given in Tables 2.59 and 2.61. Further details are given in Section 1.6.

Effect of load variations: ***

Effect of design variations: ***

Limitations: ***

Costs

Cost basis: See Section 2.6.

Capital costs: See Section 2.6 – for retrofit and for new build applications.

Annual running costs and cost factors: See Section 2.6 – for retrofit and for new build applications.

Effects of design variations on costs: ***

Effects of annual load patterns on annual running costs: ***

Process Advantages and Drawbacks

Process operated successfully in USA and Japan; high reliability; high sulphur capture efficiency (over 90% capture) can be attained; no requirement for large quantities of reagent; no large quantities of waste products; by-product potentially marketable or can be easily and safely disposed of; de-watered reaction products can be stored in open to await reprocessing (i.e. large surge capacity easily achieved); regeneration by off-site reprocessing possible.

The disadvantages are complexity of process; careful control of temperature and pH needed to avoid bisulphite formation or variable filtration characteristics; careful control of calciner off-gas composition (oxygen concentration) to avoid SO_3 and hence resulphation of regenerated MgO; need for experienced labour.

Table 2.61 Efficiency and Emission Factors–Oil-Fired Combustion System 1

Oil heat input (gross)	MWt	464
Oil fired	tonne/h	39.4
FGD plant power consumed	MWe	1.3
Equivalent oil input	tonne/h	0.3*
Fuel oil to calciner	tonne/h	3.5
Useful energy from system	GJt/h	1468
Total equiv. oil input	tonne/h	43.2(a)
Efficiency factor	GJt/tonne	34.0(a)
Emissions		
Sulphur in SO_2	kg/h	133 (222)
Nitrogen in NO_x	kg/h	127 (127)
Chlorine in HCl	kg/h	1.2 (1.2)
Particulates	kg/h	6.1 (7.1)
Emission factors (per tonne oil)		(a)
Sulphur	kg/tonne	3.08 (5.14)
Nitrogen	kg/tonne	2.94 (2.94)
Chlorine	kg/tonne	0.03 (0.03)
Particulates	kg/tonne	0.14 (0.16)

*Calculated assuming overall power generation efficiency = 0.33
(a) Based on oil fired to boiler and calciner plus oil equivalent to electric power consumed
Figures in parentheses are annual average emissions for 90% FGD plant availability

**Evaluation of Bergbau-Forschung Active Carbon FGD Process
(Process Code S51.1)**

See Section 2.2 for: outline of the basic process; its chemistry; Block diagram.

See Section 2.3 for general appraisal of the basic process.

See Section 2.3 for a list of manufacturers offering this type of equipment.

See Section 2.4 for the reason for choosing the Bergbau-Forschung (Code S51.1) basic process.

This basic process type is considered (Section 2.4) to be suitable for application in the UK to large boilers having access to reagent reprocessing facilities, and hence it is evaluated here only for Datum Combustion System 1. The process, developed by Bergbau-Forschung GmbH and Foster Wheeler Energy Corporation, is offered by Uhde GmbH (Dortmund, FRG). See Appendix 2 for details of this manufacturer.

Process Description

Figure 2.38 shows a simplified flow diagram for application of the process to an oil-fired boiler. It is assumed that the sulphur dioxide produced in the process is treated to recover elemental sulphur.

For dust and sulphur oxides removal, gas at a temperature of 120–150°C from the boiler, boosted by a Fan, enters the Adsorber containing activated carbon granules (prepared from coal in the form of extruded granules about 5 mm in length). The carbon, which is contained in louvred channels, with the gas flowing across the bed, adsorbs sulphur dioxide, oxygen and water vapour to form adsorbed sulphuric acid. Nitrogen dioxide (forming 5–10% of the total nitrogen oxides content of the gas) is also adsorbed, and the carbon bed filters out much of the particulates content of the gas. The gas temperature rises by 15–20°C across the bed; the cleaned gas passes via a Damper to Stack.

For adsorbent regeneration, the active carbon is removed continuously from the base of the Adsorber and conveyed to the Regeneration section, which can be either on- or off-site. Fines, including particulates trapped from the gas, are removed in a Classifier, and the oversize material then enters the Regenerator, where it is mixed with sand heated in a Sand Heater/Lift system to heat the carbon to 400–450°C. The adsorbed gases, sulphuric acid and nitrogen dioxide, react with the carbon, releasing sulphur dioxide, elemental nitrogen and carbon dioxide, consuming some of the carbon. The carbon remaining is screened on a Vibrating Screen to remove sand and carbon fines, cooled with air in the Carbon Cooler, and returned to the Adsorber; carbon losses are made up by adding fresh carbon. Part of the cooling air from the Carbon Cooler supplies the combustion air for the Sand Heater/Lift system.

For sulphur recovery the SO_2-rich gas from the Adsorber passes through the 'Resox' Reactor, where the SO_2 reacts with carbon to give sulphur vapour. The sulphur is condensed out in the Sulphur Condenser, and collected in the Sulphur Tank. The tail gas is returned to the inlet of the Booster Fan.

Status and Operating Experience

The status of the Bergbau-Forschung process is indicated in Section 2.3, where it is seen that five

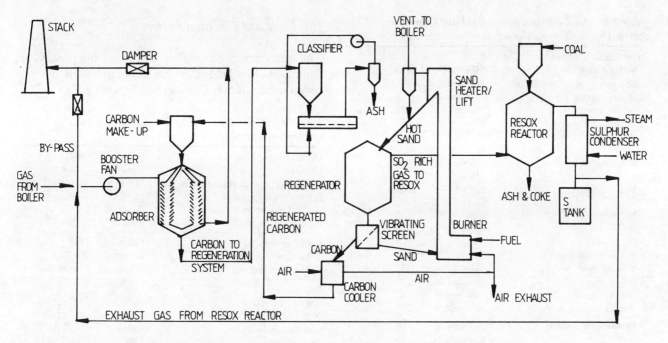

Figure 2.38 Bergbau-Forschung Active Carbon Adsorption FGD Process with Foster Wheeler Resox Sulphur Recovery

installations have been (or are being) erected, in the size range 0.3–370 MWe, totalling about 500 MWe.

Operating experience shows (see Section 2.3) that the principal problems that have been encountered have been: occurrence of hot spots in the Adsorber; poor reliability of the Vibrating Screen for separating carbon from sand; and plugging of the Sulphur Condenser of the Resox system with sulphur and carbon.

Variations and Development Potential

Design and operating variations can include:

- Simultaneous removal of SO_2 and NO_x in a two-stage Adsorber, with ammonia addition to the gas after removal of most of the SO_2 in the first stage; the carbon in the second stage catalyses the reduction of the NO [424, 435].

- Indirect heating of carbon in the Regenerator with hot combustion gases [424]

- Alternatives for sulphur recovery: as elemental sulphur by a variety of processes (Alliance process, Claus process, or by Foster-Wheeler's Resox process); as sulphuric acid (by conventional catalytic oxidation of SO_2); or as liquid SO_2.

Although complex, the process can achieve high sulphur capture without excessive reagent make-up requirements or production of waste products. The sulphur produced can be marketed as sulphuric acid, liquid SO_2 or as elemental sulphur; or it can be disposed of safely as elemental sulphur. High operating flexibility is obtainable by installing large surge capacity for regenerated and unregenerated active carbon.

Potential for process development includes improvements in process control to avoid hot spots in the Adsorber and plugging in the Resox system; improvements in efficiency of sulphur recovery in the Resox system.

Process Requirements

These are shown in Table 2.62 for the application considered: Datum Combustion System 1.

It is assumed that:

- The SO_2 content of the gas is to be reduced to 650 mg/Nm³ (dry), equivalent to 87.0% capture of SO_2.

- The Resox system has a conversion efficiency of 85%, so that 15% of the sulphur captured leaves the Resox system as various sulphur compounds which are recycled to the boiler.

- The NO_x content is reduced by 8% by adsorption of NO_2.

- The HCl content of the gas is unaffected.

- The particulates content of the gas is reduced to 15 mg/Nm³ (dry).

- The mechanical breakdown of carbon leads to a carbon make-up rate 50% higher than the theoretical.

Table 2.62 Process Requirements – Oil-Fired Combustion System 1

Inlet Gas at full load		
Volume flow	'000 Nm³/h	456
Dry gas	'000 Nm³/h	408
Water vapour	'000 Nm³/h	48
Actual volume flow	'000 m³/h	707
Temperature	°C	150
Particulates content	mg/Nm³ (dry)	39
SO_2 content	mg/Nm³ (dry)	5010
NO_x content	mg/Nm³ (dry)	1025
HCl content	mg/Nm³ (dry)	3
Exit Gas at full load		
Volume flow	'000 Nm³/h	455
Dry gas	'000 Nm³/h	407
Water vapour	'000 Nm³/h	48
Actual volume flow	'000 m³/h	729
Particulates content	mg/Nm³ (dry)	15 (17)
SO_2 content	mg/Nm³ (dry)	650 (1086)
NO_x content	mg/Nm³ (dry)	945 (953)
HCl content	mg/Nm³ (dry)	3 (3)
Reaction temperature	°C	165
Particulates removal		Simultaneous
Reagent		Active Carbon
Requirements at full load		
Active carbon	tonne/h	0.34
Natural gas	GJt/h	18.9
Air	tonne/h	98
Coal for Resox reactor*	tonne/h	0.6
Electric Power	MWe	1.2
Manpower	men/shift	***
Average load factor	%	***

*Regarded as a reagent, not as an energy source
Figures in parentheses are annual average emissions for 90% FGD plant availability

- The coal used in the Resox system has a carbon content of 60%; this coal is regarded as a reagent, not as an energy source.

By-products and Effluents

The byproduct from the system is elemental sulphur, and there is a carbon waste in addition to the captured flyash. The rates of production of the material produced are summarised in Table 2.63 for the two Datum Combustion Systems considered.

Table 2.63 Oil-Fired Combustion System 1

Rate of production:		
Sulphur (99%)	tonne/h	1.00
Active carbon fines	tonne/h	0.11
Particulates	tonne/h	0.01

Efficiency and Emission Factors

The efficiency and emission factors for the process are summarised in Table 2.64 for the application considered: Datum Combustion System 1.

Table 2.64 Efficiency and Emission Factors – Oil-Fired Combustion System 1

Oil heat input (gross)	MWt	464
Oil fired	tonne/h	39.4
FGD plant power consumed	MWe	1.2
Equivalent oil input	tonne/h	0.3*
Natural gas for sand heater	GJt/h	18.9
Equivalent oil input	tonne/h	0.4
Useful energy from system	GJt/h	1468
Total equiv. oil input	tonne/h	40.1(a)
Efficiency factor	GJt/tonne	36.6(a)
Emissions		
Sulphur in SO_2	kg/h	133 (222)
Nitrogen in NO_x	kg/h	117 (118)
Chlorine in HCl	kg/h	1.2 (1.2)
Particulates	kg/h	6.1 (7.1)
Emission factors (per tonne oil)		(a)
Sulphur	kg/tonne	3.32 (5.54)
Nitrogen	kg/tonne	2.92 (2.94)
Chlorine	kg/tonne	0.03 (0.03)
Particulates	kg/tonne	0.15 (0.18)

*Calculated assuming overall power generation efficiency = 0.33
(a) Based on oil fired to boiler plus oil equivalent to natural gas and electric power consumed
Figures in parentheses are annual average emissions for 90% FGD plant availability

In calculating the efficiency factors, it is assumed that at full load the FGD unit consumes 18.9 GJt/h of natural gas for the Sand Heater/Lift, equivalent to an increase of 0.4 tonne/h of oil; and 1.2 MWe of electric power, equivalent to the combustion of a further 0.3 tonne/h of oil at a power station, assuming an overall power generation efficiency of 33%. This additional oil is arbitrarily assumed to be included with the oil burned in the boiler for calculating the efficiency and emission factors.

Effect of plant availability: For illustration purposes, the annual average emissions and emission factors for FGD plant availabilities of 100% and 90% are given in Tables 2.62 and 2.64. Further details are given in Section 1.6.

Effect of load variations: ***

Effect of design variations: ***

Limitations: ***

Costs

Cost basis: ***

Capital costs: ***

Annual running costs and cost factors: ***

Effect of design variations on costs: ***

Effects of annual load patterns on annual running costs: ***

Process Advantages and Drawbacks

Process can be retrofitted; process readily adaptable to combined NO_x-SO_2 abatement; process also removes particulates and other air pollutants; no gas reheat needed; high sulphur capture efficiency (over 90% capture) can be attained; operating flexibility; no requirement for large quantities of reagent; no large quantities of waste products; by-product potentially marketable or can be easily and safely disposed of.

The disadvantages are complexity of process; need for experienced labour.

2.6 Costs for FGD Systems

Costs for six FGD process plants have been included in this section. Capital and operating costs for each FGD process (Datum Combustion System 1 only) were derived from the cost estimates prepared for a 2000 MWe COAL-fired plant. Details of the procedures and assumptions considered during costing are presented below and in Appendix 3.

EQUIPMENT COSTS

FGD equipment costs were prepared on the basis of the Equipment Lists presented in Appendix A of EPRI Report No. CS-3342 [95]. The EPRI Report was based on an arbitrary reference coal-fired plant of 1000 MWe (2×500 MWe) capacity but, for this study, the Equipment List was modified to represent a plant of 2000 MWe (4×500 MWe). Budget prices for items of equipment were obtained, wherever possible, from UK-based vendors and details are given in Appendix 3.

As indicated in Appendix 3, some budget costs included for supply and erection, whilst others were exclusively for supply of equipment. It was estimated that total equipment costs compiled in this way included an element for erection which amounted to, on average, 25%. The equipment costs were adjusted accordingly.

An 'installation factor' for each of the processes was derived from EPRI Report No. CS-3342 [95] by comparing the total equipment costs given in Volume 3 of the Report to the installed equipment costs, referred to as Total Process Capital and given in Volume 1 (non-regenerable processes) or Volume 2 (regenerable processes) of the Report. The installation factor, therefore, provides an indication of the relative complexity of installing each FGD system, and it was assumed that these factors would also be reflected in UK installation costs. It should be noted that equipment and installation costs for an electrostatic precipitator (and associated equipment), referred to as a Particulate Removal System, were deleted in the determination of the installation factors as this component is considered to be part of the boiler system and not of the FGD plant. For the lime spray drying process, however, an electrostatic precipitator or baghouse is an integral part of the plant and an allowance was made for this in estimating the installation factor, based on information from EPRI Report No. CS-3696 [96].

CAPITAL COSTS

Capital cost estimates for each FGD plant (2000 MWe coal-fired plant) were prepared using the procedures outlined in EPRI Report No. CS-3342 [95] for new-build plants and EPRI Report No. CS-3696 [96] for retrofit plants. In each case, the estimated total erection costs, expressed in terms of pounds sterling per kilowatt of installed boiler capacity (£/kWe), were used as the 'Base Capital Costs' for each process in the estimating procedure.

Details of the EPRI estimating procedure for new-build and retrofit plants appear in Appendix 3.

The following assumptions were made in the estimates for new or retrofit FGD plant fitted to a 2000 MWe coal-fired power station (Datum Combustion System 1):

(a) Scope Adjustments:

 – An average factor of 4% of the base capital cost has been selected (retrofit only).

(b) Process Adjustments:

 – Unit size = 500 MWe
 – Flue gas flowrate = $1.3 m^3/s$ per net MWe
 – Sulphur content = 1.6%

(c) Location:

 – Seismic and climatic considerations are ignored.
 – The soil is of poor bearing capacity and the site is near a river bank.

(d) Retrofit Adjustments:

 1) Accessibility and congestion; assumptions:
  – Limited space available, e.g. area around boiler and stack is approximately 2.8 square metres per megawatt,

- interference with existing structures or equipment which cannot be relocated,
- special designs are necessary,
- access for cranes is limited to one side,
- majority of equipment is on elevated slabs or remotely located.

(These factors add 25% to the Total Process Retrofit Capital, TPRC. Alternative conditions, as defined in EPRI Report No. CS-3696, would result in an increase of between 2% and 42%).

2) Underground obstructions; assumptions:
- more than one major obstruction such as circulating water pipe, gas main or ductbank,
- several minor obstructions, including piping, trenches and plant drainage.

(These factors add 2% to the TPRC. Alternative conditions, as defined in EPRI Report No CS-3696, would result in an increase of between 1% to 5%).

3) Ductwork length and distance from scrubber to tie-ins; assumptions:
- the scrubber is not located symmetrically with the unit and at a distance of over 90 metres from the unit,
- the tie-in with the existing chimney is from the rear.

(These factors add 12% to the TPRC. Alternative conditions, as defined in EPRI Report No. CS-3696, would result in an increase of between 2% to 12%).

(e) Escalation Adjustments:
- Cost estimates are escalated to December 1989 start-up using an escalation rate of 5%.

(f) Project Contingency:
- A single contingency factor of 10% has been selected.

(g) Process Contingency:
- A single contingency factor of 5% has been selected.

(h) General Facilities:
- A factor of 5% has been selected.

(i) Allowance for Funds During Construction:
- The duration of the engineering, procurement and construction phases is assumed to be 3 years. The EPRI allowance (3.7%) has been modified to take account of the revised escalation rate.

(j) Royalty Allowance:
- A royalty of 5% has been selected.

(k) Inventory Capital:
- The inventory Capital Costs indicated in EPRI report No. CS-3696 [96] have been adjusted to December 1986 Sterling costs.

Table 2.65 Capital Costs for FGD Plants (Newbuild)–December 1986

Process	Capital Cost for 2000 MWe* Coal-Fired Plant £ Million	Cost Index**	Capital Cost for Datum Combustion System 1 (450 t/h Steam, Oil-fired) £ Million
Limestone with forced oxidation	136	0.71	21.6
Chiyoda	126	0.81	15.4
Saarberg Hölter	124	0.65	23.0
Lime Spray Drying	133	0.65	24.7
Wellman-Lord	164	0.68	28.1
Magnesia (MgO) scrubbing	194	0.62	38.9

* See Appendix 3 (Part 4) for estimates
**Capital Cost of Plant 21 = Capital Cost of Plant 1 × (Capacity Ratio)^y where y = cost index

Table 2.66 Capital Costs for FGD Plant (Retrofit)–December 1986

Process	Capital Cost for 2000 MWe Coal-Fired Plant £ Million	Cost Index**	Capital Cost for Datum Combustion System 1 (450 t/h Steam, Oil-fired) £ Million
Limestone with forced oxidation	202	0.71	32.1
Chiyoda	186	0.81	22.8
Saarberg-Hölter	183	0.65	33.9
Lime Spray Drying	198	0.65	36.7
Wellman-Lord	242	0.68	41.5
Magnesia (MgO) Scrubbing	286	0.62	57.3

* See Appendix 3 (Part 4) for estimates
**Capital Cost of Plant 2 = Capital Cost of Plant 1 × (Capacity Ratio)^y where y = cost index

Capital cost estimates for 2000 MWe coal-fired FGD plants are outlined in Appendix 3, and summarised in Tables 2.65 (New-Build) and 2.66 (Retrofit).

To estimate capital costs for new or retrofit FGD plant fitted to a 450 t/h of steam oil-fired industrial boiler (Datum Combustion System 1), reference was made to the cost data of Samish [92]. Samish used

EPRI cost data to produce capital cost estimates for a number of FGD processes and for the following conditions:

- Two sizes of boiler plant (110.6 MWe and 1106 MWe)
- Two coals of differing sulphur content (0.5% and 4.0%)
- Two oils of differing sulphur content (0.25% and 1.2%)
- SO_2 emission levels ranging from 0.06 to 2.1 kg SO_2 per GJt fired. (N.B. 90% sulphur removal for Datum Combustion System 1 gives an emission level of 0.14 kg/GJt).

These data were used to develop 'cost indices' for estimating capital costs for FGD plants of differing sizes (see Section 2.6). Datum Case 1 oil-fired FGD plant capital costs were then derived from Datum Case 1 coal-fired (2000 MWe) capital costs using the appropriate cost index, and the capacity ratio (based on input flue gas flow rate). These capital costs are given in Tables 2.65 (New-build) and 2.66 (Retrofit).

OPERATING COSTS

Operating costs for each of the six processes were estimated using the simplified EPRI procedure in Section 5 of Ref. [96]. An escalation rate of 5% (instead of the EPRI figure of 8.5%) per annum was selected for the estimate. An outline of the procedure is included in Appendix 3, together with cost details for each of the processes for a 2000 MWe coal-fired plant. Costs were initially estimated in $/kWe-year (December 1982 prices), converted to sterling at that date and brought up-to-date (December 1986) using plant cost indices published in Process Engineering magazine. First year operating costs are summarised in Table 2.67; it has been assumed that costs for new-build and retrofit plant are identical.

Operating costs for FGD plant of two different sizes were developed by Samish [92] from EPRI cost data. This information was further developed to determine cost indices which could be used to compute operating costs for any size of plant, e.g. Datum Combustion System 1 oil-fired plant (see Section 2.6). These cost indices and the operating costs for the smaller plant are given in Table 2.67.

COST INDICES

It should be noted that cost indices were derived from capital and operating costs for the two sizes of coal-fired, rather than oil-fired, plant because the Samish data were limited to oil sulphur contents (and SO_2 reduction levels) much lower than those for Datum Combustion 1.

Table 2.67 Operating costs for FGD Plant (New-Build and Retrofit)–December 1986

Process	Operating Costs for 2000 MWe Coal-Fired Plant £/kWe-year*	Cost Index**	Operating Costs for Datum Combustion System 1 (450 t/h Steam, oil-fired) £/kWe-year
Limestone with Forced Oxidation	24.02	0.22	42.49
Chiyoda	21.88	0.14	31.45
Saarberg-Holter	22.90	0.22	40.51
Lime Spray Drying	28.82	0.25	55.10
Wellman-Lord	32.27	0.26	63.32
Magnesia (MgO) Scrubbing	25.86	0.22	45.74

* See Appendix 3 (Part 5) for estimate
**Operating costs for Plant 2 = Operating costs for Plant 1/(Capacity Ratio)^y where y = cost index

3. Nitrogen Oxides Abatement Processes (Combustion Techniques)

3.1 Classification of Combustion Techniques

3.2 Outline Descriptions

3.3 General Appraisal of Each Technique

3.4 Comparison of Techniques

3.5 Costs

3. Nitrogen Oxides Abatement Processes (Combustion Techniques)

3.1 Classification of Combustion Techniques

The general classification of NO_x abatement (deNO_x) processes adopted in this Manual involves a division into two broad groups according to whether abatement is achieved at the source of combustion, 'Combustion Techniques', or in the flue gas downstream of the boiler, 'Flue Gas Treatment'.

The first group, 'Combustion Techniques', is divided into three distinct basic categories:

- Burner design (Category N10),
- Furnace design or modification (N20),
- Furnace operating methods (N30).

Each category is further sub-divided according to specific design or operation, see Figure 3.1. Combustion techniques for NO_x abatement are dealt with in this section of the Volume.

The second group, 'Flue Gas Treatment', is dealt with in Section 4 of the Volume.

PROCESS/DESIGN CODE NUMBERS

Each major category is assigned a Category Number, e.g. Burner Design–N10. Generalised designs or operating modes within each category are characterised by a change in the last digit of the Category Number, e.g. Dual Register Burners–N11.

The basic design or process types have been assigned the Code Numbers shown in Figure 3.1.

SUPPLIERS OF DeNO_x TECHNOLOGY

Full details of all manufacturers quoted in the Manual (names, addresses, telephone and telex numbers, and names of contacts) are listed in Appendix 2.

3.2 Outline Descriptions

PRINCIPLES OF NO_x FORMATION AND ABATEMENT FOR OIL BURNING

The following mechanisms have been identified as generating NO_x during fossil fuel combustion:

- Formation of 'thermal NO_x' by fixation of atmospheric nitrogen in the combustion air.

- Formation of 'fuel NO_x' by conversion of chemically bound nitrogen in the fuel.

- Formation of 'prompt NO_x' by reactions of air nitrogen with hydrocarbon radicals.

When using light distillate oil, the formation of 'thermal NO_x' is predominant because these fuels do not contain chemically-bound nitrogen. When using residual oil and crude oil, the formation of 'fuel NO_x' may become more important than the formation of 'thermal NO_x'.

The nitrogen-containing compounds from crude oil tend to concentrate in the heavy fractions during distillation. Therefore, fuel NO_x is of importance in residual oil. The nitrogen content of residual-oil varies from 0.1 to 0.5%. The extent of oxidation of the chemically bound nitrogen during combustion is a function of the amount of the chemically bound nitrogen in the fuel.

The fundamental factors in NO_x emission control may be summarised as follows:

(a) Lowering oxygen concentration in combustion zone (for reduction of thermal and fuel NO_x).

(b) Shortening gas residence time in high temperature zone (for reduction of thermal NO_x).

(c) Reduction of fuel nitrogen.

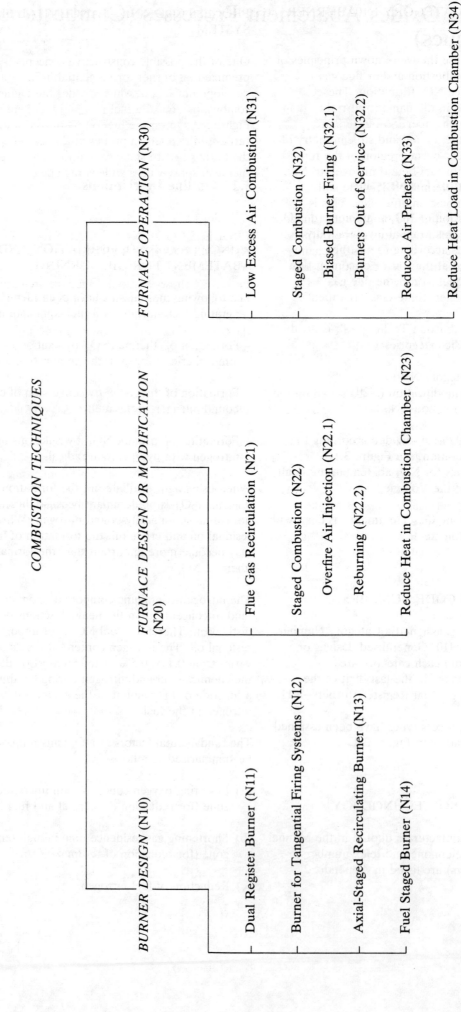

Figure 3.1 Classification of DeNO$_x$ Combustion Techniques (Oil)

TECHNIQUES IN CATEGORY N10-BURNER DESIGN

Burner designers utilise the well-known principles of off-stoichiometric combustion and/or flue gas recirculation to reduce NO_x formation. These principles are applied in the flame itself rather than in the total combustion chamber. So burners are designed to control the mixing and stoichiometry of fuel and air in the near-burner region of the furnace to retard conversion of fuel-bound nitrogen to NO_x and the formation of thermal NO_x, while still maintaining high combustion efficiency. This is accomplished by controlling the momentum, direction and quantity of fuel and air streams at the burner throat as they are injected into the furnace chamber. Various burner designs have been developed, and they may be categorised as shown in Figure 3.1.

PROCESS CODE N11-DUAL REGISTER BURNER

The low NO_x burner firing oil (and gas) based on the Dual Register Burner design is shown in Figure 3.2. The burner includes a primary gas annulus surrounding the core pipe housing the oil atomiser. Recirculated flue gas is mixed with secondary air supplied to the windbox and serves to reduce oxygen availability and reduce peak flame temperatures to control NO_x. The primary gas annulus is supplied directly with recirculated flue gas (no secondary air) which serves to blanket the base of the oil flame and further reduce oxygen availability as the oil devolatilises.

PROCESS CODE N12-TANGENTIAL FIRING SYSTEM

One of the possible constructive concepts is a premixed oil burner for tangential firing systems developed in Japan, which divides the flame into two combustion zones, a fuel-rich and fuel-lean zone. As Figure 3.3 shows, the fuel oil is sprayed into the furnace in two jets, a primary fuel-rich oil spray near the burner axis to stabilise ignition, and a secondary fuel-lean spray with a wide spray angle.

PROCESS CODE N13-AXIAL-STAGED RECIRCULATING BURNER

Figure 3.4 shows an axial-staged-recirculating burner by which the combustion air is mixed with the fuel (oil or gas) in two stages. Flue gas is recirculated to divide the combustion zone, to lower combustion temperature, and to delay the combustion process.

PROCESS CODE N14-FUEL STAGED BURNER

In the reburning (fuel-staging) process, a small part of the fuel is injected above the main combustion zone, producing a second combustion zone under reducing conditions. Downstream, this NO_x reducing zone combustion air is added to complete combustion. The principle of reburning is shown in Figure 3.5.

The process of reburning can be divided into two zones:

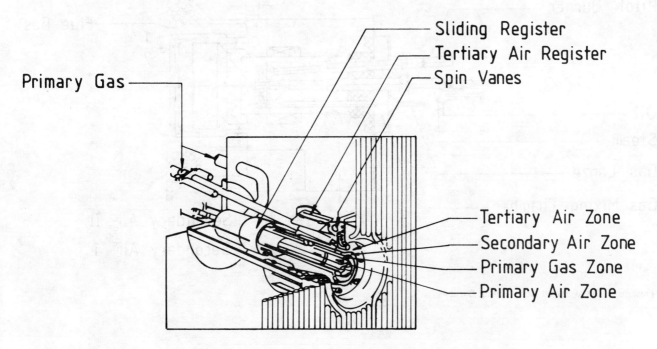

Figure 3.2 Low NO_x Burner Firing Oil and Gas Based on the Dual Register Burner Design

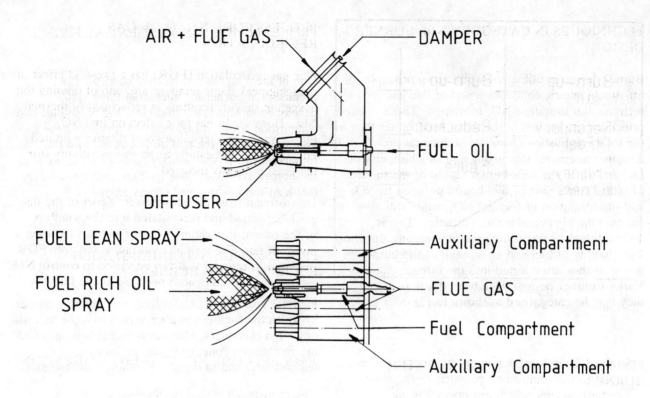

Figure 3.3 Oil-Firing Low NO$_x$ Burner for Tangential Firing Systems

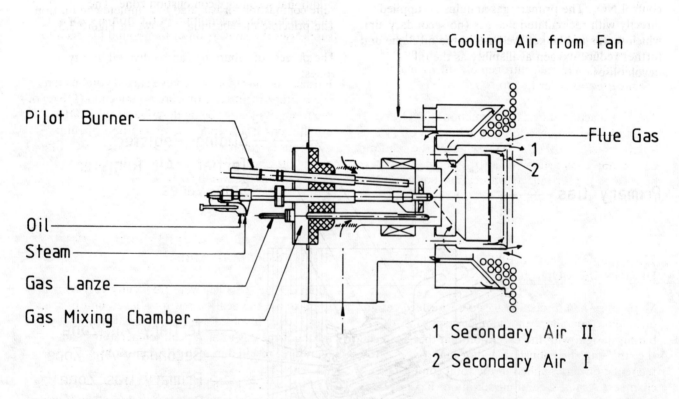

Figure 3.4 Axial-Staged Recirculating Burner

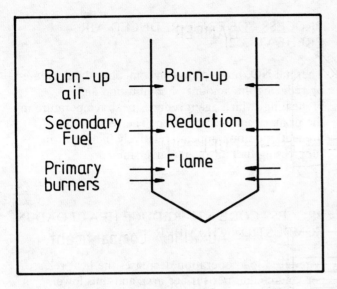

Figure 3.5 Principle of the Reburning Process

- In the NO_x reduction zone the secondary fuel is added to the combustion products of the main combustion zone which can operate using conventional low NO_x combustion modification techniques. NO_x is reduced by hydrocarbon radicals producing nitrogenous intermediates such as ammonia (NH_3) and hydrogen cyanide (HCN) as well as molecular nitrogen. Consequently, the NO_x in the exhaust from the main combustion zone is converted either to N_2, NH_3, HCN or retained as NO_x.

- In the total fixed oxidation zone air is added to ensure burnout of the secondary fuel. The NH_3, HCN, NO and char nitrogen are either mainly converted to N_2 or less to NO_x.

The last reaction shows the formation of NO_x, but the result of the total process is the reduction of approximately 50% of all NO_x entering the reburning zone in the combustion chamber. All types of fuel can be used as the secondary hydrocarbon source in a reburning process.

TECHNIQUES IN CATEGORY N20 – FURNACE DESIGN OR MODIFICATION

Methods of NO_x reduction based on furnace design or modification adopt the same principles as for burner design, with the exception that these principles (i.e. the staged combustion and/or flue gas recirculation) are applied in the total combustion chamber. Again, several modifications have been developed and categorised according to Figure 3.1.

PROCESS CODE N21 – FLUE GAS RECIRCULATION

Flue gas recirculation (FGR) has a two-fold effect of reducing peak flame temperature, and of diluting the oxygen in the air, resulting in reductions of thermal NO_x. However, it has little effect on fuel NO_x. Furthermore, this technique is of relatively limited effectiveness, particularly as an energy penalty (for the fans) is usually incurred.

Downstream from the economiser, a part of the flue gas is separated and recirculated into the windbox or the combustion chamber or the burners by using a hot draft fan. The recirculation rate is the only FGR operating parameter that can be varied to control NO_x reductions. NO_x emissions decrease as the flue gas recirculation rate is increased.

PROCESS CODE N22 – STAGED COMBUSTION

The principle of staged combustion or off-stoichiometric combustion (OSC) is based on the regulation of the oxygen content within the furnace. A primary combustion zone is created in the combustion chamber by dividing the total air, where the combustion is performed under a lack of oxygen. To ensure the total burnout at oxygen excess and relatively low temperatures, air is injected into an additional secondary combustion zone. The sub-stoichiometric condition in the primary zone suppresses the conversion of fuel-bound nitrogen to NO_x.

Furnace modification to achieve staged combustion can be implemented by overfire air injection (OFA) or reburning, the latter being described above. OFA usually consists of separate pressure port openings in the furnace above the main windboxes, with tilting nozzle tip which allow the air to be directed towards or away from the windbox.

PROCESS CODE N23 – REDUCE HEAT IN COMBUSTION CHAMBER

Older designs of boiler were frequently very compact, to maximise the heat output for a given unit of space. This usually involved using very turbulent, rapidly mixing flames, with high peak temperatures and, therefore, resulted in high NO_x emissions.

The low volumetric heat release (LVHR) criterion is based on the philosophy that a lower volumetric heat output, e.g. by reducing the intensity of mixing, could lead to a corresponding fall in NO_x emissions.

TECHNIQUES IN CATEGORY N30–FURNACE OPERATION

This is a technique where no modification is required to the existing unit, since only the operating condition of the furnace is changed and it can be implemented in several ways, as indicated in Figure 3.1.

PROCESS CODE N31–LOW EXCESS AIR COMBUSTION

Excess air is normally provided in a furnace to ensure complete combustion of the fuel. As the excess air is increased, the oxygen concentration in the flame zone increases and the temperature decreases.

Low excess air (LEA) firing causes a reduction of the local flame zone concentration of oxygen by which both the thermal and the fuel NO_x formation is reduced.

LEA can also be used in combination with other combustion modifications, such as staged combustion and operating at reduced load. However, requirements on combustion control increase if a combination of such measures is installed.

PROCESS CODE N32–STAGED COMBUSTION

The principle of off-stoichiometric combustion (OSC) or staged combustion has been described above. This principle can be applied on existing furnaces to reduce NO_x emission by making the burner operate under Biased Burner Firing (BBF) or under Burners Out Of Service (BOOS). It is mainly applicable to boilers which operate with an array of burners (e.g. utility boilers).

BBF means that by appropriate measures, a fuel-rich combustion zone is produced at the lower rows of burners and a fuel-lean combustion zone at the upper rows of burners.

BOOS means that by operating most of the burners (of a multiple burner array) under fuel-rich conditions, the remaining burners (normally the upper rows) are out of service injecting only air. This is to create a fuel-rich region, in which the fuel is devolatilised, and allowing as long a residence time within that zone as possible before the admixing of the secondary combustion air.

PROCESS CODE N33–REDUCED AIR PREHEAT [352]

Thermal NO_x production can sometimes be decreased by reducing the amount of combustion air preheating, which again reduces peak temperature in the primary combustion zone. The reduced air preheat operation appears relatively inefficient in suppressing fuel nitrogen conversion.

PROCESS CODE N34–REDUCE HEAT LOAD IN COMBUSTION CHAMBER

A reduced load operation decreases the heat release per volume and per surface area and thus lowers peak flame temperatures, resulting in lower thermal NO_x formation. Fuel/air mixing rates are also decreased and this may lower NO_x emissions from fuel-bound nitrogen.

3.3 General Appraisal of Each Technique

TECHNIQUES IN CATEGORY N10–BURNER DESIGN

PROCESS CODE N11–DUAL REGISTER BURNER

This burner is applicable to new units and can be used to retrofit directly at existing units on a plug-in basis.

Operating experience shows that NO_x reductions of 50%, relative to conventional burner designs, have been achieved in normal or staged combustion [352].

PROCESS CODE N12–TANGENTIAL FIRING SYSTEM

This system is applicable to both new units and retrofitting.

Test results with this burner at a 265 MWe existing unit indicate that the NO_x concentration in flue gas remains below 160 mg/m^3 (4% O_2) over a wide load range as fired with OFA [352].

PROCESS CODE N13–AXIAL-STAGED-RECIRCULATING BURNER

It is applicable to both new units and retrofitting.

Five units have been equipped with this burner [352]. In one, a 150 t/h industrial boiler, NO_x removal efficiencies of 38–45% (light oil) and 17–35% (heavy oil) have been achieved compared to baseline levels. Using this type of burner, NO_x emissions of 100–200 mg/m³ firing light oil and 350 mg/m³ firing heavy oil can be expected [352].

PROCESS CODE N14–FUEL-STAGED BURNER

This burner is mainly applicable to new units.

Reburning has been installed at a few oil-fired boilers in Japan (see below).

TECHNIQUES IN CATEGORY N20–FURNACE DESIGN OR MODIFICATION

PROCESS CODE N21–FLUE GAS RECIRCULATION (FGR)

This technique is suitable for new build and retrofit situations. When retrofitted, the increase in gas mass flow for a given heat output can cause boiler operating problems.

In combination with other control methods such as off-stoichiometric combustion, FGR has been effective with a limited recirculation rate.

The FGR system requires a greater investment at existing and new facilities than other combustion modifications like low NO_x burners.

There is about 1% efficiency penalty on the recirculation fan load.

PROCESS CODE N22–STAGED COMBUSTION

Process Code N22.1–Overfire Air Injection (OFA)

OFA is suitable for new build and retrofit applications. (Retrofitting to an existing boiler with OFA requires additional duct work and furnace wall penetration).

It is better applied to larger units than to small units, because total burn-out of the fuel needs a longer residence time using OFA injection. The potential gas residence time tends to be shorter in small units. This effect may cause incomplete combustion; to avoid this, an increased excess air rate through the OFA ports may be required, which leads to a decreased boiler efficiency.

A NO_x emission reduction of about 30% has been reported from Western Siberian Power Plant in the USSR when OFA has been used.

Data from the Gardanne power station, France show that NO_x reduction efficiency of 30% can be achieved by injecting 20% of the total excess air through overfire air ports.

Process Code N22.2–Reburning

This is mainly applicable to new units.

With existing boilers the retrofit of a reburning system may be difficult in most cases because of burnout problems.

Reburning has been installed at a few oil-fired boilers in Japan [352].

In one of the first applications of this technique, an 85MWt industrial boiler fired with a low-grade heavy oil containing 4% sulphur, NO_x concentrations were reduced to 400–500 mg/m³ using low-NO_x burners. NO_x concentrations of 220–260 mg/m³ were reached by the injection of 7% of the heavy oil as the secondary fuel. To operate with reburning, this particular boiler has been constructed about one metre higher than would otherwise be the case.

Further developments have taken place in Japan in utility boilers ranging up to 600 MWe with oil and oil/coal mixtures. NO_x reduction efficiencies of up to 50% have been reported [352].

PROCESS CODE N23–REDUCE HEAT IN COMBUSTION CHAMBER

This technique is suitable for new build and retrofit applications.

Operating experience is various.

TECHNIQUES IN CATEGORY N30–FURNACE OPERATION

PROCESS CODE N31–LOW EXCESS AIR COMBUSTION (LEA)

Low Excess Air (LEA) combustion can be used in combination with other combustion modifications such as staged combustion and operating at reduced load. However, requirements on combustion control increase if a combination of such measures is installed.

Operating experience shows [352] that LEA is easy to implement; there is no need to change the construction.

The boiler efficiency may increase.

The control of the combustion process has to ensure the safe operation of the boiler with LEA. This control has also to be effective at load swings and variations of fuel quality.

The use of LEA at firing installations is limited by possible negative side-effects such as increased fouling and corrosion by the reducing atmosphere, increased CO emissions and incomplete burnout.

LEA is not in widespread use as a NO_x control technique for industrial boilers.

NO_x reductions of up to 44% and efficiency increases of up to 2.5% have been reported using LEA firing.

PROCESS CODE N32–STAGED COMBUSTION

This is suitable for adoption on existing boilers; no equipment changes are needed.

With some installations, the use of staged combustion results in the boiler being de-rated, i.e. when the lower burners cannot be used to supply fuel to obtain the full boiler output. In these instances, reduction of NO_x emission by the use of staged combustion may not be a tenable solution.

A 30–40% reduction in NO_x emission can be obtained by the use of the Burners Out Of Service (BOOS) method.

PROCESS CODE N33–REDUCED AIR PREHEAT (RAP)

RAP is suitable only for new small units designed for low feed-water inlet temperatures, in which safe ignition conditions can be provided.

RAP can cause significant reductions in thermal efficiency, if no counter measures (i.e. enlarging the economiser) are applied.

PROCESS CODE N34–REDUCE HEAT LOAD IN COMBUSTION CHAMBER

This technique is suitable for new build and retrofit applications.

Operating experience is various.

3.4 Comparison of Techniques

BURNER DESIGNS

The performance of various burner designs is compared in Table 3.1. Each system is believed to have a maximum percent NO_x reduction which represents the maximum technical and economic limits for that technology. Restrictions to the application of these techniques are [352]:

- Flame stability limitations at low load operation.
- Possible increase of CO concentration in the flue gas.
- Possible increased smoke/dust emissions may require the use of an additional ESP.

Table 3.1 Comparison of Burners

Burner Type	Dual-Register Burner	Tangential Firing System	Axial-Staged Recirculation Burner	Reburning (Fuel Staging)
Code	N11	N12	N13	N14
Technique	Secondary air introduced in stages and is mixed with recirculated flue gas.	Pre-mixed burner which divides flame into two combustion zones: one fuel-rich and the other fuel-lean.	Combustion air is mixed with the fuel oil in two stages, recirculated flue gas being used to divide the combustion zone.	A small part of the fuel is injected above the main combustion zone, and additional air is added downstream to complete combustion.
Applicability*	N,R	N,R	N,R	N,R (site specific)
Operating Experience	50% NO_x reduction has been achieved–relative to conventional burner.	NO_x concentration in flue gas remained below 160 mg/m³ over a wide range of loads in tests at 265 MWe plant.	5 units have been equipped. NO_x removal efficiencies of up to 45% for light oil and up to 35% for heavy oil have been achieved.	Applied to several oil-fired boilers of up to 600 MWe in Japan. NO_x removal efficiency is about 50%.

* N = New units, R = Retrofits

FURNACE DESIGN OR MODIFICATION

Table 3.2 shows a comparison of furnace design or modification techniques.

The Flue Gas Recirculation (FGR) technique is of relatively limited effectiveness, particularly as an energy penalty is usually incurred. The Overfire Air (OFA) injection method of staged combustion is very effective for NO_x reduction and may be used with most fuels and most combustion systems; it is particularly suitable for tangentially-fired boilers.

Restrictions for these techniques are [352]:

- De-rating may occur at some units in the case of FGR.
- Flame stability limitations at low load operation.
- Possible increase of CO concentration in the flue gas.

Both FGR and OFA are usually incorporated with other forms of combustion technique, since there are limitations on the percentage NO_x reduction that can be achieved by FGR or OFA alone.

FURNACE OPERATION

Table 3.3 shows a comparison of furnace operation techniques. Low Excess Air (LEA) firing causes a reduction of the local flame zone oxygen concentration by which NO_x formation is reduced. The use of LEA firing is a simple, feasible and effective technique. There may be negative side effects such as increased fouling and corrosion by the reducing atmosphere, increased CO emissions and incomplete burnout.

Both Biased Burner Firing (BBF) and Burners Out Of Service (BOOS) techniques use the principle of Off-Stoichiometric Combustion (OSC) or Staged Combustion. They are usually applicable to boilers with multiple rows of burners. The top rows of burners may be operated under fuel-lean (BBF) or air-alone (BOOS) conditions. The advantage is that no modifications are necessary to the furnace water wall.

Most of these techniques (if being applied alone) are usually a temporary measure to reduce the NO_x emission levels on an existing furnace. They can be

Table 3.2 Comparison of Furnace Designs or Modifications

Furnace Design or Modification	Flue Gas Recirculation	Staged Combustion	Reduced Heat in Furnace
Code	N21	N22	N23
Technique	Low combustion temperature. Low oxygen concentration.	Two stage combustion.	Low volumetric heating rate. (LVHR)
NO_x level achievable	Up to 15% reduction	10–35% reduction	20–25% reduction
Applicability*	N,R	N,R	N
Problems	Cost. Flame expansion. Energy penalty.	Incomplete combustion. Flame extension. Tube erosion.	Possible boiler de-rating.

* N = New units; R = Retrofits (site specific)

Table 3.3 Comparison of Furnace Operation Techniques

Furnace Operation	Low Excess Air	Biased Burner Firing	Burners Out Of Service	De-rating
Code	N31	N32.1	N32.2	N34
Technique	Low excess air combustion, i.e. reduced oxygen concentration.	Upper row burners operate fuel-lean. Lower row operates fuel-rich.	Upper row burners are out of service, injecting only air.	Reduced heat load in furnace, thereby lowering combustion temperature.
NO_x level achievable (% reduction)	40	30–40	30–40	20–30
Applicability*	R	R	R	N,R
Problems	Unburnt carbon. CO emissions. Fouling and corrosion.		Derating of unit.	

* N = New units, R = Retrofits

Table 3.4 Estimates of Removal Efficiencies for Combustion Modifications

Technique	Reduction in NO_x Emission (%)	
	Light Oil	Heavy Oil
Low-NO_x Burners (LNB)	20–40	10–30
Reburning	30–50	30–50
Off-Stoichiometric Combustion (OSC) or Staged Combustion	10–30	10–40
Low Volume Heating Rate (LVHR)	30–40	30–40
Flue Gas Recirculation (FGR)	20–50	10–35
LVHR + LNB	44–64	37–58
LVHR + LNB + FGR	55–82	43–72
LVHR + LNB + OSC	49–74	43–78
LVHR + LNB + Reburning	60–82	56–79
LVHR + LNB + FGR + Reburning	68–91	60–86

used in combination with other combustion modifications to provide a further reduction in NO_x emission levels, although the requirements for combustion control would also increase.

COMBINATIONS OF COMBUSTION TECHNIQUES

Viewing the $deNO_x$ combustion techniques as a whole, they can be classified into LNB, OSC, FGR, LVHR and reburning processes as shown in Table 3.4. Various combinations of the processes are usually employed to achieve a lower NO_x emission level. The Electric Power Development Company (EPDC) of Japan has conducted extensive tests of combustion modifications for NO_x abatement, jointly

Table 3.5 Utility Boilers–Oil Fired (NO_x)

Source Level of Control	% Abated	Annual K Cost £/KWt	Annual O&M P/KWh	£/t Abated
UTILITY BOILERS NEW	45–50			
Low NO_x Burners				
– Oil & Gas		0.43	Negligible	7
– Oil		0.43	Negligible	7
In Furnace Reduction (IFR)		5.57		
Combustion Modifications				
O.S.C.	25–30			
L.E.A.	25–30			
LVHR	33	1.49	Negligible	29
LVHR & LNB	60	1.86	Negligible	21
LVHR & LNB & IFR	80	7.43	0.021	230
SCR: Other Countries	80–90	6.57–32.86	0.035–0.11	820–240
USA	80–90	55.71		1620–2850
SCR & LVHR & LNB	92	8.43	0.085	640
SCR & LVHR & LNB & IFR	96	13.9	0.11	760
UTILITY BOILERS EXISTING				
Combustion Modifications				
OFA–Dual Oil & Gas		0.04	0.007	336–2340
LEA–Dual Oil & Gas	20	0.07–0.5		7–135
BOOS–Dual Oil & Gas	25–50	0.2–0.6	0.007	220–960
LEA + OSC–Dual Oil & Gas		0.5–0.9		807–2435
FRG + OFA–Dual Oil & Gas	55–75	5.7	0.03	830–2235
– Oil		2–19		
NCR	40–60	2.7–3.7		114–236
LEA & OSC & NCR	70	7.5		257
SCR–Dual Oil & Gas	90	5.6		

Table 3.6 Industrial Boilers–Oil Fired (NO_x)

Source Level of Control	% Abated	Annual K Cost £/KWt	Annual O&M P/KWh	£/t Abated
INDUSTRIAL BOILERS EXISTING				
Low NO_x Burners	25–50	1.4–3.3		480–1050
Combustion Modifications				
Low Excess Air (LEA)	10	0.2		214–307
O.S.C.	20–35	0.3–3		170–1540
NCR: Japan	40–60	9		1460–2985
SCR: USA	90	47		5340–7080
INDUSTRIAL BOILERS NEW				
LEA–Dual Oil & Gas	10	0.14		107–207

with MHI, BHK, IHI, and KHI. Test results shown in Table 3.4 indicate that, by using a combination of techniques, NO_x reductions of up to 90% can be achieved.

3.5 Costs

The recent paper by Leggett [102] summarising progress on the OECD Project on Control of Major Pollutants (MAP) was used to provide the limited cost data shown in the following Tables 3.5 and 3.6. Within the MAP study, differences in data were assessed and presented as ranges rather than specific values due to substantial uncertainties regarding the starting emission level, the abatement efficiency and other similar factors.

The prices and ranges in the cost effectiveness summary tables are not intended to be indicative of the cost, or cost effectiveness, for any individual facility–rather, they attempt to represent a typical facility in that source category so that, on average, the estimated costs should be representative.

A questionnaire issued to UK burner vendors did not return any useable costing information.

Cost-effectiveness is defined in terms of US 1984 dollars per tonne ($/t) of NO_x abated (converted to £/t). The costs were derived by annualising the total capital investment required for the pollution abatement technique. Assumptions used to annualise the capital costs were made as consistent as possible, using a real discount rate of 5 per cent and equipment lifetimes that vary with the type of facility (e.g. 30 years for combustion modifications on new boilers or 15 years for retrofits; 3–5 years for catalysts for selective catalytic reduction etc). The annual charge for operation and maintenance (O & M) including by-product sales and changes in fuel consumption, is added to the annualised capital charge. The resulting net annual cost is divided by the annual tons of NO_x abated.

For further details the reader should refer to the paper by Leggett [102].

4. Nitrogen Oxides Abatement Processes (Flue Gas Treatment)

4.1 Classification of the Processes

4.2 Outline Descriptions

4.3 General Appraisal of Processes

4.4 Processes for Detailed Study

4.5 Evaluation of Selected DeNO$_x$ FGT Processes

4.6 Costs

4. Nitrogen Oxides Abatement Processes (Flue Gas Treatment)

4.1 Classification of the Processes

The general classification of NO_x abatement ($deNO_x$) processes adopted in this Volume involves a division into two broad groups according to whether abatement is achieved at the source of combustion, 'Combustion Techniques', or in the flue gas downstream of the boiler, 'Flue Gas Treatment'.

The first group, 'Combustion Techniques', is dealt with in Section 3 of this Volume.

The second group, 'Flue Gas Treatment' (FGT), is covered in this section. Figure 4.1 shows a classification of these FGT processes into two basic types:

– Dry processes (N40)
– Wet processes (N50)

Dry processes may be either catalytic or non-catalytic, whilst wet processes involve either direct absorption or gas-phase oxidation prior to absorption. One further sub-division occurs in the case of dry, non-catalytic processes, i.e. reduction or adsorption.

Only those flue gas treatment processes which selectively remove NO_x are dealt with in this section; those processes which are non-selective or have been developed specifically for simultaneous removal of NO_x and SO_2 are dealt with in Section 5.

PROCESS/DESIGN CODE NUMBERS

Each major category is assigned a Category Number, e.g. Dry Processes–N40. Generalised processes within each category are characterised by a change in the last digit of the Category Number, e.g. Selective Non-Catalytic Reduction processes–N42.

The basic design or process types have been assigned the Code Numbers shown in Figure 4.1.

SUPPLIERS OF DeNO$_x$ TECHNOLOGY

Full details of all manufacturers quoted in the Volume (names, addresses, telephone and telex numbers, and names of contacts) are listed in Appendix 2.

4.2 Outline Descriptions

DEFINITION OF NO_x

The term 'NO_x' is used to denote a mixture of the two oxides of nitrogen produced during the combustion of fuels with air; namely nitric oxide, NO, and nitrogen dioxide, NO_2. In flue gas, about 90% of NO_x is in the form of NO, a small part of which will be oxidised to NO_2 in the stack by oxygen from the excess combustion air. However, NO_x concentrations in flue gas are frequently expressed in units of $mg(NO_2)/m^3$.

CATEGORY N40

CATEGORY N41–SELECTIVE CATALYTIC REDUCTION

Outline of Processes [352]: A simplified block diagram of the process is presented in Figure 4.2. Several alternative locations for the Catalytic Reactor are possible depending on the operating temperature of the Electrostatic Precipitator and the type of FGD process installed (if any). These alternative arrangements are considered in Section 4.3. Flue gas from the Boiler enters the Catalytic Reactor at a temperature of 300°–400°C. Ammonia is injected into the flue gas immediately upstream of the Reactor and at near stoichiometric conditions (relative to the nitric oxide, NO, in the gas). The nitrogen oxides in the flue gas are converted to nitrogen and water in the presence of the catalyst, and no deleterious waste products are formed. The treated gas is discharged to the Stack via an Air Preheater and Electrostatic Precipitator. This particular arrangement is referred to as the 'High-Dust System' because dust is removed from the flue gas after denitrification.

Chemistry of Process [352, 419]: Ammonia reduces the nitrogen oxides in the flue gas to nitrogen and water according to the following equations:

$$6\,NO + 4\,NH_3 = 5\,N_2 + 6\,H_2O$$
$$6\,NO_2 + 8\,NH_3 = 7\,N_2 + 12\,H_2O$$

The reducing reaction also proceeds in the presence of oxygen, but with higher ammonia consumption:

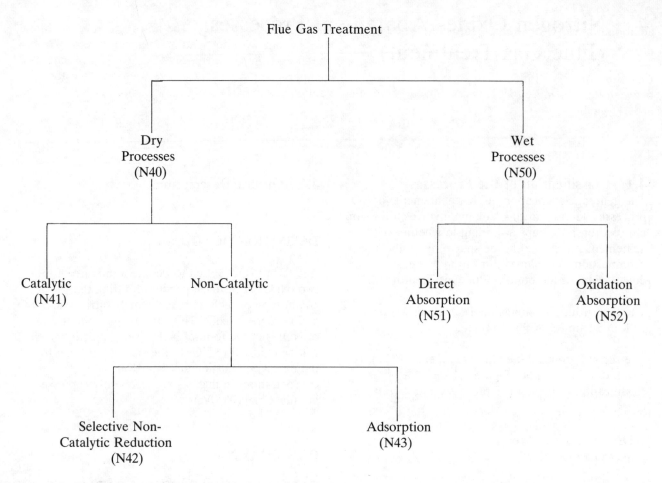

Figure 4.1 Classification of DeNO$_x$ Flue Gas Treatment Processes

$4\,NO + O_2 + 4\,NH_3 = 4\,N_2 + 6\,H_2O$
$2\,NO_2 + O_2 + 4\,NH_3 = 3\,N_2 + 6\,H_2O$

The oxidation of flue gas SO_2 to sulphur trioxide (SO_3) is also promoted by the catalyst. The SO_3 subsequently reacts with ammonia to form ammonium salts:

$SO_3 + H_2O + NH_3 = NH_4HSO_4$
$SO_3 + H_2O + 2\,NH_3 = (NH_4)_2SO_4$

The presence of SO_3 is undesirable not only for its reaction with ammonia but also because it increases the acid dewpoint of the flue gas. For this reason, catalysts used in SCR deNO$_x$ processes tend to have poor SO_2 oxidation capability.

CATEGORY N42–SELECTIVE NON-CATALYTIC REDUCTION

Outline of Process [352, 419]: A simplified block diagram of the process is presented in Figure 4.3. Denitrification of the flue gas is achieved by reaction with ammonia. As no catalyst is used in this process, much higher temperatures are required than with selective catalytic reduction processes, typically 930° to 1030°C. To achieve these temperatures, the ammonia is injected directly into the upper section of the Boiler. Optimum operating temperatures can be reduced to as low as 700°C by injecting hydrogen or natural gas with the ammonia. Exit gases from the Boiler pass to an Air Preheater and Electrostatic Precipitator prior to discharge to the Stack. No

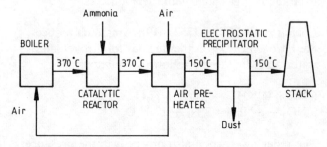

Figure 4.2 Block Diagram of Selective Catalytic Reduction Denitrification Process

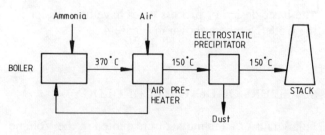

Figure 4.3 Block Diagram of Selective Non-Catalytic Reduction Denitrification Process

additional downstream equipment is required for this process.

Chemistry of Process [352]: The overall reactions between flue gas nitric oxide, excess air and ammonia are as follows:

$5O_2 + 4 NH_3 = 4 NO + 6 H_2O$
$4 NO + O_2 + 4 NH_3 = 4 N_2 + 6H_2O$

The first reaction is dominant above 1100°C, whilst the second reaction dominates in the range 800° to 1000°C.

CATEGORY N43 – DRY ADSORPTION

Outline of Process [352, 419]: A simplified block diagram of the process is presented in Figure 4.4. Adsorption processes using activated carbon were initially developed for SO_2 removal from flue gases, and later extended to simultaneous SO_2/NO_x removal (see Section 5). The optimum operating temperature for the extended process is 130°–150°C, but if selective and increased NO_x removal is possible at temperatures as low as 80°C, then an activated carbon reactor can be installed downstream of a separate FGD unit. Low levels of NO_x removal (40–60%) are reported [419] for this process, but the possibility exists to extend the use of the activated carbon bed to catalyse the reducing reaction between NO_x and added NH_3 (see above). The adsorbent is regenerated by blending with hot sand at a temperature of 650°C, whereupon nitrogen is evolved. Lower regeneration temperatures, e.g. 400°C, have been reported for simultaneous FGD/deNO$_x$ applications.

Chemistry of Process [419]: Adsorbed NO_x is converted to N_2 in the Regenerator as follows:

$2NO + C = N_2 + CO_2$
$2NO_2 + 2C = N_2 + 2CO_2$

CATEGORY N50

CATEGORY N51 – ABSORPTION OXIDATION (DIRECT ABSORPTION)

Outline of Process [419]: A simplified block diagram of the process is presented in Figure 4.5. Nitric oxide is absorbed by an aqueous salt solution containing a liquid-phase oxidising agent. Although absorption of NO is slow, due to its low solubility, the absorbed gas is rapidly converted to nitrate salts and subsequently removed from the waste water stream. The oxidising agent readily reacts with absorbed SO_2, as well as NO, so it is necessary to install a separate SO_2 Absorber upstream of the NO_x Absorber to minimise consumption of the expensive oxidising agent by the more soluble SO_2.

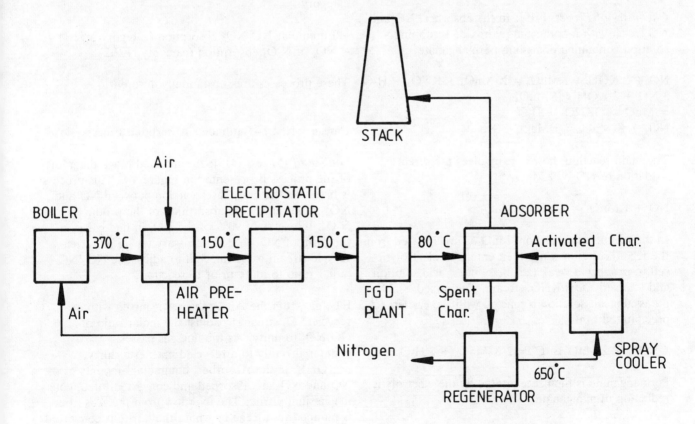

Figure 4.4 Block Diagram of Dry Adsorption Denitrification Process

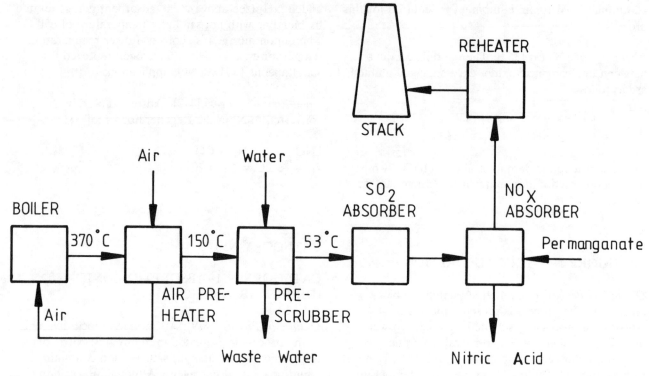

Figure 4.5 Block Diagram of Absorption-Oxidation Denitrification Process

Flue gas from the Boiler and Air Preheater enters a Prescrubber where halogen acid gases and most particulates are removed. The pre-scrubbed gases pass through an SO_2 Absorber and into the large NO_x Absorber where NO is gradually absorbed by a recirculated scrubbing agent. Gas leaving the NO_x Scrubber is reheated prior to discharge to the Stack.

Chemistry of Process [419]: In the absorber, NO and NO_2 react with a potassium hydroxide scrubbing solution containing potassium permanganate:

$NO + 2 KOH + KMnO_4 = K_2MnO_4 + KNO_2 + H_2O$
$3 NO + 2 KOH + KMnO_4$
$= MnO_2 + 3KNO_2 + H_2O$
$NO_2 + K_2MnO_4 = KMnO_4 + KNO_2$

The main reaction, however, involves the direct oxidation of NO by $KMnO_4$:

$NO + KMnO_4 = KNO_3 + MnO_2$

In the regeneration section, MnO_2 is precipitated from the KNO_3 solution which then enters an electrolytic cell to produce a weak (25–30%) nitric acid solution and a mixed KOH/KNO_3 stream. The mixed potassium salt solution is reacted with the earlier precipitated MnO_2:

$4 KOH + 2 MnO_2 + O_2 = 2 K_2MnO_4 + 2 H_2O$

Permanganate is then regenerated by the electrolytic reduction of manganate:

$2 K_2MnO_4 + 2 H_2O = 2 KMnO_4 + 2 KOH + H_2$

CATEGORY N52–OXIDATION ABSORPTION

Processes in this category include an initial gas phase oxidation stage followed by absorption in a scrubbing solution. Oxidation of the absorbed NO_x occurs to produce a by-product nitrate solution. In general, there are presently two types of process in this category:

– Equimolar $NO-NO_2$ absorption (category N52.1)
– NO_2 or N_2O_5 absorption (category N52.2)

These processes are substantially different.

Category N52.1–Equimolar Absorption Process

Outline of Process [419]: A simplified block diagram of the process is presented in Figure 4.6. This process is based on the gas-phase reaction between NO and NO_2, followed by the absorption of the product N_2O_3. In excess of 90% of the NO_x in flue gas is in the form of NO, so it is necessary for an NO_2-rich recycle stream to be injected to adjust the $NO_2:NO$ molar ratio to unity prior to reaction.

Flue gas from the Air Preheater is mixed with a recycle NO_2 stream to adjust the molar ratio of $NO:NO_2$ to unity. As the flue gas passes countercurrently to a recycled magnesia slurry, $Mg(OH)_2$, in the Absorber, equimolar amounts of NO and NO_2 are absorbed and converted to aqueous magnesium nitrite. The overflow from the Absorber is pumped to a Reactor where the nitrite is converted to the nitrate by the addition of sulphuric acid. NO

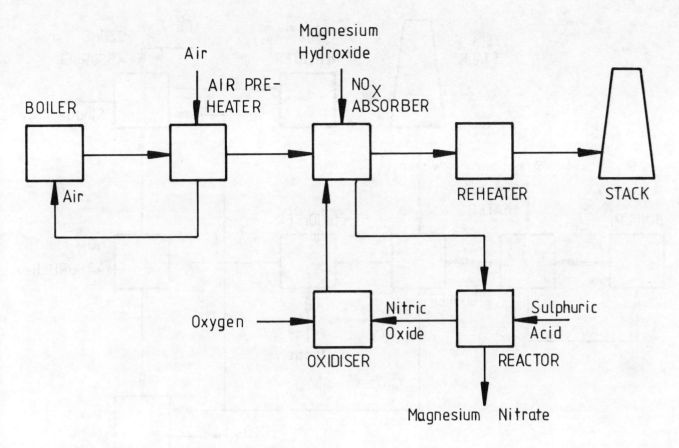

Figure 4.6 Block Diagram of Oxidation-Absorption Denitrification Process (1)

is removed from the Reactor as a gas stream and is oxidised in air to form NO_2, which is injected into the flue gas prior to the NO_x Absorber.

Chemistry of Process [419]: In the NO_x absorber, a recycle NO_2 stream is mixed with the flue gas to adjust the molar ratio of $NO:NO_2$ to unity. An $Mg(OH)_2$ slurry absorbs equimolar amounts of NO and NO_2 as follows:

$$NO + NO_2 = N_2O_3 \tag{1}$$
$$Mg(OH)_2 + N_2O_3 = Mg(NO_2)_2 + H_2O \tag{2}$$

As the NO_x concentration in the flue gas falls below 200 ppm, the reaction decreases to a negligible rate and NO_2 is absorbed:

$$2\,NO_2 = N_2O_4 \tag{3}$$
$$2\,Mg(OH)_2 + 2\,N_2O_4$$
$$= Mg(NO_3)_2 + Mg(NO_2)_2 + 2H_2O \tag{4}$$

If NO_x concentrations of less than 150–200 ppm are required, a second denitrification stage (using ozone) is necessary:

$$NO + O_3 = NO_2 + O_2 \tag{5}$$

with the NO_2 being absorbed as shown in equations 3 and 4 above.

In the reactor, sulphuric acid is added to convert the nitrite salts to nitrate:

$$3\,Mg(NO_2)_2 + 2\,H_2SO_4$$
$$= 2\,MgSO_4 + Mg(NO_3)_2 + 4\,NO + 2\,H_2O \tag{6}$$

NO is insoluble in this solution and passes to the oxidation tower, where air converts it to NO_2 prior to re-injection into the NO_x absorber:

$$2NO + O_2 = 2NO_2 \tag{7}$$

Category N52.2 – N_2O_5 Absorption Process

Outline of Process [419]: A simplified block diagram of this process is presented in Figure 4.7. This process is based on gas-phase oxidation of NO_x to N_2O_5, using excess ozone, followed by absorption in a weak (8–10%) aqueous nitric acid solution. The flue gas is adiabatically cooled and humidified prior to absorption. A small purge stream from the Absorber is concentrated to 60% acid in a Steam Evaporator. The cleaned flue gas flows to a second Absorber where it is contacted with a calcium sulphite slurry to remove the remaining ozone, prior to reheating and discharge to Stack. Make-up $CaSO_3.{}^1/_2H_2O$ is added to the Ozone Absorber and a small purge stream is removed as a waste sludge. The absence of a nitrate waste water treatment system limits the process to clean flue gas streams, i.e. free of dust or SO_x.

Chemistry of Process [419]: Flue gas NO_x rapidly and selectively reacts with ozone as follows:

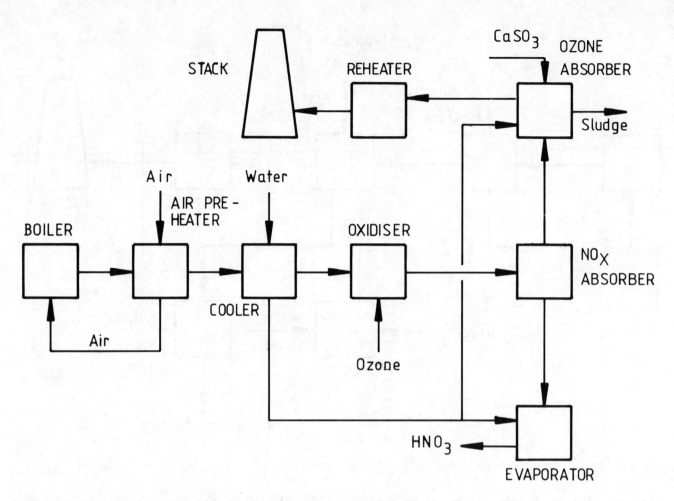

Figure 4.7 Block Diagram of Oxidation-Absorption Denitrification Process (2)

Table 4.1 Status of Selective Catalytic Reduction Plants (All Sources) [350]

Vendor	Utility Boilers		Other Plant		Additional References
	No. of Plants	Total (a) Flue Gas Vol	No. of Plants	Total (a) Flue Gas Vol	
Asahi Glass			1	70	
Babcock Hitachi & Hitachi Limited	21 (b)	19,540 (c)	8	1630	
Hitachi Zosen			9	1867	
IHI	15 (c)	18,505 (b)	1	200	
JGC			5	393	
Kawasaki HI	2	750	1	25	347, 417
Kobe Steel			1	101	
Kurabo Engineering			1	30	419
Mitsubishi HI	32	25,950	7	713	
Mitsubishi KK			10	244	
Mitsui Toatsu			13	1297	
Mitsui Engineering			12	773	
Nippon Kokan			1	1320	
Sumitomo Chemical			6	999	
Sumitomo Chemical Chem. Eng.			6	130	
Total (end 1983)	70	64,754	82	9,792	

Notes:
(a) Units of thousand Nm^3/h
(b) Total of 27 in 1984 [349]
(c) Total of 'more than' 20 in 1984 [348]

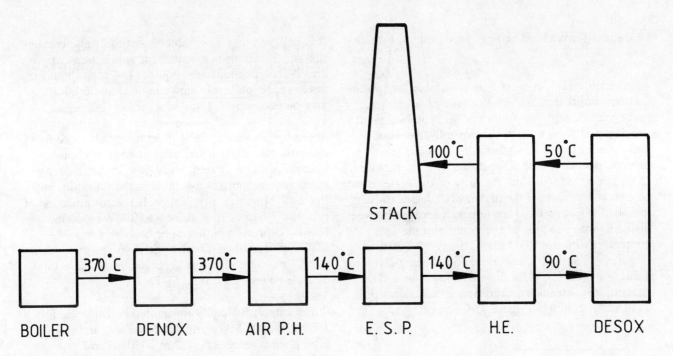

Figure 4.8 Denitrification Plant Locations–High Dust System

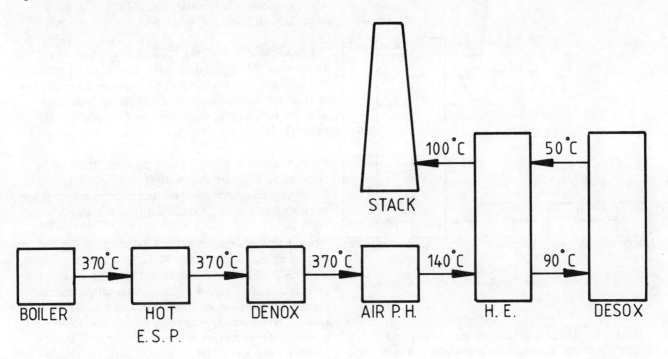

Figure 4.9 Denitrification Plant Locations–Low Dust System

$NO + O_3 = NO_2 + O_2$
$2 NO_2 + O_3 = N_2O_5 + O_2$

N_2O_5 is rapidly absorbed in the aqueous HNO_3:

$N_2O_5 + H_2O = 2 HNO_3$

4.3 General Appraisal of Processes

This section presents available information on the characteristics of the deNO$_x$ FGT processes: status, applicability, space requirements, reagent consumption, end-product or waste materials disposal requirements, typical power consumption and reductions in combustion plant efficiency, operating experience and process developments.

CATEGORY N40

CATEGORY N41–SELECTIVE CATALYTIC REDUCTION

Status: The status of this process category is indicated in Table 4.1 which gives details of plants constructed (in Japan) for coal, oil and gas-fired boilers, furnaces, etc.

Applicability: This process type is suitable for retrofit and new build applications. For retrofit situations, space may be a problem and this must be evaluated site-specifically along with the possible relocation and capacity limitations of existing fans, air preheaters, ductwork and other downstream equipment. In addition, the structural integrity of the boiler must be checked in view of the higher draft losses through the plant. SCR reactors are normally located between the economiser section of the boiler and the air preheater, with the FGD plant, if any, downstream (Figures 4.8 and 4.9). Consideration is being given to locating the deNO$_x$ plant downstream of the FGD system [349], especially where space constraints pose a particular problem (Figure 4.10).

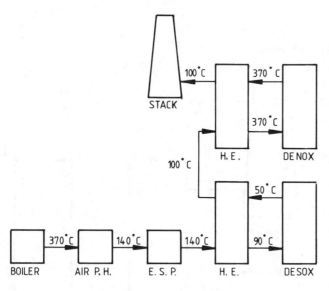

Figure 4.10 Denitrification Plant Locations-Tail Gas System

The reactor design and catalyst section depends primarily on the fuel quality, especially ash and sulphur content, and gas temperature [347]. However, current research is aimed at developing catalysts which operate at lower (e.g. air pre-heater exit) temperatures, but at the same NO$_x$ reduction efficiencies [429], in order to simplify retrofitting.

SCR plants are optimally designed for NO$_x$ reduction efficiencies of 80% [342]. However, higher efficiencies of up to 90% are feasible, but require a significant increase (30%) in catalyst volume and cost [371]. This is an important consideration, since it has been reported that catalyst costs amount to 40–60% of annual levelised operating costs [410]. Capital costs for SCR plant are dependent on unit size, fuel type, NO$_x$ reduction, location, reagent and catalyst costs, etc., but are significantly higher than for combustion techniques [345].

It is believed that a combination of SCR and combustion modifications is essential for applications where high NO$_x$ reductions are required [347], and that for lower reduction levels only combustion modifications may be necessary [341].

Space Requirements: SCR plant have only three major components: the reactor, catalyst and ammonia storage/injection system. Very little information on space requirements is available in the literature but Pruce [371] quotes a typical SCR reactor envelope of 6.1 m × 7.6 m × 18.3 m for an 80 MWe system. Kawasaki present outline specifications for a 700 MWe pulverised-coal boiler with two reactors, each of 13.4 m × 14.5 m cross-sectional area, and a total catalyst volume of 700 m³ [347].

Other catalyst volumes are quoted in Table 4.2 for 175 MWe, 500 MWe and 700 MWe oil- and coal-fired plants, being 300–407 m³, 310 m³ and 317–571 m³, respectively [352]. It can be noted that these catalyst volumes vary with NO$_x$ removal efficiency and initial flue gas NO$_x$ concentrations.

As mentioned above, the catalyst is a major cost factor and the required catalyst volume is dependent not only on the two factors previously mentioned, but also on the allowable ammonia slippage (leakage) and any redundancy necessary for fuel quality variation [347].

Consumables: In a SCR plant there are three main consumables: ammonia, catalyst and electricity. Electricity consumption, and its effect on power plant operating efficiency, is dealt with later in this section.

Typical ammonia consumption figures have been quoted as 0.7–0.8 kg/h per MWe of installed capacity for 75% NO$_x$ reduction to an output concentration of 200 mg/m³; see Table 4.2 [352]. Alternative sources express ammonia consumption in terms of the mole ratio of NH$_3$ to NO$_x$ reacted. Typically, 60 to 85 percent NO$_x$ reduction is achieved with 0.61 to 0.90 mol NH$_3$ per mol of NO$_x$ reacted, with 1–5 ppm slippage [350]. NO$_x$ reductions of 80%, with 2 ppm slippage and an outlet NO$_x$ concentration of 80 mg/m³, require a NH$_3$/NO$_x$ ratio of 0.83. This equates to 0.30 kg/h NH$_3$ per MWe of installed capacity (assuming 90% of NO$_x$ is in the form of NO).

The catalyst accounts for a high proportion of the annual levelised costs [410]. This is due to the short lifetime guarantees, normally 4–5 years, being offered for the catalyst. Catalyst replacement can be carried out in one of two ways: full volume replacement when required or partial replacement at 1–2 year intervals [347]. The latter method is normally preferred for reasons of plant economy.

For treating flue gases from oil-fired boilers, the catalyst should possess the following properties [342]:

Table 4.2 Ammonia Consumption of SCR Units [352]

A. 100 MWt plant:

Initial NO_x Concentration (mg NO_2/m^3)	Final NO_x Concentration (mg NO_2/m^3)	NO_x Reduction (%)	Ammonia Consumption (kg/h)
1800	200	89	215
1500	200	87	170
1300	200	85	145
800	200	75	80
650	200	69	60
500	200	60	40

(Consumption is 0.130–0.134 kg/h per mg NO_2/m^3 reduction)

B. 500 MWe utility boiler (hard coal-fired):

Flue gas flowrate = 1,500,000 m^3/h (4% vol. O_2)
Flue gas temperature = 400°C
Inlet NO_x concentration (as NO_2) = 800 mg/m^3
Inlet NO_2 concentration (maximum) = 40 mg/m^3
Outlet NO_x concentration (as NO_2) = 200 mg/m^3
Outlet NH_3 concentration = 3.8 mg/m^3

Ammonia consumption = 360 kg/h
Electricity consumption (for NH_3 injection) = 24 kWe
Hot water consumption (for NH_3 evaporation) = 4000 kg/h
Hot water inlet/outlet temperatures = 120°/90°C
Pressure loss (NH_3 injection + reactor) = 12 mbar
Catalyst volume = c. 310 m^3 (honeycomb/molecular sieve)

C. 700 MWe conventional boiler (oil-fired):

Flue gas flowrate (m^3/h)	2.0×10^6	1.9×10^6
Inlet NO_x concentration (mg/m^3)	400	240
NO_x removal efficiency (%)	80	80
Outlet NH_3 (ppm)	2	3
NH_3/NO_x mole ratio	0.83	0.86
Catalyst volume (m^3)	571	317
Pressure drop (mbar)	7	6
Power consumption (% generated)	0.16	0.14

D. 175 MWe wet bottom boiler (coal-fired):

Flue gas flowrate (m^3/h)	0.57×10^6	0.57×10^6
Inlet NO_x concentration (mg/m^3)	2000	2000
NO_x removal efficiency (%)	80	90
Outlet NH_3 (ppm)	3	3
NH_3/NO_x mole ratio	0.81	0.91
Catalyst volume (m^3)	300	407
Pressure drop (mbar)	10	11
Power consumption (% generated)	0.23	0.25

(i)   Resistance to toxic materials,
(ii) Mechanical strength (especially where soot blowing occurs),
(iii) Resistance to thermal cycling,   )
(iv) Resistance to oxidation of SO_2 to SO_3,) especially for high-sulphur oil-fired systems
(v)   Resistance to plugging.   )

Catalyst technology is considered proprietary by most vendors and, therefore, little information is available. Generally a base metal of iron, vanadium, chromium, manganese, cobalt, nickel, copper, barium, platinum, carbon, molybdenum or tungsten, or oxides, sulphates or combinations of these metals, is deposited on a carrier of alumina (Al_2O_3), titania (TiO_2) or silica (SiO_2) [345, 346]. Many of these materials are damaged by SO_2 and SO_3 in the flue gas, and under these conditions the most stable and most widely adopted catalyst contains vanadium compounds on a titania carrier [345].

Pressure Losses: Honeycomb and plate catalyst reactors have low pressure drops due to their high voidage. Table 4.2 indicates typical pressure drops of 6–12 mbar across the reactor and ammonia injection nozzles [352, 347].

Power Consumption: Only limited power consumption figures are available. Table 4.2 estimates overall consumption to be of the order of 0.14–0.25% of the power generated for boilers of 175 MWe and 700 MWe capacity [352]. The same reference cites 360 kWe consumption for a SCR plant located after the economiser per 100 MWe of boiler capacity, and 600 kWe per 100 MWe if installed downstream of a wet FGD plant.

Operating Experience: In Japan, all SCR reactors are installed between the boiler economiser and the air preheater, irrespective of whether hot or cold electrostatic precipitators are installed. In Europe, there is some interest in the installation of deNO$_x$ reactors downstream of the FGD plant, due to space limitations in retrofit situations [349]. There is little experience of the latter configuration at present, although six plants of this type are under construction in the FRG [28].

There is a range of catalysts for applications with all types of fuel, from those with low SO$_2$ oxidation rates to those with high activity [347]. With high-sulphur oil-fired boilers, the choice of catalyst will be a compromise between NO$_x$ reduction and SO$_2$ oxidation [348]. The oxidation of SO$_2$ to SO$_3$ and ammonia leakage is detrimental to downstream equipment due to the subsequent formation of sulphuric acid, ammonium sulphate and bisulphate (NH$_4$HSO$_4$) [347]. NH$_4$HSO$_4$ may stick to the catalyst surface causing reduced catalyst activity [348], cause fouling, erosion and reduced heat transfer in the air preheater, lead to baghouse blinding, modifications in scrubber chemistry in downstream FGD plant, and increase the disposal problems for scrubber by-products [343, 26]. Ammonium sulphate is not removed completely by electrostatic precipitators and may result in an unacceptable plume formation. In Japan, the EPDC has monitored sulphate deposition in a 250 MWe (coal-fired) power station; air preheater elements became blocked after 6 months operation, although their operational life was extended to a year by washing with water.

NH$_4$HSO$_4$ formation is temperature dependent [342, 343]. When the temperature falls below 300°C in the presence of 1 ppm SO$_3$, the potential for NH$_4$HSO$_3$ formation increases together with the possibility of deposition on the catalyst surfaces and permanent loss of catalyst activity if these low temperatures are maintained.

With higher concentrations of SO$_3$ (and NH$_3$), the minimum operating temperature to avoid sulphate formation will also increase [342]. The recommended minimum operating temperatures provide a safe margin to allow for temperature swing during load changes, the NH$_4$HSO$_4$ formation temperature being reported as 210°C in the presence of 10 ppm (36 mg/Nm^3) SO$_3$ and 10 ppm (8 mg/Nm^3) NH$_3$ in the flue gas [343]. In practice, the formation and effects of NH$_4$HSO$_3$ are minimised by the provision of an economiser flue gas bypass (to increase temperature), the reduction of SO$_3$ and NH$_3$ concentrations in the reactor outlet gases or by modifying the air preheater design; all these options increase plant costs or reduce the efficiency of power generation [343].

A maximum operating temperature is imposed by the possible dissociation of ammonia above 400°C; if temperatures of 480°C–540°C are exceeded, sintering of the catalyst may occur thereby permanently impairing its activity [342].

At 80% NO$_x$ reduction, NH$_3$ leakage is negligible. At 90% reduction, NH$_3$ leakage is 10–20 ppm or higher [371].

An outlet flue gas concentration of 5 ppm is regarded as the acceptable limit [344] and NH$_4$HSO$_3$ deposition on air preheater is avoided altogether if NH$_3$ leakage is 3 ppm (2.3 mg/Nm^3) [350]. The NH$_3$ concentration can be minimised by uniform distribution within the reactor inlet duct by means of a multi-nozzle arrangement [347].

It has been reported that there is no noticeable deterioration of titania-based catalysts caused by the presence of halogen compounds [348]. Granular and pelletised catalysts have largely been superseded by plate and honeycomb arrangements. Metal plate type catalysts with a thickness of 1 mm remain relatively dust free and usually do not need soot blowers to maintain low pressure drops through the catalyst bed [349].

Process Development: Spherical catalysts in a fixed bed are unsuitable for flue gases from oil-fired units due to possible dust deposition. This led to the development of moving catalyst beds which introduced construction complexities and abrasion of the catalyst [347]. The openness of the pipe, honeycomb and grid plate catalysts has now superseded bed catalysts for deNO$_x$ applications.

Some technical difficulties have been experienced in the formation of these configurations with some catalyst materials. Catalysts based on Al$_2$O$_3$ and ferric oxide (Fe$_2$O$_3$) are poisoned by SO$_x$ in the flue gas [350], so TiO$_2$ with a small amount of vanadium pentoxide (V$_2$O$_5$) is overwhelmingly in use for SO$_x$-containing gases.

It has been reported that V$_2$O$_5$/TiO$_2$ catalysts can be used in flue gases containing up to 5700 mg/m^3 SO$_2$ and 360 mg/m^3 SO$_3$ [352]. This catalyst combination is highly active but the V$_2$O$_5$ does promote some SO$_2$ oxidation; a low oxidation catalyst can be produced by replacing the V$_2$O$_5$ partly or fully with oxides of tungsten or molybdenum [350].

SO$_x$ poisoning is not a problem in situations where the SCR reactor is located downstream of an FGD plant [347].

For high-sulphur oil applications, vertical downward gas flow in the catalyst chamber is preferred to minimise dust deposition on the catalyst surfaces [342]. For low-sulphur oil applications, a horizontal gas flow may be acceptable. Catalyst life expectancy is quoted as 4–5 years [352] for oil-fired systems.

Extensive field trials are taking place in the FRG, on a pilot-plant scale, to investigate the effect of variable load patterns on the catalyst [28]. German boilers are considered to have more frequent load swings than those in Japan, so thermal shock on the catalyst structure is likely to be much greater. Activated coke is being considered as the catalyst in reactors located downstream of FGD plant, because it is effective at a relatively low temperature, e.g. 90°–120°C, and so less reheating of the flue gas stream is required prior to entering the deNO$_x$ reactor [28].

Another interesting development involves pilot-scale test programmes and field trials in the FRG aimed at coating the plates of a rotary regenerative-type of air preheater with catalytic material, so eliminating a separate SCR reactor [28]. In this arrangement, catalyst temperatures would typically range from 150°–350°C.

Appraisal:

1.	Information available	2
2.	Process simplicity	1
3.	Operating experience	2
4.	Operating difficulty	1
5.	Loss of power	2
6.	Reagent requirements	1
7.	Ease of end-product disposal	2
8.	Process applicability	2
	Total	13

CATEGORY N42–SELECTIVE NON-CATALYTIC REDUCTION

Status: A Selective Non-Catalytic Reduction (SNR) process was first developed by Exxon Research Engineering Corporation (ERE) in 1972, and marketed as the Thermal Denox Process [414]. A similar kind of process was independently developed by Mitsubishi (MHI) in Japan and tested at a 375 MWe oil-fired power station [24]. To date, this is the largest plant on which full-scale tests have been carried out [352].

The process has been commercially demonstrated in oil- and gas-fired utility boilers, steam boilers and process furnaces, and tests have been conducted on a municipal incinerator, oil-fired steam generator and glass melting furnace, see Table 4.3 [414].

The largest boiler equipped with SNR has a capacity of 180 MWe (equivalent) [352].

Applicability: This process type is suitable for retrofit and new build applications.

Table 4.3 Status of Selective Non-Catalytic Reduction Plants in Japan & USA Oil & Gas Fired Units

Vendor	No. of Units	Type of Plants	Dates	Range Nm^3	% DeNO$_x$ (Typical)	Reference
Exxon	6	Utility Boilers	to 1985		60–80 (b)	414, 417
Exxon	7	Industrial Boilers	to 1985		50–80 (b)	414, 417
Exxon	33	Petroleum Heaters	to 1985		75–85 (b)	414, 417
Exxon	2	Incinerators	to 1985		70–80 (b)	414, 417
Exxon	4(a)	Oil Fired Steamers	to 1985			414, 417
Exxon	1(e)	Glass Melting Furnace	to 1985		50–80 (b)	414, 417
MKK (c)	c.20	Small gas-fired boilers & heaters	1975–78	5,000–20,000	40–60	419
MHI (d)	10 min (f)	Refuse Incinerators	to 1985	30,000–100,000	40–60	419

Notes:
(a) 3 commercial, 1 demonstration. In USA only
(b) Improved technology, i.e. wall injection. Industrial boilers and furnaces in Japan with early technology: 20–60% [350]
(c) Mitsubishi Kakoki
(d) Mitsubishi Heavy Industries
(e) In USA
(f) In Japan only

The principles of SNR are similar to those of SCR except that, in the absence of a catalyst, a much higher operating temperature is required, typically 900°–1000°C, for the NH_3-NO_x reactions to take place. It is, therefore, necessary to inject the ammonia directly into the upper section of the boiler or process firebox. Operating temperatures are critical because of the narrow temperature window over which the reactions are effective and this is one of the major drawbacks of the method.

If temperatures are too high, i.e. higher than 980°C, side reactions produce NO with NO_x concentrations in the flue gas actually increasing at temperatures in excess of 1090°C [345]. If the temperature is too low, reaction rates are reduced and higher concentrations of NH_3 appear downstream.

The effective operating temperature range can be extended by adding hydrogen (H_2) or natural gas (CH_4). The optimum reaction temperature decreases with increasing H_2 (or CH_4): NH_3 ratio, with the result that at a ratio of 2.0 (and NH_3:NO_x = 1.7), the effective temperature range is 700°C ± 50°C [352].

In full-scale commercial operations, ensuring that NH_3 is injected at the correct temperature is difficult, especially during load swings. This is largely overcome by the inclusion of multiple NH_3 injection points and the controlled use of H_2, as indicated above.

SNR systems installed in early commercial boilers in Japan achieved 50–60% NO_x reduction. The recent development of wall injectors (see later) has extended the system's capabilities to about 80% [352], although higher efficiencies (90%) have been claimed to be possible [409].

While this method is harder to control and generally less efficient than SCR techniques, it is much less expensive to install, especially on existing boilers and heaters [345]. In addition it is free from the problems associated with catalysts, namely blinding/plugging and SO_2 oxidation to SO_3 [345].

Higher NH_3 consumption and the use of H_2 (or CH_4) have an effect on operating costs, although these must be compared to the comparatively high cost of the catalyst in an SCR system [410].

Problems with locating early designs of the NH_3 injection system, especially in retrofit situations, have been noted [409]. It is claimed, however, that the new wall injector design can be easily retrofitted to existing boilers with minimal impact on the boiler structure, and without the need to relocate ducting, air preheaters, stacks, etc. [409]. Proper design and location require the use of a detailed kinetic model of the system [409].

To date, most SNR applications have been on flue gas containing virtually no SO_3 [409]. Unreacted NH_3 will combine with SO_3 to produce ammonium sulphate and bisulphate, with the possibility that the bisulphate will cause fouling of downstream equipment.

Space Requirements: No details of space requirements are reported in the literature, but this is expected to be insignificant, i.e. the space required for the ammonia storage, mixing and injection system. Compared to the SCR system, major space savings are to be expected because of the absence of the catalytic reactor [352].

Consumables: Typical NH_3 consumption figures have been presented by Exxon [416] for a sample 83t/h of steam oil- or gas-fired industrial boiler; see Table 4.4.

Ammonia consumption is also expressed in terms of an NH_3:NO_x molar ratio. Quoted ratios are variable (Table 4.5), but are generally of the order of 2.0 for

Table 4.4 Ammonia Consumption in SNR plant

Design Conditions:	83t/h
	42 bar
	370°C
Fuel:	OIL OR GAS
Initial NO_x Concentration	200 ppm (vol)
NH_3:NO_x Ratio:	2.0:1
NH_3 Vaporiser:	Electric Element; direct contact
NH_3 Storage Capacity:	38m³ (30 days)
Annual NH_3 Consumption (a):	164 tonnes
Annual Electric Consumption:	55 MWh (a)

Note:
(a) 65% load factor

Table 4.5 Typical NH_3:NO_x Molar Ratios for SNR

Mol. NH_3 / Mol. NO_x	% deNO$_x$	Other Plant Details	Ref (Year)
0.3–0.5	10–20	Oil Refinery/Petrochemical Plant Inlet NO_x = 400 mg/m³	352 (1986)
0.8–2.0	30	Wall Injectors (new design)	350 (1985)
1.0	30–40	Unreacted NH_3 = 10–15 mg/m³	350 (1985)
1.0–2.0	45–55	Oil Refinery/Petrochemical Plant	352 (1986)
1.0–2.0	48–63	235 MWe Utility Boiler Injection Grid (old design)	414 (1985)
1.5	35–45	375 MWe Plant, 0.2%S oil. Unreacted NH_3 = 13–23 mg/m³	352 (1986)
1.5–2.0	50–60	Commercial Boilers	345 (1981)
1.5–2.5 (optimum)	c. 90	Oil Fired Steam Generator	416 (1982)
3.0	70		343 (1980)
4.0	60		352 (1986)

NO_x reductions of about 60%. These compare with a molar ratio of 0.61 for 60% $deNO_x$ with SCR [350].

Hydrogen (or natural gas) consumption is dependent upon temperature variations in the boiler. For example, with a NH_3/NO_x ratio of 1.7 and no hydrogen, optimum flue gas temperatures are in the region of 970°C [352]. As the H_2/NO_x molar ratio increases to 0.5, 1.3 or 2.4, the optimum flue gas temperature falls to 825°C, 750°C and 700°C, respectively. It is claimed that with the new wall injector design, the injection of hydrogen may be unnecessary [414].

Pressure Losses: No information on typical pressure losses is quoted in the literature. Pressure drops through SCR systems have been reported as 6–12 mbar across the ammonia injection nozzles and catalyst reactor, so SNR systems are expected to sustain significantly lower losses than these. In addition, the newer wall injector systems will reduce the pressure losses still further.

Power Consumption: Limited information on power requirements in SNR plant is available. Reference to Table 4.4 suggests that the annual consumption of 55 MWh for a 91t/h industrial boiler approximates to about 0.05% of the equivalent power output [416].

Operating Experience: The main factors affecting the injection of ammonia into an SNR process are [352]:

- Temperature,
- Residence time at temperature,
- Temperature profile,
- initial NO_x concentration,
- NH_3/NO_x molar ratio,
- H_2/NH_3 molar ratio,
- Mixing conditions.

Early commercial or test facilities involved the location of one or more injection nozzles in the flue gas stream at the appropriate temperature, so that a mixture of NH_3 and its carrier gas (steam or air) could be injected [352]. The carrier gas is usually required to prevent overheating of the grid [409]. In more recent applications, simpler and less expensive wall injectors have largely superseded the injection grids. The wall injectors can be easily retrofitted to an existing boiler with minimum impact on the structure of the boiler and without the need to relocate ducting, air preheaters, stacks, etc. [409].

The ammonia must be injected within a specific and narrow temperature range which varies according to boiler load. To accommodate these requirements the wall injectors are arranged in two or more zones, and the amount of NH_3 is controlled. It is necessary to inject varying quantities of hydrogen into the system with the NH_3 in order to obtain good NO_x removal rates at low boiler loads. This adds to the cost and complexity of the operation [409]. Variations in temperature (and flue gas composition) within the large-size ducts, pose additional operating difficulties, thereby limiting the extent to which NO_x reduction may be achieved with SNR techniques [371].

Other difficulties associated with SNR may be:

(a) The leakage and subsequent emission of ammonia (generally below 45 mg/m^3) and by-products. In this aspect, SNR differs little from SCR although, with higher operating NH_3/NO_x molar ratios, leakage rates are expected to be higher.

(b) The potential fouling of downstream equipment, especially the air preheater, with ammonium bisulphate. In the high temperature flue gas, sulphur oxides from the fuel do not interfere with the SNR reaction chemistry [414, 417]. In addition, the injected NH_3 does not promote the oxidation of SO_2 to SO_3: this occurs in SCR due to the presence of the catalyst. However, unreacted ammonia reacts with SO_3 and H_2O in the flue gases, as the temperature falls, to form ammonium bisulphate and, at still lower temperatures, ammonium sulphate. The sulphate is a dry solid which creates neither corrosion nor unacceptable fouling problems in the air preheater [417], whereas the bisulphate is a sticky, corrosive liquid at preheater temperatures. Exxon research [414, 417] has predicted that ammonium bisulphate formation can be minimised by one or more methods:

- Maintaining an $NH_3:SO_3$ molar ratio above 2.0 and providing sufficient reaction time for the formation of sulphate in preference to the bisulphate,

- Limiting NH_3 leakage to 5 ppm (4 mg/Nm^3) or less,

- Maintaining the air preheater outlet flue gas temperature above 204°C,

- Maintaining the temperature at the NH_3 injection point lower than the sulphate formation temperature.

The first method is considered the most practical for SNR processes since NH_3 leakage rates (c. 50 ppm or 38 mg/Nm^3) are generally twice those of SO_3 [417]. In addition, extensive testing in Japanese oil-fired boilers indicates that sulphate/bisulphate deposits can easily be removed by water-washing at intervals [417]. Ammonium sulphate creates a 'blue haze' problem at the stack [409].

(c) Thermal damage to the NH_3 injection nozzles and the difficulties associated with keeping them cool. Corrosion of the injection grid is a further potential problem with oil-fired boilers [409].

These problems are avoided with the improved wall injectors.

The SNR process has been demonstrated commercially in oil- and gas-fired boilers. Operation at a 375 MWe oil-fired (0.2%S) utility boiler in Japan has been virtually trouble free, except for problems associated with ammonium sulphate and bisulphate deposits in the air preheater and plume formation at the stack [352].

Process Development: The development of wall injectors has eliminated many of the disadvantages associated with injection grids and reportedly improved NO_x reduction levels [409, 414]. The proprietary designed injectors are located at or near the boundary walls of the injection zone, and can be readily retrofitted to existing boilers. Optimum design and location of the simple, large jets does require the use of a detailed kinetic model of the unit to be fitted.

Wall injectors can be located at the optimum flue gas temperature, even within the combustion zone of the boiler [414]. Two sets of injectors may be used to accommodate load variations with the result that hydrogen injection, to promote low flue gas temperature performance, may be unnecessary.

The first commercial unit installed with wall injectors started operation in late 1984, although NO_x reduction levels were kept relatively low (30%) [350].

Exxon have developed a fundamental kinetic model for SNR which has broadened the understanding of the process, allowing them to predict accurately the performance and NH_3 leakage rate for any type of fired unit [414].

In the late 1970s tests were carried out in Japan and the USA to increase $deNO_x$ efficiency by including a small amount of SCR catalyst with an SNR system in oil-fired utility boilers [350]. NO_x removal efficiencies of 50–60% and NH_3 leakage rates below 10 ppm (8 mg/Nm^3) were targeted. The tests were not successful and the SNR facilities were removed.

Appraisal:

1.	Information available	1
2.	Process simplicity	2
3.	Operating experience	1
4.	Operating difficulty	0
5.	Loss of power	2
6.	Reagent requirements	1
7.	Ease of end-product disposal	2
8.	Process applicability	2
	Total	11

CATEGORY N43–DRY ADSORPTION

Activated carbon (coke or char) is capable of catalysing the reaction between NO_x and NH_3 at significantly lower temperatures than for SCR [352, 4]. The ability of activated carbon to adsorb SO_2 from the flue gases has led to the development of this process for simultaneous removal of NO_x and SO_2 (see Section 5). However, there is considerable interest in the FRG in the use of activated coke as a catalyst in cold-side SCR applications, i.e. downstream of the FGD plant [28]. Operating temperatures would be in the region of 90° to 120°C, thereby requiring significantly less reheat of the saturated flue gases than would be the case with conventional cold-side SCR catalysts. In addition, with reduced SO_3 concentrations in the flue gas at the NH_3 injection point, the deposition of ammonium bisulphate and consumption (and leakage) of NH_3 would be minimised [352].

Disadvantages of the process include high carbon losses and a low NO_x removal efficiency of 40–60% [343]. Higher $deNO_x$ efficiencies (80%) have been reported at a 90 MWt demonstration plant for simultaneous NO_x/SO_2 removal in Japan, with normal flue gas temperatures of 120°–155°C [24]. It is recognised that for commercial applications a low-cost activated coke is required to counter the considerable coke consumption rate [350].

Appraisal (for NO_x reduction only):

1.	Information available	0
2.	Process simplicity	1
3.	Operating experience	0
4.	Operating difficulty	0
5.	Loss of power	2
6.	Reagent requirements	1
7.	Ease of end-product disposal	2
8.	Process applicability	2
	Total	8

CATEGORY N50–FLUE GAS TREATMENT–WET PROCESSES

Although several wet processes have been developed or advocated for the removal of NO_x from flue gases, they are all limited by the low solubility of NO in water. The general approach is either to oxidise the NO to the more soluble NO_2 in the gas phase prior to absorption or absorb the NO_x directly in the absorption solution. In addition, the absorbed NO_x can either be reduced or oxidised in solution. Therefore, four major categories of wet NO_x-removal process are recognised:

(i) Gas-phase oxidation, absorption and liquid-phase reduction (oxidation-absorption-reduction)

(ii) Gas-phase oxidation, absorption and liquid-phase oxidation (oxidation-absorption)
(iii) Direct absorption and liquid-phase reduction (absorption-reduction)
(iv) Direct absorption and liquid-phase oxidation (absorption-oxidation)

Most scrubbing solutions for wet NO_x removal also remove SO_x, so current developments are aimed at simultaneous NO_x-SO_x processes, which are particularly suitable for coal or high-sulphur oil applications [345].

Oxidation-absorption-reduction processes are, in most cases, developments of commercially available FGD techniques [345], and absorption-reduction processes are being developed specifically for the simultaneous removal of NO_x and SO_x to avoid the use of the expensive gas-phase oxidant [419]. Of the remaining two wet $deNO_x$ categories, absorption-oxidation processes were originally developed to treat nitric acid plant tail gases and have a major problem with the high consumption of the expensive liquid-phase oxidant [343]. This problem is caused by the oxidant reacting with absorbed SO_2 as well as NO, and so a separate (and more conventional) SO_2 absorber is necessary. This factor, plus the need for an extremely large NO_x absorber to ensure adequate NO absorption rates, makes the absorption-oxidation processes unattractive and expensive when compared to other NO_x and simultaneous NO_x-SO_x processes [343].

The oxidation-absorption category encompasses a number of processes that have no common mechanism other than a gas phase oxidation stage followed by an absorption stage. The two main processes in this category involve liquid phase (oxidation) reactions which convert the NO_x to nitrate salts or nitric acid, respectively. Of all the wet processes, (ozone) oxidation processes are expected to exhibit the highest NO_x removal efficiencies [347]. The main disadvantages are a very high oxidant consumption rate and the need to scrub the flue gas of SO_2 prior to the NO_x removal stage [343]. The latter problem makes the process more complicated than similar wet processes where SO_2 and NO_x are simultaneously removed in a single absorber.

Small units have been built in Japan to test wet $deNO_x$ technologies, but it has been reported that these processes will be of little importance for some time [4]. This is primarily due to the waste water problems, caused by the presence of nitrite, nitrate and nitrogen-sulphur compounds, which make wet $deNO_x$ processes of little importance compared to dry techniques [350, 352].

Kawasaki Heavy Industries in Japan have developed wet and dry $deNO_x$ processes since 1970 [347]. Owing to economic and technical difficulties, their effort since 1977 has been concentrated solely on the development of the dry SCR process.

Appraisal (for NO_x reduction only):

1.	Information available	0
2.	Process simplicity	0
3.	Operating experience	0
4.	Operating difficulty	0
5.	Loss of power	1*
6.	Reagent requirements	0
7.	Ease of end-product disposal	0
8.	Process applicability	2
	Total	3

*Assumed in absence of data

4.4 Processes for Detailed Study

The selection of processes for detailed study in this Volume has been based upon their suitability for application in the UK for the three datum combustion systems (Section 1.4) considered:

– Large (450 tonne steam/h) industrial boiler (Datum system 1).

– Small (25 tonne steam/h) factory boiler (Datum system 3).

In principle, all of the processes listed in Section 4.1 can be applied to all combustion plant, but the attraction of many processes diminishes with factors such as decrease in plant operating scale, and increases in $deNO_x$ process complexity, reagent costs, and end-product disposal difficulty.

Appraisal of Processes
To evaluate some of these factors, a rough appraisal of each process type has been made in Section 4.3 by assigning 'merit points' for a number of features; merit points have been awarded according to the scale:

0 Below average merit

1 Average merit

2 Above average merit

3 Outstandingly above average merit

The features to which these points have been assigned are described in Section 1.5; they are briefly:

1 Information available

2 Process simplicity

3 Operating experience – extent and difficulties encountered

4 Operating difficulty – availability, reliability

5 Loss of power sent out – by installation of the FGD process

6   Reagent requirements – quantities
7   Ease of end-product disposal
8   Process applicability – e.g. for retrofit

All of the processes listed in Section 4.1 and outlined in Section 3.2 are appraised in Section 4.3. The merit points assigned to the process types for each of the above features are summarised in Table 4.6. It should be noted that the number of points in the merit point system adopted in other sections of the Manual are not strictly comparable with those considered here.

Table 4.6   Summary of deNO$_x$ Process Appraisals

Code No.	Name	Merit Points for Feature No:								Total Points
		1	2	3	4	5	6	7	8	
N41	Selective Catalytic Reduction	2	1	2	1	2	1	2	2	13
N42	Selective Non-Catalytic Reduction	1	2	1	0	2	1	2	2	11
N43	Dry Adsorption	0	1	0	0	2	1	2	2	8
N50	Wet Processes	0	0	0	0	1	0	0	2	3

Features: 1. Information available
2. Process simplicity
3. Operating experience
4. Operating difficulty
5. Loss of power
6. Reagent requirements
7. Ease of end-product disposal
8. Process applicability

Processes suitable for the UK
The principal purpose of assigning merit points to each of the processes was to aid in the selection of processes that could be considered suitable for application in the UK. It was arbitrarily assumed that suitable deNO$_x$ processes would be those having a total of 10 or more merit points. It is worthy of note that one process, Dry Adsorption, was assigned 8 points, differing from SCR and SNR processes primarily in the lack of information (for deNO$_x$ alone) and operating experience. Although the process has recognised potential, it is being developed primarily as a simultaneous NO$_x$-SO$_2$ abatement process and is discussed further in Section 5.

Selection of processes for detailed study
No flue gas treatment processes are considered to be suitable for the smallest operating scale dealt with in this Manual (Datum System 2), primarily because the lower NO$_x$ reduction efficiencies required to meet expected NO$_x$ emission standards can be readily, and less expensively, attained by combustion modifications alone.

For utilities and large industrial boilers, two processes for consideration in the UK are:

– Selective Catalytic Reduction (SCR) process (Process Code N41; merit rating 12 points). SCR is a well-established process with commercial experience in oil- and gas-fired units dating back to the early 1970s. It is a relatively simple process and can achieve NO$_x$ reduction efficiencies of over 90%.

– Selective Non-Catalytic Reduction (SNR) process (Process Code N42; merit rating 11 points). SNR has been commercially applied to oil- and gas-fired units. It is a very simple and relatively inexpensive process, capable of achieving NO$_x$ reduction efficiencies of 70–80% for utility and industrial boiler applications.

4.5   Evaluation of Selected DeNO$_x$ FGT Processes

EVALUATION OF SELECTIVE CATALYTIC REDUCTION PROCESS
(Process Code N41)

See Section 4.2 for: outline of basic process, chemistry and schematic diagram.

See Section 4.3 for a list of manufacturers offering this type of equipment and for a general appraisal of the basic process.

See Section 4.4 for the reason for choosing the Selective Catalytic Reduction DeNO$_x$ process.

This basic process type is considered to be suitable for application in the UK only to large industrial boilers. The prohibitive cost of the catalyst and the scale of operation make SCR unsuitable for application with small factory boilers, where combustion techniques would be more cost effective. Hence, the SCR process is evaluated here only for Datum Combustion System 1.

Process Description

Figure 4.11 shows a simplified flow diagram for application of the process to a heavy fuel oil-fired water tube boiler.

A mixture of ammonia gas and air is injected into the flue gas upstream of a catalytic reaction chamber which, in a typical arrangement, is located between the outlet of the economiser section of the boiler and the flue gas inlet of the air preheater. The flue gas mixture flows vertically downwards (or horizontally) through the Reactor where the ammonia reacts with NO$_x$ to form nitrogen gas and water vapour. The Reactor consists of several levels of active catalyst modules. A vertical flow path through the reactor would be selected for high-sulphur oil applications

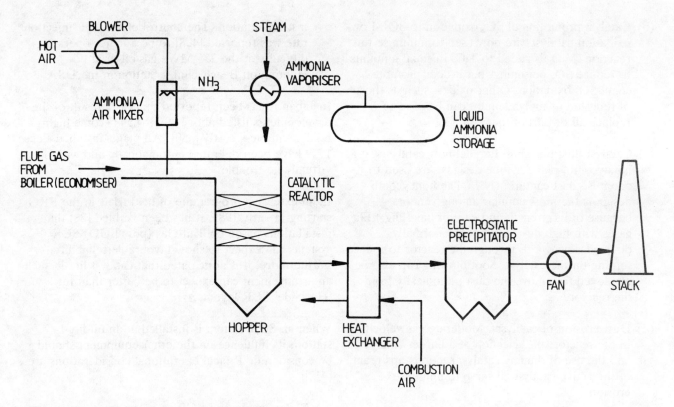

Figure 4.11 Selective Catalytic Reduction Process

because it minimises the amount of dust settling on the surface of the catalyst. Some dust is collected in Hoppers at the bottom of the Reactor chamber.

Flue gas leaving the Reactor is cooled in a Heat Exchanger prior to entering an Electrostatic Precipitator (if installed) at a temperature of about 150°C. The gas is discharged to the Stack via an induced-draught Fan.

There are several possible locations for an SCR reactor. The system described above is termed a 'High Dust' system (Figure 4.8) because the flue gas contains a high dust loading when it is introduced to the SCR reactor. If an FGD plant is also installed, it would be located downstream of the Electrostatic Precipitator, possibly after being further cooled in a Heat Exchanger. An alternative arrangement is for the SCR reactor to be located downstream of a high temperature electrostatic precipitator in order to minimise deposition in the reactor. This is termed a 'Low Dust' system. (Figure 4.9.)

A third arrangement, which is favoured by some German companies, locates the SCR reactor downstream of an FGD plant, the so-called 'Tail Gas' system (Figure 4.10). This alleviates many of the problems associated with retrofitting deNO$_x$ technology to boiler plant already fitted with an FGD process. In addition to the benefits of the low dust system, this arrangement significantly reduces the SO$_2$/SO$_3$ concentrations in the SCR reactor, thereby limiting the problems caused by ammonium bisulphate on downstream equipment, such as the Air Preheater and minimising wastage of ammonia. The major disadvantage of this arrangement is the need to reheat the saturated flue gases from the FGD plant (typically 50°C) to a temperature that is suitable for SCR, e.g. 370°C. The use of low temperature catalysts such as activated carbon may ultimately lead to low temperature reheat requirements.

Status and Operating Experience

The status of the SCR process is indicated in Section 4.3 and Table 4.1, where it can be seen that in the region of 200 SCR plants have been installed.

Despite some potential problems (see later) SCR is the most widely utilised and developed process for NO$_x$ reduction. It is finding widespread application in the FRG because the stringent deNO$_x$ regulations call for reductions beyond the capabilities of combustion modifications, i.e. at least 50–70%, and because of the early deadlines (late 1980s) for compliance.

In its early stages of development, a number of operational problems were identified and solved. These were:

(a) SO$_x$ poisoning of the catalyst. TiO$_2$-based catalysts have largely replaced those based on Al$_2$O$_3$ and Fe$_2$O$_3$ to minimise damage caused by SO$_2$ and SO$_3$. The 'Tail Gas' arrangement will also minimise the presence of sulphur oxides in the flue gases entering the SCR reactor.

(b) Catalyst promotion of SO_2 oxidation to SO_3. Low-oxidation catalysts are now used to minimise this reaction. V_2O_5 is added to TiO_2 in small amounts to reduce SO_x poisoning, but it does promote some SO_2 oxidation. Other oxides, such as those of tungsten or molybdenum, can be used to replace all or part of the V_2O_5 [350].

(c) Catalyst dust plugging. Parallel flow catalysts, e.g. honeycomb, plate or tube designs, are used to minimise dust capture. (*N.B.* The honeycomb design is the most popular among vendors because of its strength and ease of handling.) Flue gases with high dust loadings are normally directed downwards through the reactor to reduce settling on the catalyst. Soot blowing can also be employed to remove the dust periodically from the reactor.

(d) Dust erosion of catalysts. Moderate gas velocities in the reactor, the selection of a harder catalyst, and the use of dummy catalyst sections upstream of the active catalyst all serve to minimise erosion.

(e) Ammonium bisulphate deposition in the catalyst. Maintaining flue gas temperatures above 330°C in the reactor inhibits the reactions which produce the bisulphate. This can be achieved by installing a hot gas by-pass upstream of the economiser. Temperature control is especially important during periods of load swing.

(f) Ammonium bisulphate deposition in the air preheater. The potential for bisulphate formation increases, with decreasing temperature, making the air preheater (and induced-draught fan) vulnerable to this problem. The main countermeasure is to minimise the concentration of the main reactants, NH_3 and SO_3, in the gas stream leaving the reactor. NH_3 leakage from the reactor should, typically, be less than 5 ppm (4 mg/Nm^3). Alternative measures are the installation of soot blowers or design changes to the air preheater plates.

Solutions for minimising the effects of bisulphates either increase costs or reduce the efficiency of power generation. It should be noted, however, that EPDC field tests in Japan have shown that without these measures, air preheater elements could become blocked in only 6 months of operation; with water washing 12 months operation could be achieved [343].

(g) Ammonia contamination of fly-ash and FGD waste-water. In Japan, NO_x removal efficiencies are normally limited to 80% to avoid significant NH_3 leakage to downstream equipment.

(h) Maintenance and accuracy of NO_x and NH_3 instrumentation. The control of the NH_3-injection rate was reported [408] to be a major operating difficulty at the 215 MWe gas/oil-fired Huntington Beach Unit 2 Station in the USA.

In Japan it has been reported that, as a result of the developments listed above, SCR has become highly reliable and less costly [350]. All plants installed since 1979 have been operated automatically and with virtually no trouble.

In an extensive programme of field trials in the FRG starting in early 1985, it has been verified [28] that installations based on Tail-Gas (post FGD) SCR systems can expect to show favourable reliability. Furthermore, it is anticipated that catalyst life in such an arrangement can expect to be better than for 'Hot-Side' SCR layouts.

When an SCR system is installed at an oil-fired station, its influence on the other equipment should be considered. Typical operational considerations are [413]:

(a) Boiler–control of SCR inlet temperature (if necessary).

(b) Air preheater/fan–increase of draught loss due to SCR, and prevention of air preheater plugging and corrosion.

(c) Electrostatic precipitator–prevention of corrosion.

(d) FGD plant–prevention of plugging and corrosion of a gas-gas heater (if installed), and influence of NH_3 in waste-water.

Variations and Development Potential

NO_x is reduced by NH_3 at a temperature of 150°–450°C in the presence of a catalyst. The optimum temperature is in the range 300°–400°C; at lower temperatures the reaction is too slow, whilst at higher temperatures the activity of the catalyst tends to be lowered by thermal effects [350]. These temperatures are readily achieved by locating the SCR reactor between the economiser section of the boiler and the air preheater, so that no flue gas reheat is required. Low temperature catalysts for use at 150°–250°C and high temperature catalysts for use at 400°–550°C have been developed for special gases [350] and there is much interest in the FRG in activated coke as a low-temperature (90°–120°C) in Tail Gas SCR systems [28].

Ammonia is a suitable reducing agent for SCR applications because it selectively reacts with NO_x. Other reducing agents such as hydrogen, carbon monoxide and methane could be used as alternatives, but

Table 4.7 Features of High-Dust & Low-Dust SCR Systems [24]

	High-Dust System	Low-Dust System
SCR Reactor	1. Catalyst should have open shape to prevent plugging by coarse dust.	No special measures against dust are necessary.
Air Preheater	Ammonium bisulphate deposition tends to be large. To prevent plugging, it is necessary to select suitable element shape, operate suitable soot blower and control NH_3 leakage to 5 ppm (4 mg/Nm^3).	Ammonium bisulphate deposition tends to be large. To prevent plugging, it is necessary to select suitable element shape, operate soot blower and control NH_3 leakage to 5 ppm (4 mg/Nm^3).
Electrostatic Precipitator	1. Performance is low in cases of low sulphur oil combustion. Otherwise, performance is stable. 2. Lower cost.	1. Performance variation due to oil composition is small. 2. Radiation losses from the casing affects the boiler efficiency. 3. Higher flue gas volumes may be handled if gas is recirculated back to the boiler from downstream of ESP. 4. Higher cost.
FGD plant	SCR leakage NH_3 has no effect on FGD; NH_3 reacts with SO_3 and the products are collected in ESP. NH_3 concentration in waste-water from FGD is kept low.	SCR leakage NH_3 reacts with SO_3 and products are collected in FGD plant. The NH_3 concentration in waste-water from FGD is higher.

Table 4.8 Advantages & Disadvantages of Parallel-Flow Catalysts [352]

Main Advantages	Main Disadvantages
1. No plugging by dust, even at high dust loadings.	1. Shutdown required to replace catalyst (compared to moving-bed designs).
2. Smaller frontal area; granular catalysts must have thin beds and large frontal areas to avoid high pressure drop.	2. Higher cost per unit surface area.
3. Higher gas velocity, thus improving transfer to the catalyst surface.	3. Larger reactor volume because of space velocity.
4. Minimum attrition compared to moving-bed design.	4. Higher cost for installing the catalyst, compared to dumping the granular type.
5. Less pore blinding.	

Table 4.9 Advantages of Molecular Sieve Catalyst [352]

1. Low catalyst volume because of high inner surfaces. Only small reactors required.

2. No oxidation of SO_2 to SO_3 at temperatures below 450°C.

3. NH_3 is stored in the catalyst causing highly flexible behaviour at load variations.

4. Low NH_3 leakage, e.g. below 5 ppm, even at high removal efficiencies.

5. No poisoning of the catalyst can occur, so lifetime is limited only by mechanical erosion.

6. The used catalyst can be used as a raw material in the ceramic industry.

consumption is generally high due to reaction with the oxygen present in the flue gas.

As seen previously, apart from the Tail-Gas arrangement, where the SCR reactor is located downstream of an FGD plant, there are two basic 'hot-side' arrangements, e.g. High-Dust and Low-Dust systems. A comparison of the features of these two latter systems is given in Table 4.7.

The most popular catalyst material at present is V_2O_5 on a TiO_2 base. V_2O_5 is highly active with a comparatively long life, whilst TiO_2 is affected very little by SO_3. Other catalyst materials that can be considered are the oxides of iron (ferric), molybdenum, tungsten and chromium. An alternative base material is silica.

In addition to the selected catalyst material, NO_x removal performance will be affected by such criteria as the molar ratio of NH_3 to NO_x, and the volume, temperature, NO_x concentration and O_2 concentration of the flue gas entering the reactor.

The design of the catalyst bed is determined by the dust and SO_3 concentration in the flue gas and the need to minimise pressure drop. Granular or pellet type catalyst have been largely superseded by the parallel-flow catalysts which take the shape of a honeycomb, plate or ring arrangement. The honeycomb shape appears to give the best combination of specific surface (area per volume) and pressure drop [352]. The advantages and disadvantages of the parallel-flow catalysts compared to granular types are listed in Table 4.8. A special catalyst that is totally based on ceramics is the molecular sieve catalyst. The reported advantages of this type of catalyst are given in Table 4.9. A zeolite type catalyst has been licensed for use in three small SCR plants in the FRG using the 'Tail-Gas' arrangement [28].

This is a well-established and proven design for gas- and oil-fired systems. It is capable of achieving high NO_x reduction efficiencies (90% or more), which are beyond individual combustion techniques and, possibly, the selective non-catalytic reduction method.

Potential for Process Development: Japanese experience has indicated actual catalyst life to be 4–5 years for oil-fired boilers [352]. It is anticipated that the future will see some effort being made by catalyst manufacturers to extend the lifetime of the catalyst. It is further expected that low SO_2-oxidation catalysts will be developed for application with flue gases having high SO_2 concentrations [2].

Low-temperature SCR systems, with activated coke as the catalyst, are being tested in the FRG. Such a system is of particular interest in 'Tail-Gas' arrangements because it needs little or no preheating, experiences no catalyst poisoning, and the catalyst can either be regenerated or disposed of as a fuel [361].

One of the most interesting developments in the field of SCR is also taking place in the FRG. This is the installation of catalyst-coated plates on the surfaces of a rotary regenerative air preheater [28]. The catalyst operates over a temperature range of 150°C to 350°C, and the NH_3 is injected into the flue gas behind the economiser. The idea is attractive insofar as the catalyst surface area required for a moderate $deNO_x$ efficiency is comparable to that available in typically installed regenerative heat exchangers. In this way, a separate SCR reactor would not be required, saving space and cost. NO_x removal efficiencies in excess of 60% have been reported in small-scale field tests, but early test on larger scale (150 MWe) in late 1985 produced lower efficiencies [28]. An alternative source of information on this development indicates that $deNO_x$ efficiencies of 90% with low-sulphur (0.2%) fuels decreased rapidly when higher sulphur fuels were tested [352]. The potential for this arrangement seems to be as a low-cost $deNO_x$ method, used in addition to combustion modifications.

A great deal of research and field testing has been carried out in Japan and the FRG to establish the design of air preheaters for conventional SCR applications and gas-gas heaters in a 'Tail-Gas' SCR arrangement. Important design criteria are as follows [28]:

(a) Heating surfaces must have a configuration that can be easily cleaned by soot blowing.

(b) The cross-over of heating-element layers should be avoided in areas most likely to be affected by ammonium bisulphate deposition.

(c) Enamelled heating elements should be used, especially where SO_3 concentrations exceed 2 ppm (7 mg/Nm³).

(d) Highly efficient soot-blowing devices should be used to limit outages.

(e) Provision for frequent off-line water washing of the air preheater should be considered in cases where SO_3 concentrations in excess of 2 ppm are encountered.

Process Requirements for Each Application Considered

Process requirements are shown in Table 4.10 for the one application considered: Datum Combustion System 1.

It is assumed that:

– The NO_x content of the gas is to be reduced to 200 mg/Nm³, corresponding to a NO_x reduction of about 80%.

Table 4.10 Process Requirements

Datum Combustion System		1
Gas at full load:		
Volume flow	'000 Nm³/h	456
Dry gas	'000 Nm³/h	408
Water vapour	'000 Nm³/h	48
Actual volume flow	'000 m³/h	1075
Temperature	°C	370
Particulates content	mg/Nm³ (dry)	39
SO_2 content	mg/Nm³ (dry)	5010
HCl content	mg/Nm³ (dry)	3
Inlet gas at full load:		
NO_x content (as NO_2)	mg/Nm³ (dry)	1025
Outlet gas at full load:		
NO_x content (as NO_2)	mg/Nm³ (dry)	200 (282)
NH_3 content (a)	ppm	2
Reagent		Ammonia
NH_3:NO_x Molar Ratio		0.82
Requirements at full load:		
Ammonia	kg/h	166
Hot water (120°–90°C)	tonnes/h	1.8
Electric power	MWe	0.8
Manpower	men/shift	0
Average load factor	%	***

(a) 1 ppm NH_3 = 0.76 mg/Nm³
Figures in parentheses are annual average emissions for 90% $deNO_x$ plant availability

- $NH_3:NO_x$ molar ratio is 0.82.

- The SCR reactor is located immediately downstream of the boiler.

- NO_x consists of 90% NO and 10% NO_2 by volume.

- Particulate settling in the reactor is negligible.

- Reactions between NH_3 and SO_2 can be ignored.

By-products and Effluents

The reaction between NO_x and ammonia produces nitrogen gas and water vapour, so essentially no problem waste material is formed. Side-reactions between NH_3 and SO_3 in the flue gas result in the formation of ammonium bisulphate/sulphate which cause fouling problems on downstream equipment; these reactions are inhibited by the use of low SO_2-oxidation catalysts and by limiting NH_3 leakage to less than 5 ppm (4 mg/Nm³).

Where an FGD plant is located downstream of the SCR reactor, leakage NH_3 and ammonium salts will be flushed from the gas stream by the FGD scrubbing liquor and discharged in the waste water stream. The presence of these gas-borne wastes will affect the FGD reagent consumption and the nature of the waste water. Typically, an NH_3 leakage rate of 5 ppm (4 mg/Nm³) from an SCR unit fitted to a 2000 MWe utility boiler is equivalent to 23 kg/h of NH_3 to be neutralised in the FGD plant.

Efficiency and Emission Factors

The efficiency and emission factors for the process are summarised in Table 4.11 for the application considered: Datum Combustion System 1.

Table 4.11 Efficiency and Emission Factors

Datum Combustion System		1
Oil heat input (gross)	MWt	464
Oil fired	tonnes/h	39.4
SCR plant power consumed	MWe	0.8
Equivalent oil input	tonne/h	0.21 (b)
Useful energy from system	GJt/h	1467
Total equiv. coal input	tonnes/h	39.61
Efficiency factor	GJt/tonne	37.0
Nitrogen emission (in NO_x)	kg/h	25 (35)
NO_x emission factor (a)	kg/tonne	0.63 (0.89)

(a) Emission factor expressed as kg nitrogen per tonne of oil
(b) Calculated assuming overall power generation efficiency = 0.33
Figures in parentheses are annual average emission for 90% deNO$_x$ plant availability

In calculating the efficiency factors, it is assumed that at full load:

- The performance without the incorporation of the SCR plant would be as shown in Table 1.4.

- The SCR plant overall power consumption is 0.18% of the energy generated (including hot water for ammonia evaporation), i.e. 0.8 MWe. This is taken to be equivalent to the combustion of 0.21 tonne/h of oil at a power station, assuming an overall power generation efficiency of 33%. This additional oil is arbitrarily assumed to be included with the oil burned in the boiler for calculating the efficiency and emission factors.

The efficiency of the SCR process is primarily determined by the activity of the catalyst, although it is influenced by reaction temperature, $NH_3:NO_x$ molar ratio and the flue gas space velocity through the catalyst [352].

Effect of plant availability: For illustration purposes, the annual average emissions and emission factors for deNO$_x$ plant availabilities of 100% and 90% are given in Tables 4.10, 4.11 and 4.12. Further details are given in Section 1.6.

Table 4.12 Efficiency and Emission Factors for Tail Gas System

Datum Combustion System		1
Gas at full load:		
Volume flow (dry)	'000 Nm³/h	407
Temperature	°C	370
NO_x concentrations:		
At inlet	mg/Nm³	945
At exit	mg/Nm³	200 (274)
NO_x removal efficiency	%	79
Nitrogen emission (in NO_x)	kg/h	25 (35)
NO_x emission factor (a)	kg/tonne	0.63 (0.89)

(a) Emission factor expressed as kg of nitrogen per tonne of oil. Figures in parentheses are annual average emissions for 90% deNO$_x$ plant availability

Effect of load variations: Reduced boiler loads are accompanied by drops in the temperature of the flue gas leaving the boiler economiser and, hence, entering the SCR reactor. Lower operating temperatures in the reactor promote the formation of ammonium bisulphate (and sulphate) which cause fouling problems in downstream equipment. This factor, and the potential damage caused by thermal shock to the catalyst by frequent cycling (as is common in German power stations and is being rigorously investigated there), can be countered by installing a by-pass around the economiser to maintain acceptable temperature levels. This leads to a reduced plant efficiency factor, although no data are available on the precise effort.

Effect of design variations: Two major design variations have an effect on NO_x emission factors: 'Tail Gas' SCR arrangements with conventional and activated carbon catalysts, respectively. The latter arrangement is in the development stage and will not be considered further. Estimated emission factors for the former arrangement, located downstream of a limestone-gypsum process, are presented in Table 4.12. It is assumed that the process conditions leaving the FGD plant are as given in Table 2.41, and that the gases leaving the FGD absorption tower are heated to the SCR reactor operating temperature by regenerative-type heat exchangers (see Figure 4.8). Emission factors without the SCR plant would be as indicated in Table 2.43 and efficiency factors for this and the more conventional (hot-side) arrangement are expected to be as presented in Table 4.11.

Limitations: Selective Catalytic Reduction is limited in its application by its low cost effectiveness in comparison to combustion techniques. It is most likely to be applied where high NO_x reduction efficiencies (80–90%) are required, although at the higher end of the range there remains the possibility that increased leakage rates of NH_3 will result in fouling and waste water problems downstream. These higher efficiencies can be attained by a combination of combustion techniques and by the more recent developments in the field of selective non-catalytic reduction, although the latter still has problems with temperature control.

Costs

Capital Costs: See Section 4.6.

Annual running costs and cost factors: See Section 4.6

Effect of design variations on cost: ***

Effect of annual load patterns on annual running costs: ***

Process Advantages and Drawbacks

The advantages are:

1. NO_x removal efficiencies in excess of 90% can be achieved.
2. The process has been commercially applied to flue gas from oil-fired boilers since 1980.
3. Low leakage NH_3 rates, i.e. less than 5 ppm (4 mg/Nm³), are experienced for NO_x removal efficiencies of up to 80%.

The disadvantages are:

1. Catalyst lifetime is limited to 4–5 years, accounting for a high proportion of the annualised operating costs.

EVALUATION OF SELECTIVE NON-CATALYTIC REDUCTION PROCESS
(Process Code N42)

See Section 4.2 for outline of basic process, chemistry and schematic diagram.

See Section 4.3 for a list of manufacturers offering this type of equipment and for a general appraisal of the basic process.

See Section 4.4 for the reason for choosing the selective non-catalytic reduction $DeNO_x$ process.

This basic process type is considered to be suitable for application in the UK only to large industrial boilers. The comparatively lower NO_x reduction efficiencies required for small factory oil-fired boilers, to attain the same NO_x emission concentrations as for larger boilers, can be achieved more cost effectively using combustion techniques. Hence, the SNR process is evaluated here only for Datum Combustion System 1.

Process Description

The selective non-catalytic reduction process (SNR), sometimes referred to as Thermal $DeNO_x$, is based on the reaction between flue gas NO_x and injected ammonia gas to produce nitrogen and water vapour. The reaction is very sensitive to temperature with peak NO_x reduction rates occurring over a limited temperature range of about 900° to 1000°C. Generally, the NH_3 is injected into the flue gas stream by means of an air stream carrier gas at a location (or locations) specifically selected to provide an optimum reaction temperature and residence time. Hydrogen (or natural gas) can be injected with the ammonia to extend the effective NO_x reduction reactions down to temperatures of about 700°C (for $H_2:NH_3$ ratio of 2:1) with the result that lower optimum reaction temperatures are achieved with higher H_2 injection rates.

The major factors affecting the process are:

– Temperature,
– Residence time at temperature,
– Temperature profile,
– Initial NO_x concentration,
– $NH_3:NO_x$ molar ratio,
– $H_2:NH_3$ molar ratio,
– Mixing condition,
– Interaction of flue gas constituents, especially O_2, H_2O and free radicals.

To achieve the temperatures required for the reaction, the NH_3 (and H_2) is normally injected into the upper section of the boiler, either upstream or downstream of the superheater, between two adjacent superheater sections, or at two or more of these locations. Each boiler needs to be considered separately for optimum location of the NH₃ injection system.

A schematic diagram of a typical system is depicted in Figure 4.12.

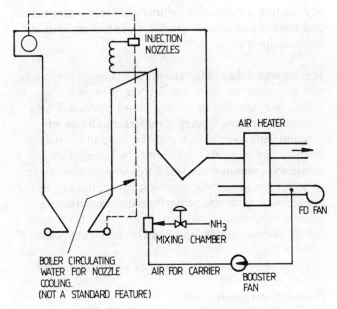

Figure 4.12 Schematic Diagram of NH₃ Injection System

Status and Operating Experience

The status of the SNR process is indicated in Section 4.3 and Table 4.3, where it can be seen that more than 80 commercial SNR plants have been installed in Japan and the USA. All are gas- or oil-fired units, of which thirteen are known to have been fitted to utility and industrial boilers.

The largest boiler equipped with SNR has an equivalent capacity of 180 MWe, whilst the largest oil-fired plants on which test work has been carried out is 375 MWe.

Early applications of this deNO$_x$ method involved the location of an injection grid within the flue gas stream, at the appropriate temperature, for the injection of NH₃ and carrier gas (air or steam). Sometimes, multiple grids were used (at several locations) or hydrogen injected with the NH₃ to account for changes in temperature, especially during load swings. The injection grid was subject to high temperature, corrosive fouling conditions, whilst multiple grids and H₂ injection increased the cost and complexity of the system.

The injection grids have now been largely superseded by wall injectors, which consist of large jets located at or near the boundary walls of the injection zone. The advantages of wall injectors in an SCR application have been listed [414] as:

– Higher performance,
– Lower investment cost,
– Better load following (without hydrogen),
– Lower fouling tendency,
– No cooling requirement,
– Simple installation,
– Easily retrofitted,
– Greater cost effectiveness.

Despite their simplicity and lower cost, wall injectors require careful design and location to achieve the optimum performance. It is now possible to locate wall injectors at the optimum flue gas temperature, and even within the combustion zone of the boiler. Two or more sets of injectors can be installed, at little extra cost, to allow for temperature fluctuations due to load changes. As such, the injection of H₂ will be minimised, thereby reducing operating costs.

Wall injectors can be retrofitted to existing boilers without the need to relocate ducting or other downstream equipment, and with minimal impact of the boiler structure. If actual and anticipated flue gas temperatures in the region of the wall injectors differ significantly, it is relatively easy to relocate the injectors.

The injection grid system was first applied in Japan in early 1974 where NO$_x$ reduction efficiencies of up to 65% were achieved. Wall injector technology has been applied since autumn 1984 to a 200 t/h boiler generating steam and power in a Japanese chemical plant. NO$_x$ reduction efficiencies of 70% were achieved at a cost estimated to be 20% of that of a comparable SCR facility [414]. Efficiencies of up to 80% for industrial and utility boilers (and up to 90% for other applications) have been achieved with this technology [416].

SNR, like SCR, has experienced downstream ammonium bisulphate fouling problems caused by the low temperature reaction between leakage NH₃ and the SO₃ present in the flue gas stream. SNR NH₃ leakage rates are somewhat higher (c. 50 ppm or 38 mg/Nm³) than for SCR, but flue gas SO₂ oxidation to SO₃ is reduced due to the absence of a promoting catalyst. It is claimed that the problems of ammonium bisulphate fouling can be minimised by satisfying one or more of the following criteria [414]:

1. Maintaining an NH₃:SO₃ molar ratio above 2.0, and providing sufficient reaction time for the formation of ammonium sulphate rather than the bisulphate.

2. Limiting NH₃ leakage to 5 ppm (4 mg/Nm³) or less.

3. Maintaining the air preheater flue gas outlet temperature above 200°C.

4. Maintaining the temperature at the NH₃ injection

point lower than the sulphate formation temperature.

Method '1' is the most practical method for SNR processes because typical NH_3 leakage rates are generally twice those of SO_3 [417].

NH_3 leakage in SNR systems is extremely complicated because it is dependent upon the interaction of numerous factors, including the concentration of flue gas components, time–temperature relationships, NH_3 mixing efficiency, and the design of downstream equipment. The amount of NH_3 leakage can be determined for particular fired systems, permitting Exxon to guarantee leakage rates as low as 5 ppm (4 mg/Nm³) where favourable conditions exist [414].

Variations and Development Potential

Although the older injection grid technology is still installed in a number of oil- and gas-fired units, it has been largely superseded by the newer wall injection system. These developments, plus the use of a fundamental kinetic model of the process chemistry and a three-dimensional flow model, form the current state of the art for SNR.

The use of multiple injection nozzles allows for fluctuations in flue gas temperature away from the optimum NH_3/NO_x reaction temperature range of 900°–1000°C. The injection of a readily oxidisable gas, such as H_2 or CH_4, with the NH_3 lowers the optimum temperature range progressively to about 700°C ± 50°C for a $H_2:NH_3$ ratio of about 2:1 [352].

SNR is a flue gas $deNO_x$ treatment process that has, in the last few years, achieved high NO_x reduction efficiencies comparable to those of SCR. It is claimed that the process is up to five times as cost effective as SCR [414] primarily due to the relative ease of installation and lack of catalyst replacement costs. Its potential appears to be in the middle ground between the low cost, lower efficiency combustion techniques and the high cost, high efficiency SCR methods.

Whilst the new wall injector nozzles appear to have eliminated many of the disadvantages associated with early grid designs, their proper design and location requires the use of a detailed kinetic model of the system to be fitted. This recently developed kinetic model is based on the fundamental chemistry of the process, and it is claimed to predict accurately the performance and NH_3 leakage rates for any type of fired unit [414].

The use of wall injectors has also promoted the development of three-dimensional, turbulent flow modelling techniques to ensure adequate mixing of NH_3 and flue gas NO_x, especially in larger boilers. This technique also plays a part in predicting $deNO_x$ performance (and NH_3 leakage) for specific applications.

Process Requirements for Each Application Considered

Process requirements are shown in Table 4.13 for the application considered: Datum Combustion System 1.

It is assumed that for System 1:

– The NO_x content of the gas is to be reduced to 300 mg/Nm³, corresponding to a NO_x reduction of about 71%.

– $NH_3:NO_x$ molar ratio is 2.5.

– H_2 is not required for performance control.

– NO_x consists of 90% NO and 10% NO_2 by volume.

– Reactions between NH_3 and SO_3 can be ignored.

Table 4.13 Process Requirements

Datum Combustion System		1
Gas at full load:		
Volume flow	'000 Nm³/h	456
Dry gas	'000 Nm³/h	408
Water vapour	'000 Nm³/h	48
Actual volume flow	'000 m³/h	2043
Temperature	°C	950
Particulates content	mg/Nm³ (dry)	39
SO_2 content	mg/Nm³ (dry)	5010
HCl content	mg/Nm³ (dry)	3
Inlet gas at full load:		
NO_x content (as NO_2)	mg/Nm³ (dry)	1025
Outlet gas at full load:		
NO_x content (as NO_2)	mg/Nm³ (dry)	300 (372)
NH_3 content (a)	ppm	50
Reagent		Ammonia
$NH_3:NO_x$ molar ratio		2.5
Requirements at full load:		
Ammonia	kg/h	447
Hot water (120°C–90°C)	tonnes/h	49
Electric power	MW	0.23
Manpower	men/shift	Nil
Average load factor	%	***

***Data not available
(a) 1 ppm NH_3 = 0.76 mg/Nm³
Figures in parentheses are annual average emissions for 90% $deNO_x$ plant availability

NO_x reduction efficiencies of the order of 70% have been selected, being an average over the range of efficiencies to be expected with utility and industrial boilers fitted with the latest technology [417].

By-Products and Effluents

The reaction between NO_x and ammonia produces nitrogen gas and water vapour, so essentially no waste material is formed. Side reactions between NH_3 and SO_3 in the flue gas result in the formation of ammonium bisulphate which causes fouling problems on downstream equipment; these reactions are normally inhibited in SNR systems by maintaining $NH_3:SO_3$ molar ratios above 2.0, so that the less troublesome ammonium sulphate is formed in preference to the bisulphate.

Where an FGD plant is located downstream of the SNR reactor, leakage NH_3 and ammonium salts will be flushed from the gas stream by the FGD scrubbing liquor and discharged in the waste water stream. The presence of these gas-borne wastes will affect FGD reagent consumption and the nature of the waste water.

Efficiency and Emission Factors

The efficiency and emission factors for the process are summarised in Table 4.14 for the application considered: Datum Combustion System 1.

In calculating the efficiency factors, it is assumed that at full load:

- The performance without the incorporation of the SNR plant would be as shown in Table 1.4.

Table 4.14 Efficiency and Emission Factors

Datum Combustion System		1
Oil heat input (gross)	MWt	464
Oil fired	tonnes/h	39.4
SNR plant power consumed	MWe	0.23
Equivalent oil input	tonne/h	0.06 (b)
Useful energy from system	GJt/h	1469
Total equiv. coal input	tonnes/h	39.46
Efficiency factor	GJt/tonne	37.2
Nitrogen emission (NO_x)	kg/h	37 (46)
NO_x emission factor (a)	kg/tonne	0.94 (1.17)

(a) Emission factor expressed as kg nitrogen per tonne of oil
(b) Calculated assuming overall power generating efficiency = 0.33
Figures in parentheses are annual average emissions for 90% $deNO_x$ plant availability.

- The SNR plant overall power consumption is 0.05% of the energy generated (including hot water for NH_3 evaporation), i.e. 232 kWe. This is taken to be equivalent to the combustion of 60 kg/h of oil at a power station, assuming an overall power generation efficiency of 33%. This additional oil is arbitrarily assumed to be included with the oil burned in the boiler for calculating the efficiency and emission factors.

Exxon have indicated ranges of expected NO_x reduction efficiencies for different types of fired equipment, but stress that the SNR process is plant specific and would need to be evaluated for each application [417].

Effect of plant availability: For illustration purposes, the annual average emissions and emission factors for $deNO_x$ plant availabilities of 100% and 90% are given in Tables 4.13 and 4.14. Further details are given in Section 1.6.

Effect of Load Variations: Reduced boiler loads are accompanied by flue gas temperature fluctuations in the upper sections of the boiler. This would be accompanied by a loss of NO_x reduction efficiency if it were not for the multiple location of NH_3 injection nozzles and the facility to inject H_2 into the flue gas (with NH_3) to reduce the optimum reaction temperature. It is anticipated, therefore, that emission factors would change very little at reduced loads, but possibly at the expense of increased H_2 consumption.

Effect of Design Variations: Systems fitted with the older injection grid layouts would be expected to exhibit lower emission factors than for the newer wall injection arrangement. However, it is expected that only the latest technology will be available commercially.

Limitations: Selective non-catalytic reduction is limited in its application by its lower cost effectiveness in comparison to combustion techniques, and by its lower NO_x reduction efficiency in comparison to selective catalytic reduction techniques [416]. In both cases, however, it is beginning to approach the required performance. It may find applications in combined low-NO_x burners and SNR arrangements where the high $deNO_x$ levels attainable by SCR are required, but at a substantially lower investment cost. SNR is more difficult to control than SCR, and the comparatively high leakage rates of NH_3 present potential downstream fouling and waste water problems.

Costs

Capital Costs: See Section 4.6

Annual Running Costs and Cost Factor: See Section 4.6

Effect of Design Variations on Cost: ***

Effects of Annual Load Patterns on Annual Running Costs: ***

Process Advantages and Drawbacks

The advantages are:

1. Catalysts and specific reactors are not required, thereby avoiding plugging problems and SO_2 oxidation.

2. Preheating is unnecessary because the reaction takes place in a high temperature zone.

3. Space for a catalytic reactor is unnecessary.

4. Much less expensive to install than SCR, especially on existing boilers.

5. Capable of dealing with particulate-laden flue gas streams.

The disadvantages are:

1. Considerable amounts of NH_3 leakage occur, which could result in the formation of ammonium bisulphate and subsequent fouling of downstream equipment.

2. NH_3 consumption is high, and H_2 is sometimes required to maintain $deNO_x$ efficiencies.

3. Only a narrow temperature range exists over which

Table 4.15 Utility Boilers–Oil Fired (NO_x)

Source Level of Control	% Abated	Annual K Cost £/kWt	Annual O&M P/kWh	£/t Abated
UTILITY BOILERS NEW	45–50			
Low NO_x Burners				
– Oil & Gas		0.43	Negligible	7
– Oil		0.43	Negligible	7
In Furnace Reduction (IFR)		5.57		
Combustion Modifications				
O.S.C.	25–30			
L.E.A.	25–30			
LVHR	33	1.49	Negligible	29
LVHR & LNB	60	1.86	Negligible	21
LVHR & LNB & IFR	80	7.43	0.021	230
SCR: Other Countries	80–90	6.57–32.86	0.035–0.11	820–240
USA	80–90	55.71		1620–2850
SCR & LVHR & LNB	92	8.43	0.085	640
SCR & LVHR & LNB & IFR	96	13.9	0.11	760
UTILITY BOILERS EXISTING				
Combustion Modifications				
OFA–Dual Oil & Gas		0.04	0.007	336–2340
LEA–Dual Oil & Gas	20	0.07–0.5		7–135
BOOS–Dual Oil & Gas	25–50	0.2–0.6	0.007	220–960
LEA & OSC–Dual Oil & Gas		0.5–0.9		807–2435
FRG & OFA–Dual Oil & Gas	55–75	5.7	0.03	830–2235
–Oil		2–19		
NCR	40–60	2.7–3.7		114–236
LEA & OSC & NCR	70	7.5		257
SCR–Dual Oil & Gas	90	5.6		

Table 4.16 Industrial Boilers–Oil Fired (NO_x)

Source Level of Control	% Abated	Annual K Cost £/kWt	Annual O&M P/kWh	£/t Abated
INDUSTRIAL BOILERS EXISTING				
Low NO_x Burners	25–50	1.4–3.3		480–1050
Combustion Modifications				
Low Excess Air (LEA)	10	0.2		214–307
O.S.C.	20–35	0.3–3		170–1540
NCR: Japan	40–60	9		1460–2985
SCR: USA	90	47		5340–7080
INDUSTRIAL BOILERS NEW				
LEA–Dual Oil & Gas	10	0.14		107–207

optimum deNO$_x$ efficiencies can be achieved. Hence, control can sometimes be difficult.

4. Lower NO$_x$ reduction efficiencies are possible, compared to SCR.

5. Variations in flue gas temperature and composition within the dust can present operational problems, especially where duct sizes are large.

4.6 Costs

The recent paper by Leggett [102] summarising progress on the OECD Project on Control of Major Pollutants (MAP) was used to provide the limited cost data shown in the Tables 4.15 and 4.16. Within the MAP study, differences in data were assessed and presented as ranges rather than specific values due to substantial uncertainties regarding the starting emission level, the abatement efficiency and other similar factors.

The prices and ranges in the cost effectiveness summary tables are not intended to be indicative of the cost, or cost effectiveness, for any individual facility–rather, they attempt to represent a typical facility in that source category so that, on average, the estimated costs should be representative.

A questionnaire issued to UK burner vendors did not return any usable costing information.

Cost-effectiveness is defined in terms of US 1984 dollars per tonne ($/t) of NO$_x$ abated (converted to £/t). The costs were derived by annualising the total capital investment required for the pollution abatement technique. Assumptions used to annualise the capital costs were made as consistent as possible, using a real discount rate of 5 per cent and equipment lifetimes that vary with the type of facility (e.g. 30 years for combustion modifications on new boilers or 15 years for retrofits; 3–5 years for catalysts for selective catalytic reduction, etc.). The annual charge for operation and maintenance (O & M) including by-product sales and changes in fuel consumption, is added to the annualised capital charge. The resulting net annual cost is divided by the annual tons of NO$_x$ abated.

For further details the reader should refer to the paper by Leggett [102].

5. Combined SO_2-NO_x Abatement Processes

5.1 Classification of the Processes

5.2 Outline Description of Combined $SO_2 - NO_x$ Abatement Processes

5.3 General Appraisal of Processes

5.4 Processes for Detailed Study

5.5 Evaluation of Selected Combined Abatement Processes

5. Combined SO_2-NO_x Abatement Processes

5.1 Classification of the Processes

The general classification of processes for the simultaneous abatement of SO_2 and NO_x emission adopted in this Manual is presented in Figure 5.1. The processes are divided into four categories (NS10–NS40) according to the scheme:

- First division: distinction between dry processes (NS10 and NS20) and wet processes (NS30 and NS40).

- Second division: for dry processes, distinction between catalytic (NS10) and non-catalytic (NS20); for wet processes, distinction between direct absorption (NS30) and oxidation-absorption (NS40).

- Third division: for dry catalytic processes, distinction between regenerable (NS11) and non-regenerable (NS12); for dry non-catalytic processes, distinction between those using burner modifications (NS21), radiation (NS22) and absorption (NS23); for wet processes (both direct absorption and oxidation absorption) distinction between liquid-phase NO_x oxidation (NS31 and NS41) and liquid-phase NO_x reduction (NS32 and NS42). Note that the LIMB process (NS21.1) is not a process but is essentially a modified combustion technique.

Some of the processes yielding a dry end-product involve spray-drying of solutions or slurries as an integral feature of the process. Such processes are frequently described in the literature as 'dry', 'semi-dry' or 'wet-dry' processes. These terms are regarded as misleading for such processes, which are referred to in this Manual as solution-based or slurry-based (as appropriate) absorption processes with dry end-products. In the Manual, the terms 'wet' and 'dry' are used to describe the state of the reagent contacted with the gas; the terms 'semi-dry' and 'wet-dry' are not used.

Process Code Numbers: A number of basic process types occur within each process category. Basic process types are assigned a Code Number comprising the relevant Category Number followed by a Type Number; e.g. the electron beam radiation processes fall in Category NS22 and have been assigned the Code No. NS22.1.

The Code Numbers assigned to the basic process types are shown in Table 5.1.

Table 5.1 Classification of Combined SO_2/NO_x Abatement Processes

Category NS10: Dry processes–Catalytic
Category NS11: Regenerable catalyst
NS11.1 Active carbon adsorption/selective catalytic reduction (SCR)
NS11.2 Copper oxide
Category NS12: Non-regenerable catalyst
NS12.1 Non-selective catalytic reduction of NO_x and SO_2
NS12.2 Catalytic reduction of NO_x and catalytic oxidation of SO_2
Category NS20: Dry processes–Non-catalytic
Category NS21: Burner modifications
NS21.1 Limestone Injection into Multistage Burner (LIMB)
Category NS22: Radiation
NS22.1 Electron beam
Category NS23: Adsorption
NS23.1 Sodium carbonate process
Category NS30: Wet processes–Direct absorption
Category NS31: Liquid phase NO_x oxidation
NS31.1 Lime spray dryer
Category NS32: Liquid-phase NO_x reduction
NS32.1 Asahi chemical process
NS32.2 Limestone scrubbing with chelating compound
NS32.3 Sulf-X process
Category NS40: Wet processes–Oxidation absorption
Category NS41: Liquid phase NO_x oxidation
NS41.2 Oxidation plus ammonia scrubbing
NS41.2 Kawasaki process
Category NS42: Liquid phase NO_x reduction
NS42.1 Oxidation plus limestone slurry scrubbing–IHI Process
NS42.2 Oxidation plus limestone slurry scrubbing–Moretana Process

5.2 Outline Descriptions of Combined SO_2-NO_x Abatement Processes

Process Code NS11.1–Active Carbon Adsorption/SCR

Outline of Process [352, 424]: A simplified block diagram of the process is presented in Figure 5.2. Gas from the Boiler at up to 150°C is cooled, if

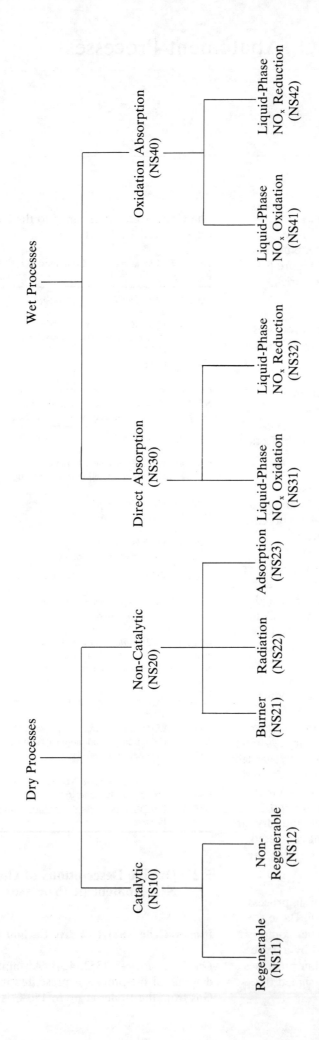

Figure 5.1 Simultaneous SO_2/NO_x Abatement Processes

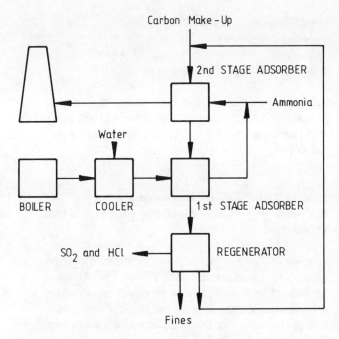

Figure 5.2 Block Diagram of Carbon Adsorption/SCR Combined Abatement Process

required, to about 120°C by injection of water. It then flows through the First Stage Adsorber containing a moving bed of activated carbon pellets sized about 10 mm, or of coke, from the Second Stage Adsorber. The carbon adsorbs SO_2, together with NO_2 which forms 5–10% of the NO_x content of the gas. The carbon catalyses the oxidation of SO_2 to SO_3, which reacts with water vapour also adsorbed from the gas, forming sulphuric acid. Ammonia is added to the gas leaving the First Stage Adsorber, and the gas then enters the Second Stage Adsorber, where the carbon catalyses the reaction of ammonia with NO to form nitrogen and water; it behaves as a selective catalytic reduction (SCR) catalyst. Some reaction also occurs between adsorbed ammonia and SO_2 adsorbed in the Second Stage to form ammonium sulphate and bisulphate. There is a temperature rise of 15–20°C in the Adsorbers, and the cleaned gas is exhausted to stack. Carbon leaving the base of the First Stage Adsorber is transferred to the Regenerator, where it is heated in tubes to 400–450°C by hot combustion gases outside the tubes.
The sulphuric acid and ammonium salts are desorbed and decomposed by carbon to produce SO_2, H_2O, CO_2, oxygen and nitrogen. The SO_2 is treated: by liquefaction for sale as liquid SO_2; by oxidation to SO_3 for the manufacture of sulphuric acid; or by reduction to elemental sulphur for sale or safe disposal. The regenerated carbon is cooled and returned to the Second Stage Adsorber, together with fresh carbon to replace that lost in regeneration. The carbon bed also acts as a panel bed filter, capturing particulates which are screened out before regeneration. Further, it adsorbs halogen acid gases, which are released during regeneration; they are separated from the regenerator tail gas before treatment of the SO_2.

Chemistry of the Process [352, 424]: In the First Stage Adsorber, SO_2 and water vapour are adsorbed by the activated carbon, and they react on the carbon surface with oxygen in the gas. The reaction can be represented by:

$$2\ SO_2 + O_2 + 2\ H_2O = 2\ H_2SO_4$$

The sulphuric acid formed remains adsorbed in the pores of the carbon. The reaction is exothermic, and there is a 15–20°C rise in gas temperature.

At the higher temperature (400–450°C) in the Regenerator, the sulphuric acid decomposes to sulphur trioxide and water; the trioxide is reduced by the carbon to form the dioxide:

Decomposition: $H_2SO_4 = SO_3 + H_2O$

Reduction: $2\ SO_3 + C = 2\ SO_2 + CO_2$

The reduction involves transient formation of a surface carbon-oxygen species, C...O:

$$2\ H_2SO_4 + 2\ C = 2\ SO_2 + 2\ H_2O + 2\ C...O$$
$$2\ C...O \quad\quad = C + CO_2$$

In the Second Stage Absorber, the carbon catalyses the reduction of NO by NH_3:

NO Reduction:
$6\ NO + 4\ NH_3 = 5\ N_2 + 6\ H_2O$
$4\ NO + 4\ NH_3 + O_2 = 4\ N_2 + 6\ H_2O$

Reaction also occurs between ammonia and residual SO_2 adsorbed in the Second Stage Adsorber to form ammonium sulphate and bisulphate:

$$2\ SO_2 + O_2 + 2\ H_2O = 2\ H_2SO_4$$
Sulphate $\quad H_2SO_4 + 2\ NH_3 = (NH_4)_2SO_4$
Bisulphate $\quad H_2SO_4 + NH_3 = NH_4HSO_4$

In regeneration, these reactions are reversed, and the ammonia released is decomposed by reaction with the surface carbon-oxygen species, C...O; this involves no loss of carbon:

NH_3 Decomposition
$2\ NH_3 + 3\ C...O = N_2 + 3\ H_2O + 3\ C$

A carbon make-up is required to replace that consumed in the reduction of H_2SO_4 formed in the First Stage Adsorber, which can account for 70–80% of the total sulphur dioxide captured in the process (0.09 kg/kg SO_2 removed in the First Stage). The loss of carbon during reduction increases porosity, and hence the internal surface of the carbon remaining, which is therefore further activated. However, the reduction weakens the carbon particles, so that there are carbon break-down losses which also have to be replaced, resulting in a carbon make-up rate (typically 0.12–0.18 kg/kg SO_2 removed) that is higher than the theoretical.

Process Code NS11.2 – Copper Oxide Absorption/SCR

Outline of Process [145, 352]: A simplified block diagram of the process is presented in Figure 5.3. The process operates at a temperature of about 400°C, and the Reactors are therefore located after the Boiler Convection Passes, and upstream of the Air Heater and Electrostatic Precipitator (ESP). There are two Reactors, containing copper oxide supported on an alumina base, and operated cyclically: one Reactor is on stream, with the gas flowing over (not through) the copper oxide, whilst the other is being regenerated by passing hydrogen over the sulphated copper oxide, reducing it to metallic copper and giving an SO_2-rich off-gas. When the Reactor is put back on stream, the copper is oxidised by oxygen in the gas, reforming copper oxide for further reaction.

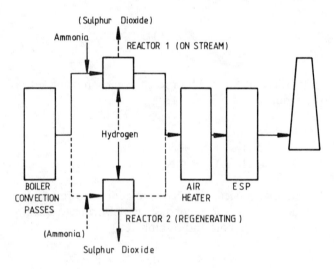

Figure 5.3 Block Diagram of Copper Oxide Combined Abatement Process

Chemistry of Process [145, 352]: The SO_2 absorption reaction is:

Absorption $\qquad CuO + SO_2 + O_2 = 2\ CuSO_4$

In the regeneration phase, hydrogen reduces the copper sulphate to metallic copper. The copper is re-oxidised when the absorption phase is resumed. The reactions are:

Reduction
$CuSO_4 + 2\ H_2 = Cu + SO_2 + 2\ H_2O$

Oxidation $\qquad 2\ Cu + O_2 = 2\ CuO$

Ammonia reacts with nitrogen oxides in the presence of copper oxide and sulphate which act as catalysts for the selective reduction of NO_x; the copper sulphate has the greater catalytic activity. The reduction reaction is:

$6\ NO + 4\ NH_3 = 5\ N_2 + 6\ H_2O$

All of the reactions occur at about 400°C, and the absence of temperature cycling between the absorption and regeneration phases avoids thermal stresses on the copper oxide.

Process Code NS12.1 – Non-Selective Catalytic Reduction of SO_2 and NO_x (Ralph M. Parsons Co. Process)

Outline of Process [419]: A simplified block diagram of the process is presented in Figure 5.4. A reducing agent such as natural gas or producer gas is added to the combustion gas leaving the boiler Economiser, and particulates are removed in a Hot Electrostatic Precipitator (ESP). The gas then enters a Reactor containing a non-noble metal non-selective reduction catalyst which catalyses the reduction of SO_2 to H_2S and of NO_x to elemental N_2. The gas is then cooled by passage through the Air Heater, and the H_2S is converted to elemental sulphur in a Stretford Absorber. The gas is reheated in the Reheater before being exhausted to stack.

Chemistry of Process [419]: The reduction agents are hydrogen and carbon monoxide; when natural gas is used, H_2 and CO can result from reforming of methane, e.g. by the reaction:

Reforming $\qquad CH_4 + H_2O = 3\ H_2 + CO$

The reduction reactions, yielding H_2S and N_2, can be represented as follows:

Reduction of SO_2 $\quad SO_2 + 3\ H_2 = H_2S + 2\ H_2O$

Reduction of NO $\quad 2\ NO + 2\ CO = 2\ CO_2 + N_2$

Reduction of NO_2 $\quad 2\ NO_2 + 4\ CO = 4\ CO_2 + N_2$

In the Stretford Absorber, H_2S is absorbed by sodium carbonate solution; this can be simplified as:

H_2S Absorption
$H_2S + Na_2CO_3 = NaHS + NaHCO_3$

The Stretford absorber solution, which also contains soluble vanadium compounds, is then oxidised by air to precipitate sulphur and regenerate the sodium carbonate; simplified as:

S Production
$2\ NaHS + 2\ NaHCO_3 + O_2$
$= S_2 + 2\ Na_2CO_3 + 2\ H_2O$

The vanadium acts as the oxygen carrier, being

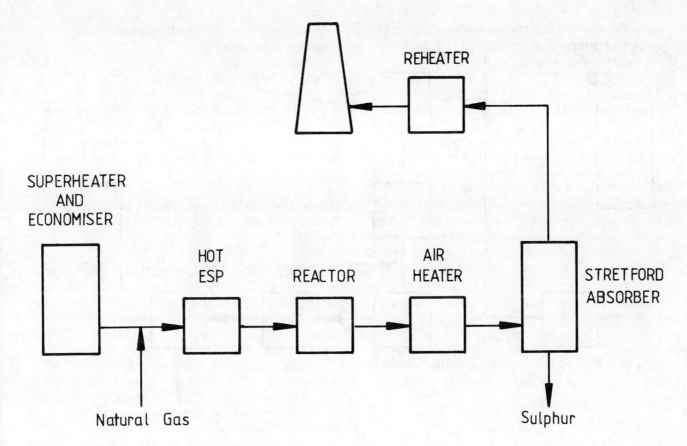

Figure 5.4 Block Diagram of Combined Abatement Process–Non-Selective Catalytic Reduction of NO_x and SO_2

oxidised to the pentavalent state by the oxidising air, and reduced to the tetravalent state in the reaction with NaHS.

Process Code NS12.2 – Catalytic Reduction of NO_x and Catalytic Oxidation of SO_2

Outline of Process [352]: A simplified block diagram of the process is presented in Figure 5.5. Gas leaving the Superheater and Economiser section of the boiler at a temperature of 450°C passes through a Hot Electrostatic Precipitator (ESP) to remove particulates. Natural gas is injected into the gas stream to reduce the oxygen content before it is contacted with the First Catalyst, and to reduce nitrogen oxides to nitrogen in contact with the First Catalyst. The gas is then cooled to about 360°C by passing over further economiser heat transfer surface in the Cooler. Oxidising conditions are restored by the addition of air, and the gas is then contacted with the Second Catalyst to oxidise sulphur dioxide to trioxide. The cleaned gas passes through the boiler Air Heater and the air-cooled Condenser before being exhausted to stack. The condensate by-product is 93% sulphuric acid.

Chemistry of the Process [352]: The addition of natural gas results in establishing reducing conditions for reaction with the First Catalyst, and nitrogen oxides are reduced:

$$4\ NO + CH_4 = CO_2 + 2\ H_2O + 2\ N_2$$

After restoring oxidising conditions by the addition of air, the oxidation of SO_2 to SO_3, catalysed by e.g. vanadium pentoxide, and subsequent formation of sulphuric acid, occurs as in the Cat-Ox FGD process (see Section 2.2).

Oxidation $\qquad 2\ SO_2 + O_2 = 2\ SO_3$

Acid formation $\qquad SO_3 + H_2O = H_2SO_4$

Process Code N21.1 – Limestone Injection into Multi-Stage Burner (LIMB)

Outline of Process [431]: Figure 5.6 is a schematic diagram of a Limestone Injection/Multi-staged Burner (LIMB).

The technology is a combination of low-NO_x combustion techniques (see Section 3.2) and the injection of dry limestone into the furnace for SO_x control (see Section 2.2). In the LIMB design, however, the finely ground limestone is injected into the furnace either through the burner itself or in close proximity to it. The reaction products and any unreacted limestone leaving the furnace are removed with the fly-ash in an electrostatic precipitator or baghouse.

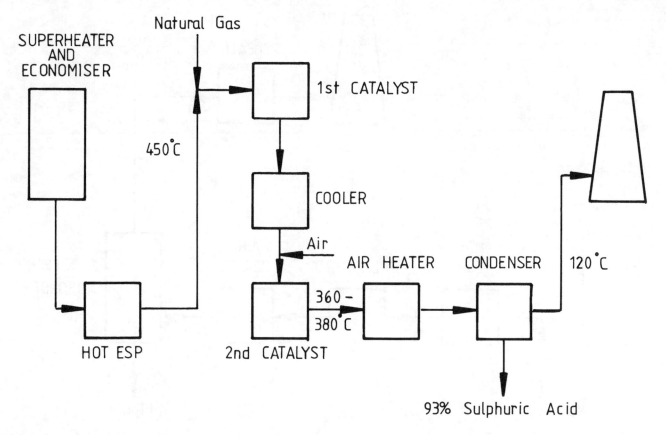

Figure 5.5 Block Diagram of Catalytic NO$_x$ Reduction/Catalytic SO$_2$ Oxidation Combined Abatement Process

Chemistry of Process [297]: When finely ground limestone is injected into a flame, it undergoes a calcination reaction:

$$CaCO_3 = CaO + CO_2$$

Experimentally confirmed thermodynamic considerations indicate that calcination starts at about 800°C. With pulverised coal flame temperatures being significantly higher than this, rapid calcination is expected to take place thus allowing reaction with SO$_2$:

$$2\ CaO + 2\ SO_2 + O_2 = 2\ CaSO_4$$

Intermediate stages in this overall reaction involve the formation of calcium sulphite (CaSO$_3$) or the direct reaction between CaO and SO$_3$.

At high temperatures, CaSO$_4$ becomes thermally unstable and the above reaction may become reversible.

Sulphur released as H$_2$S in the reducing zone of the burner is captured by the calcined limestone but not by direct reaction with the limestone.

Process Code NS22.1–Electron Beam Radiation Process

Outline of Process [352, 419]: A simplified block diagram of the process is presented in Figure 5.7. Gas from the Boiler Electrostatic Precipitator (ESP) or Baghouse is cooled and humidified in the Cooler by injection of water; ammonia is also added, and the gas passes through a Reactor in which it is subjected to an intense field from an electron beam. The field brings about reactions between ammonia, sulphur dioxide and nitrogen oxides to form solid ammonium nitrate and ammonium nitrate sulphate. The solids are removed in a second ESP or Baghouse, and are potentially saleable as a fertiliser by-product. The gas is exhausted to stack.

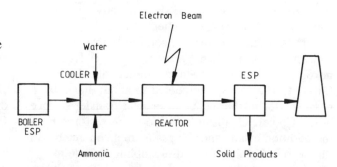

Figure 5.7 Block Diagram of Electron Beam Radiation Combined Abatement Process

Chemistry of the Process [352, 419]: Under the influence of the electron beam radiation, SO$_2$ and NO$_x$ react with ammonia in the presence of water vapour to form ammonium sulphate and ammonium nitrate-sulphate:

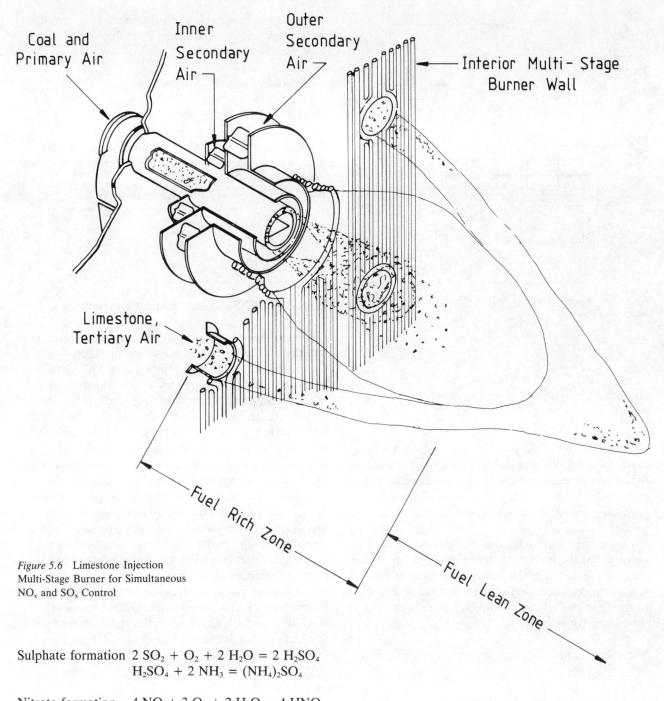

Figure 5.6 Limestone Injection Multi-Stage Burner for Simultaneous NO_x and SO_x Control

Sulphate formation $2 SO_2 + O_2 + 2 H_2O = 2 H_2SO_4$
$H_2SO_4 + 2 NH_3 = (NH_4)_2SO_4$

Nitrate formation $4 NO + 3 O_2 + 2 H_2O = 4 HNO_3$
$NH_3 + HNO_3 = NH_4NO_3$

Process Code NS23.1–Sodium Carbonate Adsorption Process (NO_xSO Process)

Outline of Process [352, 434]: A simplified block diagram of the process is presented in Figure 5.8. Gas leaving the Boiler Air Preheater at a temperature of about 120°C enters the Adsorber. Here it fluidises a bed of sodium carbonate deposited on gamma-alumina, which absorbs SO_2 and NO_x. The clean gas then passes via an Electrostatic Precipitator (ESP) to stack. The sorbent flows continuously from the Absorber to the Sorbent Heater, where it is fluidised and heated to 550–600°C by a stream of air and combustion products from an Air Heater fired with coal or oil. The nitrogen oxides are desorbed and are returned, with the heating air, to the boiler where chemical equilibrium controls the NO_x concentrations; hence, there is no accumulation of NO_x in the gas entering the Adsorber. From the Sorbent Heater, the sorbent passes to the moving-bed Regenerator where it is contacted with a reducing gas–hydrogen, carbon monoxide or hydrogen sulphide. The sorbent is then treated with steam in the moving-bed Steam Treatment vessel to remove residual sulphur as H_2S and to reactivate the sorbent. The sorbent is cooled with air in the fluidised bed Cooler, and is then conveyed pneumatically to the Adsorber. The cooling air passes to the Air Heater en route to the Sorbent Heater and eventually to the Boiler.

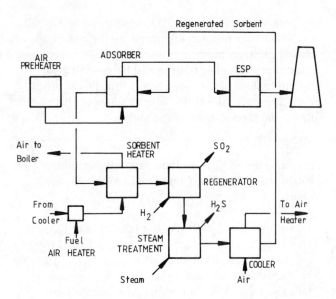

Figure 5.8 Block Diagram of Sodium Carbonate Combined Abatement Process

Chemistry of the Process [352, 434]: The sodium carbonate and the gamma-alumina substrate react to form sodium aluminate, which is the actual sorbent:

$Na_2CO_3 + Al_2O_3 = 2 NaAlO_2 + CO_2$

Nitrogen oxides are adsorbed by the sorbent, and are desorbed at the higher temperature prevailing in the Sorbent Heater. The desorption is accompanied by some disproportionation of the adsorbed NO_x, resulting in the release of nitrous oxide, N_2O, and dinitrogen trioxide, N_2O_3.

The sodium aluminate reacts with water vapour and SO_2 as follows:

$2 NaAlO_2 + H_2O = 2 NaOH + Al_2O_3$

$2 NaOH + SO_2 = Na_2SO_3 + H_2O$

The formation of the more alkaline NaOH results in increased reactivity to the SO_2. In the presence of NO_x, the sodium sulphite is oxidised by oxygen in the gas to sodium sulphate:

$2 Na_2SO_3 + O_2 = 2 Na_2SO_4$

In the Regenerator, the sulphate is reduced by hydrogen, hydrogen sulphide, carbon monoxide or other reducing gases; for example, with H_2 and H_2S the reactions are:
$Na_2SO_4 + Al_2O_3 + 4 H_2 = 2 NaAlO_2 + H_2S + 3 H_2O$
$Na_2SO_4 + Al_2O_3 + 3 H_2S$
$= 2 NaAlO_2 + (4/x) S_x + 3 H_2O$

With carbon monoxide the sulphur reaction product is carbonyl sulphide, which is hydrolysed on further treatment with steam:
$Na_2SO_4 + Al_2O_3 + 4 CO$
$= 2 NaAlO_2 + COS + 3 CO_2$

$COS + H_2O = H_2S + CO_2$

Some reduction to sulphide occurs in the Regenerator, and regeneration has to be completed by steam treatment:
$Na_2S + Al_2O_3 + H_2O = H_2S + 2 NaAlO_2$

Process Code NS31.1–Lime Spray Dryer Process

Outline of Process [352, 432]: A simplified block diagram of the process is presented in Figure 5.9. Gas from the Boiler, at a temperature 120–160°C, enters the top of a Spray Dryer into which a slurry of lime containing an additive, e.g. sodium hydroxide, is sprayed. The lime, which is hydrated and slurried in the Slaker, is pumped to the slurry atomiser nozzles via a Feed Tank, where it is mixed with the additive. The gas passes down the Spray Dryer co-current with the slurry spray droplets; the sulphur oxides, nitrogen oxides and acid halides in the gas react with the lime and alkali, and water is evaporated from the slurry droplets to produce a dry product which also contains the particulates present in the in-going gas. Coarse particles of product are collected at the base of the Spray Dryer, and fine particles are separated from the gas in a Baghouse or Electrostatic Precipitator (ESP). The gas leaves the system at 90–100°C, and is exhausted to the stack. Part of the dry end-product is recycled to the Feed Tank to increase the conversion of lime.

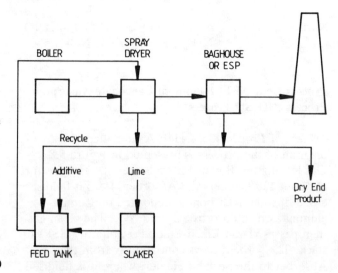

Figure 5.9 Block Diagram of Lime Spray Dryer Combined Abatement Process

The dry end-product contains calcium and sodium sulphite, sulphate, nitrite, nitrate and halides, together with some carbonate, unreacted lime and ash. It is usual to recycle part of the dry end-product to the Slurry Preparation stage to increase the lime utilisation.

Chemistry of Process [122]: The sulphur oxides, SO_2 and SO_3, acid halides, HF, HCl and HBr (represented below as 'HHa') and nitrogen oxides react in the Spray Dryer with the lime and alkali in the slurry as by reactions including the following:

$$Ca(OH)_2 + SO_2 = CaSO_3 + H_2O$$

$$Ca(OH)_2 + SO_3 = CaSO_4 + H_2O$$

$$Ca(OH)_2 + 2\,HHa = CaHa_2 + 2\,H_2O$$
$$2\,NaOH + 2\,NO + O_2 = NaNO_2 + NaNO_3 + H_2O$$
$$2\,NaOH + 2\,NO_2 = NaNO_2 + NaNO_3 + H_2O$$

The absorption of NO_x does not occur in the absence of SO_2, and the efficiency of NO_x abatement increases with increase in the molar ratio of SO_2/NO_x. For example [352] the abatement is negligible at a zero molar ratio, about 30% at a molar ratio of 0.3 and about 70% at a molar ratio of 1. However, it has been shown [444] that NO_x absorption at low SO_2/NO_x molar ratios is enhanced by the use of an ionising electron beam; see also above. The reaction temperature is higher than would be desirable for sulphur capture alone (see above), but this is offset by the presence of NaOH, which enhances the sulphur capture activity of the slurry.

Process Code NS32.1–Sodium Sulphite Scrubbing Process with Chelating Compound (Asahi Chemical Process)

Outline of Process [419]: A simplified block diagram of the process is presented in Figure 5.10. Gas from the Boiler passes through the Prescrubber to remove particulates and acid halides and to cool the gas, and then enters the Absorber where SO_2 and NO are absorbed by a solution of sodium sulphite containing the ferrous salt of ethylene diamine tetracetic acid (Fe^{++}.EDTA). The clean gas is heated in the Reheater and exhausted to the Stack. The solution leaving the Absorber is recirculated (together with sodium carbonate make-up) via a Reducer, where the nitric oxide chelate complex of the Fe^{++}.EDTA reacts with sodium sulphite to form sodium sulphate, imidodisulphonate and dithionate, and to release elemental nitrogen. The sulphate and dithionate are

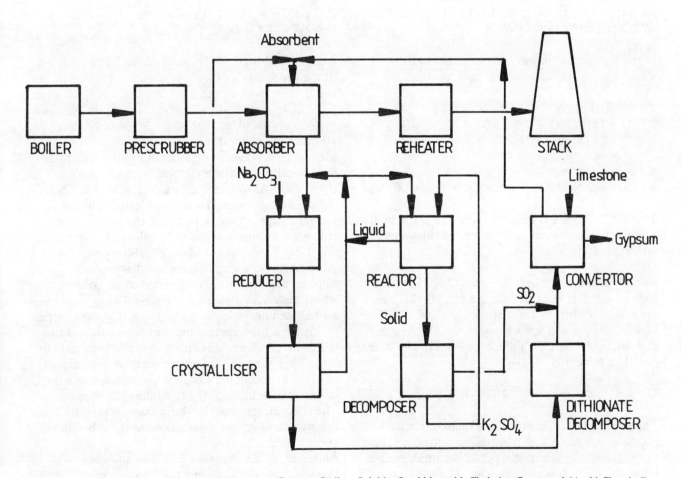

Figure 5.10 Block Diagram of Combined Abatement Process: Sodium Sulphite Scrubbing with Chelating Compound (Asahi Chemical)

separated from a side stream of the circulating solution in the Crystalliser system and sent to the Dithionate Decomposer; the mother liquor is treated with potassium sulphate in the Reactor to precipitate potassium imidodisulphonate which is separated from the solution and heated in the Decomposer to yield potassium sulphate (recycled to the Reactor) and sulphite together with SO_2 and N_2. The dithionate is thermally decomposed in the Dithionate Decomposer to sulphate, with release of SO_2. The sulphate, together with SO_2 from the Decomposer and Dithionate Decomposer, water and limestone, are fed to the Converter system where sodium sulphite absorbent is recovered and recirculated to the Absorber, and gypsum is produced.

Chemistry of the Process [419]: The absorbent is a solution of sodium sulphite and the ferrous salt of ethylene diamine tetracetic acid (Fe^{++}.EDTA) at a pH of 6.3. The principal reactions in the Absorber from sodium bisulphite and dithionate (with SO_2), and a ferrous chelate complex (with NO):

Bisulphite $SO_2 + Na_2SO_3 + H_2O = 2\ NaHSO_3$

Dithionate $4\ NaHSO_3 + O_2 = 2\ Na_2S_2O_6 + 2\ H_2O$

Chelate $NO + Fe^{++}.EDTA = Fe^{++}.EDTA.NO$

Oxygen in the gas causes some oxidation of the ferrous EDTA to the ferric EDTA, Fe^{+++}.EDTA, and of sulphite to sulphate:

In the Reducer the main reactions restore the absorbent for recirculation to the Absorber:

Chelate reduction
$2\ Fe^{++}.EDTA.NO + 2\ Na_2SO_3$
$= 2\ Fe^{++}.EDTA + 2\ Na_2SO_4 + N_2$

Bisulphite reaction
$Na_2CO_3 + 2\ NaHSO_3 = 2\ Na_2SO_3 + H_2O + CO_2$

The Fe^{+++}.EDTA is reduced to Fe^{++}.EDTA by the sulphite ion, and some sodium imidodisulphonate, $NH(SO_3Na)_2$ is formed.

Sodium dithionate dihydrate, $Na_2S_2O_6.2\ H_2O$, and sodium sulphate decahydrate, $Na_2SO_4.10\ H_2O$, are crystallised out from a side-stream in the Crystalliser system, dehydrated and treated in the Dithionate Decomposer where the dithionate is thermally decomposed to sulphate:

Dithionate decomposition $Na_2S_2O_6 = Na_2SO_4 + SO_2$

The mother liquor, containing sodium imidodisulphonate, is treated with potassium sulphate solution in the reactor to precipitate potassium imidodisulphonate, which is centrifuged out. The concentrate is returned to the Reducer, and the solid is thermally decomposed in the Decomposer to give K_2SO_4 which is recycled to the Reactor:

Precipitation
$NH(SO_3Na)_2 + K_2SO_4 = NH(SO_3K)_2 + Na_2SO_4$

Decomposition
$2\ NH(SO_3K)_2$
$= 2\ SO_2 + K_2SO_4 + K_2SO_3 + H_2O + N_2$

The SO_2 released by decomposition of dithionate and imidodisulphonate are sent to the Converter system, where sodium sulphate from the Dithionate Decomposer, together with a feed of water and limestone, are used to produce sodium sulphite for recycle to the Absorber, and gypsum, $CaSO_4.2\ H_2O$. A complex series of operations occurs in the Converter system; the reactions can be simplified as the formation of calcium sulphite hemihydrate, and its conversion to gypsum:

Sulphite
$2\ CaCO_3 + 2\ SO_2 + H_2O$
$= 2\ CaSO_3.(1/2)H_2O + 2\ CO_2$

Gypsum
$2\ CaSO_3.(1/2)H_2O + 2\ Na_2SO_4 + 3\ H_2O$
$= 2\ Na_2SO_3 + 2\ CaSO_4.2\ H_2O$

Process Code NS32.2 – Limestone Slurry Scrubbing Process with Chelating Compound

Outline of Process [352]: A simplified block diagram of the process is presented in Figure 5.11. Gas from the Boiler passes through the Prescrubber to remove particulates and acid halides and to cool the gas, and then enters the Absorber where SO_2 and NO are absorbed by a limestone slurry (prepared in the Mixing Tank) containing a chelating compound – the ferrous salt of ethylene diamine tetracetic acid (Fe^{++}.EDTA) or of nitrilo triacetic acid (Fe^{++}.NTA). The chelating compound forms a ferrous chelate complex with nitric oxide, and in the presence of chloride ions this is reduced by calcium bisulphite in the slurry to nitrogen and oxygen. Slurry from the Absorber, containing SO_2 captured as calcium sulphite hemihydrate, is circulated to an Oxidiser (in which the calcium sulphite is oxidised to gypsum by air bubbled into the slurry) and to a Regenerator (in which any of the chelate that has been oxidised to the ferric condition is reduced to the ferrous state by a reducing agent, e.g. ascorbic acid). A side stream of the slurry from the Oxidiser is circulated to a thickener and/or centrifuge for separating the gypsum. The concentrate, and the slurry from the Oxidiser and the Regenerator, are recirculated to the Absorber.

Chemistry of the Process [352]: The absorbent is a limestone slurry containing the ferrous salt of ethylene diamine tetra-acetic acid (Fe^{++}.EDTA) or

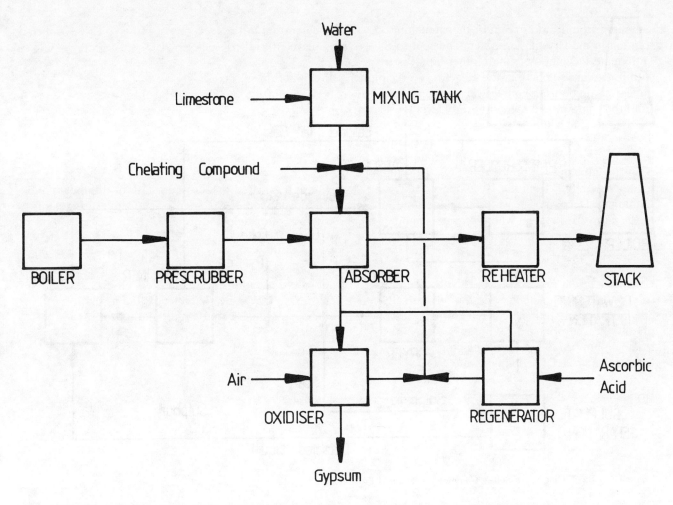

Figure 5.11 Block Diagram of Combined Abatement Process: Limestone Slurry Scrubbing with Chelating Compound

of nitrilo triacetic acid (Fe^{++}.NTA). The principal absorption reactions form: calcium sulphite and bisulphite (with SO_2); and a ferrous chelate complex (with NO):

$SO_2 + CaCO_3 = CaSO_3 + CO_2$

$CaSO_3 + SO_2 + H_2O = Ca(HSO_3)_2$

Fe^{++}.EDTA + NO = Fe^{++}.EDTA.NO
Fe^{++}.NTA + NO = Fe^{++}.NTA.NO

The ferrous chelate complex is reduced by bisulphite present in the slurry in the presence of chloride ions but the ferrous EDTA and NTA become oxidised to the ferric state:

4 Fe^{++}.EDTA.NO + Ca(HSO$_3$)$_2$
 = 2 N$_2$ + O$_2$ + Ca(HSO$_4$)$_2$ + 4 Fe^{+++}.EDTA

4 Fe^{++}.NTA.NO + Ca(HSO$_3$)$_2$
 = 2 N$_2$ + O$_2$ + Ca(HSO$_4$)$_2$ + 4 Fe^{+++}.NTA

They are restored to the ferrous state by reduction with, e.g., ascorbic acid in the Regenerator.
In the combined abatement modification of the Saarberg–Hölter–Lurgi FGD process (Section 2.5) the oxidation of sulphite to gypsum occurs in the Absorber, and formic acid is added to the absorbent slurry to accelerate SO_2 absorption and gypsum formation.

Process Code NS32.3–Sulf-X Process

Outline of Process [145, 175] (see Section 2.2): A simplified block diagram of the process is presented in Figure 5.12. Gas from the Boiler first passes through a Prescrubber to remove acid halides and particulates, and then into the Absorber where it is scrubbed with a slurry containing a mixture of iron sulphides in an aqueous sodium sulphide solution. The reactions with sulphur dioxide produce a range of insoluble iron/sulphur compounds, some soluble iron sulphate, and trace amounts of elemental sulphur. Reduction of NO_x to elemental nitrogen also occurs. The solid reaction products are separated from the solution in the Dewatering and Filtration system, dried in a steam-heated Dryer, and roasted at 650–750°C in an indirectly fired Calciner to which coal or coke is fed as a reducing agent, regenerating the iron sulphides and releasing sulphur. Part of the solution from the Filtration stage is treated in a

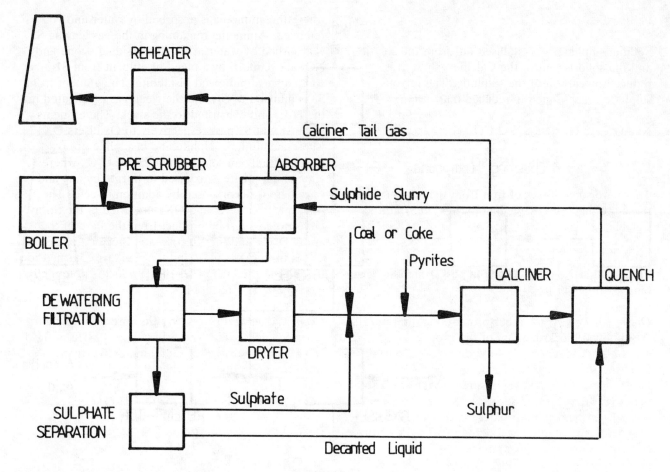

Figure 5.12 Block Diagram of Sulf-X Combined Abatement Process

Sulphate Separation system to crystallise out sodium sulphate which is transferred to the Calciner. The solids from the Calciner pass to the Quench, where they are mixed with liquid from the Filtration and Sulphate Separation stages to reform the sulphide slurry sent to the Absorber.

Sulphur vapour in the tail gas from the Calciner is condensed for marketing or safe disposal, and the gas is returned to the Prescrubber inlet.

Chemistry of Process [95]: The sorbent slurry is a complex mixture of iron compounds in an aqueous sodium sulphide solution, with a pH controlled at 6.0–6.4 by the buffering action of $Fe(OH)_2$. The main absorbing compound is FeS. The sulphur capture reactions occurring in the Absorber include:

$FeS + H_2O = Fe^{++} + HS^- + OH^-$

$SO_2 + H_2O = H^+ + HSO_3^-$

$2\ HS^- + 2\ HSO_3^- + O_2 = 2\ S_2O_3^{--} + 2\ H_2O$

$S_2O_3^{--} + H^+ = HSO_3^- + S$

$2\ HSO_3^- + O_2 = 2\ SO_4 + 2\ H^+$

The overall reaction can be expressed by the following simplification (note that the equation does not balance):

$FeS + SO_2 + O_2 = Fe^{++} + SO_4^{--} + H^+$

Nitrogen oxides are reduced by ferrous sulphide, releasing elemental nitrogen and forming ferrous sulphate:

$4\ NO + FeS = 2\ N_2 + FeSO_4$
$2\ NO_2 + FeS = N_2 + FeSO_4$

These reactions result in the formation of complex mixtures of iron sulphides, Fe_xS_y, together with $FeSO_4$, $Fe(OH)_2$ and Na_2SO_4. The key controlling factors are maintenance of the correct pH, the correct Na_2S concentration, and the buffering action of the $Fe(OH)_2$.

The main regeneration reaction is the thermal decomposition of the iron sulphides to give the sorbent FeS and sulphur, e.g. for FeS_2:

$FeS_2 = FeS + S$

Sodium sulphate is crystallised out from the solution, and the solid is sent to the Calciner where it is reduced by carbon to the sulphide. The ferrous sulphate is subsequently reduced in solution:

$Na_2SO_4 + 2\ C = Na_2S + 2\ CO_2$

$FeSO_4 + Na_2S = FeS + Na_2SO_4$

There are some losses of iron from the system, and these are made up by feeding pyrites, FeS_2, to the Calciner.

Process Code NS41.1 – Oxidation plus Ammonia Scrubbing (Walther Process)

Outline of Process [Walther promotional literature]: A simplified block diagram of the process is presented in Figure 5.13. Ammonia is added to gas from the Boiler, and the gas is cooled in a Heat Exchanger (H.E.) before passing through three scrubbers (S1, S2 and S3) in series. In S1, the ammonia is absorbed, and sulphur oxides and acid halides removed, with ammoniacal effluent from S3. Ozone is then added to the gas to oxidise NO to NO_2, and more ammonia is also added before the gas enters S2, where the ammonia is absorbed by water and NO_2 is removed. Ammonia remaining in the gas leaving S2 is absorbed by water in S3, and residual ozone and NO_2 are reduced by a bleed of effluent from S1 containing sulphite and bisulphite. The effluent from S3 is pumped to S1, and the clean gas is reheated in the H.E. and exhausted to the stack. The effluents from S1 and S2 pass respectively to Oxidisers OX1, where ammonium sulphite and bisulphite are oxidised to sulphate by air and ammonia, and OX2, where ammonium nitrite is oxidised to nitrate by air, and carbonate is decomposed by addition of acid. The solutions from OX1 and OX2 are dried in the Spray Dryer to give a solid product from the base of the Spray Dryer and the Electrostatic Precipitator (ESP). Gases from OX1, OX2 and the ESP are returned to the inlet to the H.E. The solid product is potentially marketable as a fertiliser.

Chemistry of the Process [Walther promotional literature]: Sulphur oxides and acid halides ('HHa' = HCl and HF) are absorbed by ammonium hydroxide solution in Scrubber S1, where the pH = 6:

$SO_2 + 2\ NH_4OH = (NH_4)_2SO_3 + H_2O$

$(NH_4)_2SO_3 + SO_2 + H_2O = 2\ NH_4HSO_3$

$SO_3 + 2\ NH_4OH = (NH_4)_2SO_4 + H_2O$

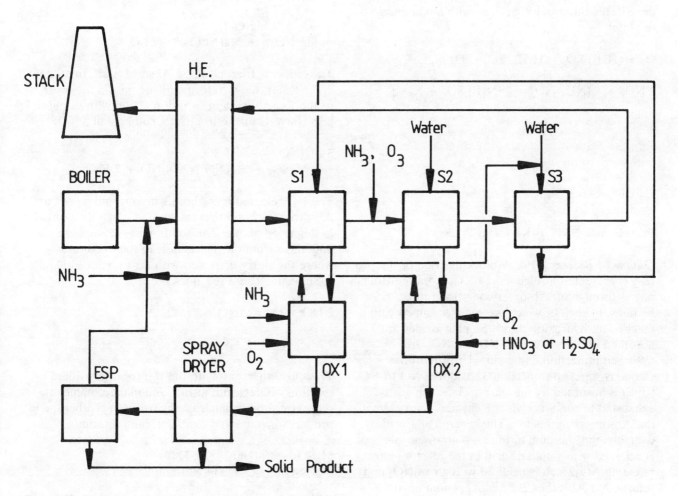

Figure 5.13 Block Diagram of Walther Combined Abatement Process

HHa + NH$_4$OH = NH$_4$Ha + H$_2$O

In Oxidiser OX1 at a pH of 4.0, sulphite and bisulphite are oxidised to sulphate:

NH$_4$HSO$_3$ + NH$_3$ = (NH$_4$)$_2$SO$_3$

2 (NH$_4$)$_2$SO$_3$ + O$_2$ = 2 (NH$_4$)$_2$SO$_4$

Ozone oxidises NO to NO$_2$, and this is absorbed by ammonium hydroxide in S2 at pH = 8–9 to form ammonium nitrite and nitrate:

NO + O$_3$ = NO$_2$ + O$_2$

2 NO$_2$ + 2 NH$_3$ + H$_2$O = NH$_4$NO$_2$ + NH$_4$NO$_3$

The nitrite is oxidised to nitrate in OX2, and ammonium carbonate formed in S2 is decomposed at pH = 3–4 by nitric or sulphuric acid addition:

2 NH$_4$NO$_2$ + O$_2$ = 2NH$_4$NO$_3$

(NH$_4$)$_2$CO$_3$ + 2 HNO$_3$ = 2NH$_4$NO$_3$ + CO$_2$ + H$_2$O
(NH$_4$)$_2$CO$_3$ + H$_2$SO$_4$ = (NH$_4$)$_2$SO$_4$ + CO$_2$ + H$_2$O

In Scrubber S3, residual ammonia in the gas is absorbed by water, and residual ozone and NO$_2$ are reduced by ammonium sulphite and bisulphite from the S1 effluent at pH = 5; thus, with ammonium sulphite:

O$_3$ + (NH$_4$)$_2$SO$_3$ = (NH$_4$)$_2$SO$_4$ + O$_2$

2 NO$_2$ + 4 (NH$_4$)$_2$SO$_3$ = 4 (NH$_4$)$_2$SO$_4$ + N$_2$

Process code NS41.2 – Kawasaki Process

Outline of Process [419]: A block diagram of the process is presented in Figure 5.14. Gas from the Boiler passes through an Absorber comprising three sections, in each of which the gas is scrubbed with a magnesium hydroxide slurry. Sulphur oxides are absorbed in Section 1, and NO and NO$_2$ are absorbed in Section 2 in equimolar proportions. Ozone is supplied to Section 3, oxidising NO to NO$_2$ which is absorbed by the slurry. The clean gas is exhausted to the Stack via a Reheater. Slurry leaving the Absorbers passes to a Thickener. The overflow, containing magnesium nitrite and nitrate in solution, is acidified with sulphuric acid in the reactor system, releasing NO which is oxidised with air to NO$_2$ and returned to Absorber 2. The magnesium nitrite, converted to nitrate and sulphate in the Reactor, passes together with the underflow from the Thickener, containing magnesium sulphite and sulphate, to the Oxidiser, where sulphite is oxidised to sulphate by treatment of the slurry with air. In the Regenerator system, the magnesium hydroxide is regenerated by reaction between calcium nitrate, to produce gypsum, followed by treatment with lime to produce calcium nitrate solution, some of which is bled off.

Chemistry of the Process [419]: The gas is cooled and humidified in Absorber 1, and sulphur dioxide is absorbed by the magnesium hydroxide slurry to form the sulphite hexahydrate:

SO$_2$ + Mg(OH)$_2$ + 5 H$_2$O = MgSO$_3$.6H$_2$O

In Absorber 2, NO and NO$_2$ (including NO$_2$ from the Reactor) react together and are absorbed by the slurry to form magnesium nitrite:

NO + NO$_2$ + Mg(OH)$_2$ = Mg(NO$_2$)$_2$ + H$_2$O

This reaction ceases when the NO$_x$ concentration falls below about 200 ppm, and the gas then has to be treated with ozone in Absorber 3, where NO is oxidised to NO$_2$ which is absorbed by the slurry to form magnesium nitrate and nitrite:

NO + O$_3$ = NO$_2$ + O$_2$

4 NO$_2$ + 2 Mg(OH)$_2$
= Mg(NO$_3$)$_2$ + Mg(NO$_2$)$_2$ + 2 H$_2$O

The effluents from all three Absorbers are treated in a Thickener; the nitrate and nitrite pass in the overflow to the Reactor system, where reaction with sulphuric acid converts nitrite to nitrate, sulphate and NO:

3 Mg(NO$_2$)$_2$ + 2 H$_2$SO$_4$
= 2 MgSO$_4$ + Mg(NO$_3$)$_2$ + 4 NO + 2 H$_2$O

The NO released is oxidised with air, and passes to Absorber 2 where it reacts with NO and is absorbed as described above. The solution passes with the underflow from the Thickener, to the Oxidiser, where the slurry is treated with air to oxidise sulphite to sulphate. These reactions are:

2 NO + O$_2$ = 2 NO$_2$

2 MgSO$_3$.6H$_2$O + O$_2$ = 2 MgSO$_4$ + 6 H$_2$O

The slurry then passes to the Regenerator, where reaction with calcium nitrate and make-up lime regenerates the magnesium hydroxide absorbent, and produces gypsum and excess calcium nitrate:

MgSO$_4$ + Ca(NO$_3$)$_2$ + 2 H$_2$O
= CaSO$_4$.2H$_2$O + Mg(NO$_3$)$_2$

Mg(NO$_3$)$_2$ + Ca(OH)$_2$ = Mg(OH)$_2$ + Ca(NO$_3$)$_2$

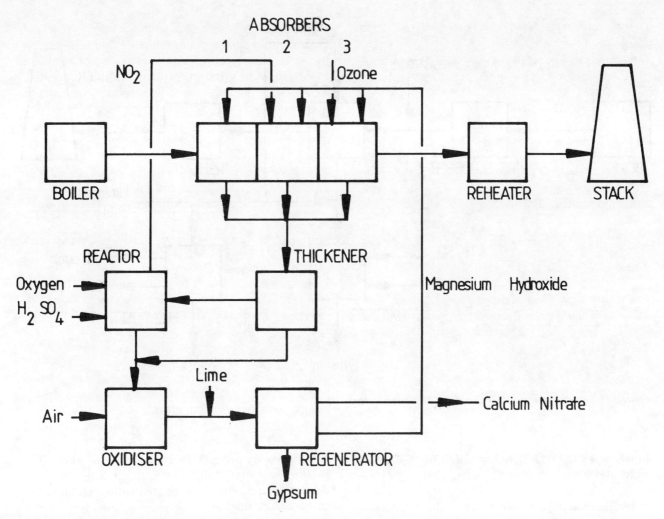

Figure 5.14 Block Diagram of Kawasaki Combined Abatement Process

Process NS42.1–IHI Process

Outline of Process [419]: A simplified block diagram of the process is presented in Figure 5.15. Gas from the Boiler first enters a Prescrubber to remove acid halides and particulates, and to cool and humidify the gas. Ozone is mixed with the gas, to oxidise NO to NO_2, and the gas is then scrubbed in a Turbulent Contact Absorber (TCA) through which a slurry of lime or limestone is circulated. The slurry, which absorbs SO_2, contains copper and sodium chloride catalysts to promote absorption and reduction of NO_2. The clean gas is exhausted to stack via the Reheater. A side stream of the circulating slurry is treated in the Oxidiser system; the pH is reduced by sulphuric acid addition, and the calcium sulphite hemihydrate is oxidised with air to gypsum, which is centrifuged off. Fresh slurry is added to neutralise the centrate, in the Decomposer system; this system includes a thickener for separation of slurry (returned to the TCA), an evaporator for the overflow, and a thermal decomposer for the evaporator residue. The decomposition stage is needed to eliminate nitrogen-sulphur compounds of calcium, and it produces further gypsum contaminated by copper and calcium chloride.

Chemistry of the Process [419]: SO_2 is absorbed by the slurry in the TCA to form calcium sulphite hemihydrate and bisulphite:

Sulphite
$$2\ SO_2 + 2\ CaCO_3 + H_2O = 2\ CaSO_3.(1/2)H_2O + 2\ CO_2$$

Bisulphite
$$2\ SO_2 + 2\ CaSO_3.(1/2)H_2O + H_2O = 2\ Ca(HSO_3)_2$$

Nitric oxide is oxidised to NO_2 by ozone, and the NO_2 is absorbed by the slurry and reduced to nitrogen by calcium sulphite and bisulphite to produce gypsum, calcium sulphamates and calcium nitrite and nitrate; the reactions include:

$$NO + O_3 = NO_2 + O_2$$

$$2\ NO_2 + 4\ CaSO_3.(1/2)H_2O + 6\ H_2O = N_2 + 4\ CaSO_4.2H_2O$$

$$4\ NO_2 + 2\ CA(HSO_3)_2 + 10\ CaSO_3.(1/2)H_2O + 17\ H_2O = 2\ Ca(NH_2SO_3)_2 + 10\ CaSO_4.2H_2O$$

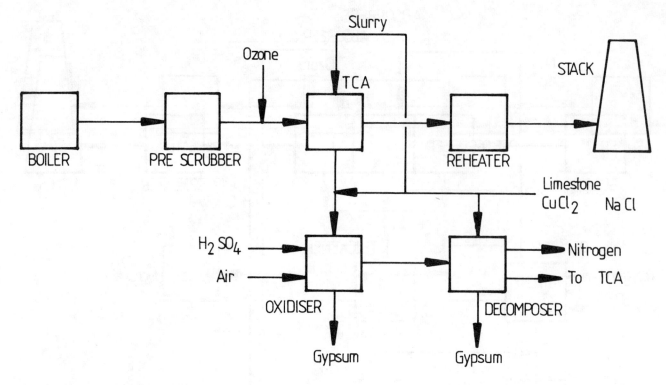

Figure 5.15 Block Diagram of IHI Combined Abatement Process

$4 NO_2 + 4 CaSO_3.(1/2)H_2O = Ca(NO_2)_2 + Ca(NO_3)_2 + 2 Ca(HSO_3)_2$

In the thermal decomposition stage, the sulphamates, nitrite and nitrate are decomposed by reactions including:

Sulphamate: The following equation, reproduced from Reference [419], does not balance:

$2 (NH_2SO_3)_2Ca + 4 CaCO_3 + 2 Ca(HSO_4)_2 + 3 O_2 + 8 H_2O = 2 N_2 + 8 CaSO_4.2 H_2O + 4 CO_2$

$Ca(NO_2)_2 + Ca(NO_3)_2 + 2 Ca(HSO_4)_2 = 2 N_2 + 4 CaSO_4.2 H_2O + 2 H_2O$

Process Code NS42.2–Moretana Calcium Process

Outline of Process [419]: A simplified block diagram of the process is presented in Figure 5.16. Gas from the Boiler is passed through a Prescrubber to remove acid halides and particulates, and to cool and humidify the gas. Chlorine dioxide is added to the gas, and it is then scrubbed with a limestone slurry, containing a catalyst, which is circulated through a plate tower Absorber. The clean gas is exhausted to stack via a Reheater. A side stream of the circulating slurry is centrifuged to remove gypsum, and the centrate is split into two streams; the larger passes to a Slurry Tank, where fresh limestone and catalyst is added for return to the Absorber, and the remainder

is sent to a By-product Treatment system, where it is either evaporated to dryness, to produce calcium chloride and nitrate, or is treated with ammonium carbonate or sulphate to produce a liquid fertiliser.

Chemistry of the Process [419]: Chlorine dioxide oxidises NO to NO_2 and nitric acid:

$2 NO + ClO_2 + H_2O = NO_2 + HNO_3 + HCl$

In the Absorber, the SO_2 is absorbed by limestone to form calcium sulphite and bisulphite, and NO_2, HNO_3 and HCl are absorbed and undergo reactions with calcium sulphite and bisulphite (reduction of NO_2 in the presence of a catalyst), and with calcium carbonate:

$2 NO_2 + CaSO_3 + CaCO_3 + 2 H_2O = Ca(NO_2)_2 + CaSO_4.2H_2O + CO_2$

Bisulphite: The following equation, reproduced from Reference [419], does not balance:

$Ca(NO_2)_2 + Ca(HSO_3)_2 + H_2O = N_2 + 2 CaSO_4.2 H_2O$

$CaCO_3 + 2 HNO_3 = Ca(NO_3)_2 + CO_2 + H_2O$

$CaCO_3 + 2 HCl = CaCl_2 + CO_2 + H_2O$

In the By-product Treatment system, the centrate, containing calcium nitrate and chloride, is either eva-

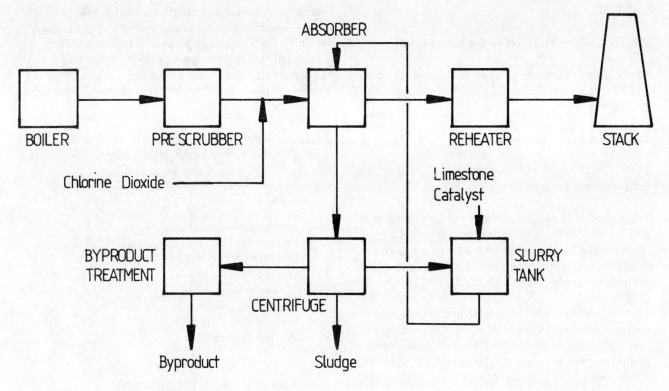

Figure 5.16 Block Diagram of Moretana Calcium Combined Abatement Process

porated to dryness or treated with ammonium carbonate to produce calcium carbonate, together with a liquid fertiliser solution of ammonium chloride and nitrate:

$$CaCl_2 + (NH_4)_2CO_3 = 2\ NH_4Cl + CaCO_3$$

$$Ca(NO_3)_2 + (NH_4)_2CO_3 = 2\ NH_4NO_3 + CaCO_3$$

5.3 General Appraisal of Processes

This Section presents available information on the characteristics of the combined SO_2-NO_x abatement process types: status, applicability, space and typical land area requirements, fresh water and water treatment requirements, reagent consumption, end-product or waste materials disposal requirements, typical power consumptions and reductions in combustion plant efficiency, operating experience and process developments; see Section 1.5.

The status of Category NS11 processes is indicated in Table 5.2, which presents information on the size of plants built, year of completion, and percentage abatement of SO_2 and NO_x.

Table 5.2 Status of Combined Abatement Processes–Dry Regenerable Catalyst

Code No.	Vendor	No. of Plants	Plant size Range MWe	Total MWe	Dates	% Abatement SO_2	NO_x	Ref.
NS11.1	Mitsui/BF/Uhde	4	30–1110*	1506*	1984–87	90–98	60	Q, 352
	Sumitomo	3	2–300*	312*	1984**	95	80–90	423, 34
NS11.2	Shell	1	–	–	–	–	–	34
		+1	0.6	0.6	1980	88	50–65	137

* Gas flow, thousand Nm³/h
** For 300 thousand Nm³/h plant
– Indicates information unavailable
Q Response from Mitsui Miike Engineering Corporation to Questionnaire

CATEGORY NS11

Process Code NS11.1 – Active Carbon Adsorption/SCR

Applicability: This process type is suitable for retrofit and new build applications to plant in the size range 30–700 MWe.

Typical land area requirements: ***

Typical fresh water and water treatment requirements: Estimates are presented in Table 5.3.

Consumption of reagents: Hydrogen consumption is 2.6–3.2 mol/mol sulphur captured; ammonia consumption is 0.5–1.4 mol per mol in-going NO_x [656]. Further information in the form of published estimates is presented in Table 5.3.

End-product or waste disposal requirements: End product is elemental sulphur (99.9% purity), sulphuric acid (98% purity) or liquid sulphur dioxide; if elemental sulphur is produced the quantity is equivalent to 90–93% of the sulphur captured. Estimates are presented in Table 5.3.

Table 5.3 Estimated Requirements

Reference		Q	352
MWe		60	460
Gas flow	'000 Nm^3/h	200	1200
Gas: Inlet SO_2 content	ppmv	373	1000
Inlet NO_x content	ppmv	200	240
Abatement: SO_2	%	90	98
NO_x	%	50	90
Water requirement per MWh:			
Process water		Nil	Nil
Cooling water	tonne	0.17	–
Consumption of reagents per MWh:			
Activated coke	kg	0.73	1.18
Ammonia	kg	0.42	0.67
Heavy oil	kg	–	0.65
Coke (for reduction)	kg	–	6.30
Nitrogen	Nm^3	0.22	–
Instrument air	Nm^3	0.50	–
End-product or waste disposal requirements per MWh:			
Activated coke fines	kg	–	0.49
Coal ash	kg	–	0.63
Sulphur	kg	1.3	3.41

– Indicates information unavailable
Q Response from Mitsui Miike Engineering Corporation to Questionnaire

Typical pressure losses: ***

Typical power requirements: ***

Typical reductions in combustion plant efficiency: ***

Operating experience: A demonstration plant (130,000 Nm^3/h from a coal-fired utility boiler) has operated for at least 16,000 hours, and a small commercial unit (30,000 Nm^3/h) was reported in 1986 to have operated trouble-free since October 1984 [352]. Estimated availability, based on experience of the 30,000 Nm^3/h plant, exceeds 95%, or equal to 100% if maintenance work on the abatement plant is carried out during scheduled boiler shut-down periods. It has been reported [656] that fouling by fly-ash in coal-fired systems is not a problem, but that there is some emission of ammonia.

Process developments: Development and use of lower-cost activated carbon is the most likely development to reduce the process operating cost.

Appraisal:

1.	Information available	1
2.	Process simplicity	2
3.	Operating experience	0
4.	Operating difficulty	1
5.	Loss of power	2*
6.	Reagent requirements	2
7.	Ease of end-product disposal	2
8.	Process applicability	2
	Total	12

*Assumed in absence of information

Process Code NS11.2 – Copper Oxide Absorption/SCR

Applicability: This process type is suitable for new build applications only.

Typical land area requirements: ***

Typical fresh water and water treatment requirements: ***

Consumption of reagents: Hydrogen consumption [282] 0.2 kg/kg sulphur captured. Ammonia consumption 0.52–0.57 kg/kg NO_x (as NO_2) abated (for 50.65% abatement) [137].

End-product or waste disposal requirements: End product is SO_2 which can be processed to produce sulphuric acid, elemental sulphur or liquid SO_2.

Typical pressure losses: Probably somewhat in excess of 20 mbar, the value (see Section 2.3) for a plant handling 125,000 Nm^3/h, 2500 ppmv SO_2, 90% sulphur capture.

Typical power requirements: ***

Typical reductions in combustion plant efficiency: ***

Operating experience: The process was applied to an industrial boiler in Japan for several years to remove 90% of the SO_2 and 40% of the NO_x, but the plant is no longer operated. The US Environmental Protection Agency operated a pilot plant for 90 days on a side-stream (1600 m^3/h equivalent to 0.6 MWe)

Table 5.4 Status of Combined Abatement Process–Dry Non-Regenerable Catalysts

Code No.	Vendor	No. of Plants	Plant size Range MWe	Total MWe	Dates	% Abatement SO$_2$	NO$_x$	Ref.
NS12.1	Parsons	1	0.2*	0.2*	–	–	–	419
NS12.2		–	–	–	–	–	–	352

* Thousand Nm^3/h
– Indicates information unavailable

from a 400 MWe coal-fired boiler at the Big Bend station, North Ruskin, Florida; sulphur capture was 90% and NO$_x$ abatement exceeded 70% [137, 352].

Process developments: The need for hydrogen makes this process more suitable for application at petroleum refineries than for utility or industrial boilers. Shell are no longer licensing the process.

Appraisal:

1. Information available   1
2. Process simplicity   1
3. Operating experience   1
4. Operating difficulty   0*
5. Loss of power   1*
6. Reagent requirements   2
7. Ease of end-product disposal 1
8. Process applicability   1
              Total   8

*Assumed in absence of information

The status of Category NS12 processes is indicated in Table 5.4, which presents information on the size of plants built, year of completion, and percentage abatement of SO$_2$ and NO$_x$.

CATEGORY NS12

Process Code NS12.1–Non-Selective Catalytic Reduction of SO$_2$ and NO$_x$ (Ralph M. Parsons Co. Process)

Applicability: This process type is suitable for new build applications only.

Typical land area requirements: ***

Typical fresh water and water treatment requirements: ***

Consumption of reagents: ***

End-product or waste disposal requirements: ***

Typical pressure losses: ***

Typical power requirements: ***

Typical reductions in combustion plant efficiency: ***

Operating experience: A 200 m^3/h pilot plant has been operated in FRG [352].

Process developments: ***

Appraisal:

1. Information   0
2. Process simplicity   2*
3. Operating experience   0
4. Operating difficulty   0*
5. Loss of power   2*
6. Reagent requirements   2*
7. Ease of end-product disposal 2*
8. Process applicability   1
              Total   9

*Assumed in absence of information

Process Code NS12.2–Catalytic Reduction of NO$_x$ and Catalytic Oxidation of SO$_2$

Applicability: This process is suitable only for new build applications.

Total land area requirements: ***

Typical fresh water and water treatment requirements: Nil.

Consumption of reagents: Natural gas is needed to remove oxygen from the gas and to reduce the nitric oxide. Air is required to restore oxidising conditions for catalytic oxidation of SO$_2$.

End-product or waste disposal requirements: The end-product is concentrated (93%) sulphuric acid, equivalent in quantity to the SO$_2$ removed; this can be sold.

Typical pressure losses: ***

Typical power requirements: ***

Typical reductions in combustion plant efficiency: ***

Operating experience [352]: A 200 Nm^3/h pilot plant was reported in 1986 to be in operation to test the process.

Process development [352]: It was reported in 1986 that installation of a demonstration plant was planned.

Appraisal:

1.	Information available	0
2.	Process simplicity	1
3.	Operating experience	0
4.	Operating difficulty	0*
5.	Loss of power	2*
6.	Reagent requirements	1
7.	Ease of end-product disposal	2
8.	Process applicability	1
	Total	7

*Assumed in absence of information

CATEGORY NS21

Combustion techniques in Category NS21 processes are still at the development stage. The US EPA have recently initiated a full-scale demonstration of the LIMB technique, which is expected to last for 4½ years [351], whilst the pilot-scale development of a slagging burner was reported by Rockwell International in 1983 [427].

Process Code NS21.1–Limestone Injection into Multi-stage Burner (LIMB)

Applicability: This technique is suitable for retrofit and new build applications, although highest levels of NO_x and SO_x removal are to be expected from new boilers [431]. At present, the effects of the sorbent on the performance of an existing particulate collection system have not been determined, so upgrading of this equipment may be required in retrofit situations.

Space requirements: No information is available, but this technique requires sufficient space for limestone storage, preparation and injection, and for the disposal of the solid waste products and unreacted limestone.

Typical land area requirements: ***

Consumption of reagents: Limestone is normally used for SO_x absorption although alternative sorbents have been used in trials, e.g. calcite, dolomite or argonite [431]. Fine grinding of the limestone is necessary to improve the reactivity (surface area) of the sorbent and increase its utilisation. The Peabody Coal Co. found that limestone utilisation improved from 6.3 to 5.9 when the maximum particle size of the sorbent was reduced from 149 μm to 74 μm [431]. Calcium/sulphur molar ratios of 2.0 are typically required for SO_x reductions of 50%.

End-product or waste disposal requirements: ***

Typical pressure losses: ***

Typical power requirements: ***

Typical reductions in combustion plant efficiency: ***

Operating experience: The forerunner of LIMB was the injection of dry limestone into the furnace above the flame zone. In early tests carried out for the EPA, low levels of SO_2 control were achieved. These levels were especially low when the limestone was added to the coal: e.g. less than 8% SO_2 reduction was reported by Babcock and Wilcox with this arrangement [431]. It was subsequently found that at high peak temperatures, e.g. in excess of 1250°C, dead-burning of the limestone occurred leading to loss of reactivity.

With the development of low-NO_x coal burner designs, and their inherently lower operating temperatures, interest in limestone injection was revived. The EPA in the USA and several European and US companies renewed their efforts to achieve acceptable SO_2 control levels by sorbent injection. For example, EPA-sponsored trials by Energy and Environmental Research Corporation in 1978 achieved 73% SO_2 removal with limestone injection (Ca:S ratio = 2) through an advanced, but experimental, 1.8 MWt low-NO_x burner. Two coals of 0.7% and 3.8% sulphur content were used in these trials and the limestone was mixed with the coal. With the lower sulphur coal only, SO_2 control rates of 50% and 88% were achieved for Ca:S ratios of 1.0 and 3.0, respectively [431]. These tests were aimed at minimising NO_x emission without optimising SO_2 removal.

Other companies which are known to be developing limestone injection techniques in conjunction with low-NO_x burner designs include the International Flame Research Foundation in the Netherlands [297], L. and C. Steinmuller in FRG [431], Foster Wheeler Energy Corporation (FWEC) [421, 368], Riley Stoker, TRW Inc. and Rockwell International [296].

Process development: The limited pilot-scale testing in the USA indicated that direct limestone injection into the furnace could not compete with other FGD processes because of its limited sulphur capture capabilities. There has, however, been an incentive to develop the LIMB technology as a combined SO_x-NO_x system because of its apparently high cost effectiveness where levels of control of the order of 70% for NO_x and 50% for SO_x are adequate [431].

The EPA have announced that their pilot unit testing of LIMB is essentially complete and a full-scale demonstration initiated in association with Ohio Edison, at the latter's 150 MWe Edgewater Station [351]. The programme is expected to last 4.5 years and has the prime objectives of demonstration NO_x and SO_x reductions of 50–60% and determining the impact of the technique on the operation of the boiler.

Appraisal:

1.	Information	0
2.	Process simplicity	2
3.	Operating experience	0
4.	Operating difficulty	0
5.	Loss of power	2
6.	Reagent requirements	1
7.	Ease of end-product disposal	1
8.	Process applicability	2
	Total	8

CATEGORY NS22

The status of Category NS22 processes is indicated in Table 5.5, which presents information on the size of plants built, year of completion, and percentage abatement of SO_2 and NO_x.

Process Code NS22.1 – Electron Beam Radiation Process

Applicability: This process type is suitable for retrofit and new build applications.

Typical land area requirements: ***

Typical fresh water and water treatment requirements: ***

Consumption of reagents: Ammonia: 0.53 kg/kg SO_2 removed; 0.74 kg/kg NO_x (as NO_2) abated.

End-product or waste disposal requirements: About 3% of equivalent electrical power generation of combustion system [34, 419].

Typical reductions in combustion plant efficiency: ***

Operating experience: Appears to be confined to: a 6000 Nm^3/h pilot plant at the Paducah, Kentucky, station of TVA [678]; and to a 10,000 m^3/h pilot plant in Japan, where over 95% sulphur capture and 80% NO_x abatement were obtained with gas containing 570 mg/m^3 SO_2 and 360 mg/m^3 NO_x [352].

Process developments: The Pittsburgh Energy Technology Center of the US Department of Energy was reported [145, 685] to be co-sponsoring a development programme with Indianapolis Power & Light.

Appraisal:

1.	Information available	0
2.	Process simplicity	2
3.	Operating experience	0
4.	Operating difficulty	0*
5.	Loss of power	1
6.	Reagent requirements	2
7.	Ease of end-product disposal	2
8.	Process applicability	2
	Total	9

*Assumed in absence of information

CATEGORY NS23

The status of Category NS23 processes is indicated in Table 5.6, which presents information on the size of plants built, year of completion, and percentage abatement of SO_2 and NO_x.

Process Code NS23.1 – Sodium Carbonate Adsorption Process (NOXSO Process)

Applicability: This process type is suitable for retrofit and new build applications.

Space requirements: The overall space requirements for a 125 MWe module are estimated [434] to be approximately: area 7.4 m^2; height 24.5 m.

Typical land area requirements: ***

Table 5.5 Status of Combined Abatement Processes – Radiation Processes

Code No.	Vendor	No. of Plants	Plant size Range MWe	Total MWe	Dates	% Abatement SO_2	% Abatement NO_x	Ref.
NS22.1	Ebara/JAERI	1	100	100	–	over 90	80	145, 34
		+2	6*–10*	16*	–	–	–	352, 678

* Thousand Nm^3/h
– Indicates information unavailable

Table 5.6 Status of Combined Abatement Processes – Adsorption Processes

Code No.	Vendor	No. of Plants	Plant size Range*	Total *	Date	% Abatement SO_2	% Abatement NO_x	Ref.
NS23.1	NOXSO		60	60	1982	90	90	434, 673

*Gas flow, Nm^3/h

Typical fresh water and water treatment requirements: ***

Consumption of reagents: A bed inventory loss of 0.02% per hour has been reported [673].

End-product or waste disposal requirements: SO_2 and H_2S streams could be treated to produce elemental sulphur.

Typical pressure losses: ***

Typical power requirements: ***

Typical reductions in combustion plant efficiency: ***

Operating experience: Not yet tested beyond the bench scale, in which over 50 sorbent regenerations have been achieved [673].

Process developments: The Pittsburgh Energy Technology Center of the US Department of Energy conducted development work on the process [685]. Testing on a large scale is needed.

Appraisal:

1.	Information available	0
2.	Process simplicity	0
3.	Operating experience	0
4.	Operating difficulty	0*
5.	Loss of power	1
6.	Reagent requirements	1
7.	Ease of end-product disposal	2
8.	Process applicability	2
	Total	6

*Assumed in absence of information

CATEGORY NS31

The status of Category NS31 processes is indicated in Table 5.7, which presents information on the size of plants built, year of completion, and percentage abatement of SO_2 and NO_x.

Process Code NS31.1 – Lime Spray Dryer Process

Applicability: This process type is suitable for retrofit and new build applications.

Typical land area requirements: ***

Typical fresh water and water treatment requirements: Water to prepare a slurry containing 20% suspended lime absorbent (see below). No waste water from the process.

Consumption of reagents [432]: Lime at a molar ratio of $Ca/(SO_2 + NO_x)$ of about 1.5. Sodium hydroxide at a Na/Ca molar ratio of about 0.56.

End-product or waste disposal requirements [352]: Solid product for disposal containing calcium and sodium sulphite, sulphate, nitrate, carbonate and hydroxide, with captured particulates.

Typical pressure losses: ***

Typical power requirements: ***

Typical reductions in combustion plant efficiency: ***

Operating experience: This process has been operated in short-term tests at the two plants indicated in Table 5.7, and in the laboratory. The possibility of emission of NO_2, formed by oxidation of NO, has been reported [685].

Process developments: The Pittsburg Energy Technology Center of the US Department of Energy has conducted development work on the process [685]. Further full-scale testing is needed.

Appraisal:

1.	Information available	1
2.	Process simplicity	2
3.	Operating experience	0
4.	Operating difficulty	0
5.	Loss of power	2*
6.	Reagent requirements	1
7.	Ease of end-product disposal	1
8.	Process applicability	2
	Total	9

*Assumed in absence of information

CATEGORY NS32

The status of Category NS32 processes is indicated in Table 5.8, which presents information on the size of plants built, year of completion, and percentage abatement of SO_2 and NO_x.

Table 5.7 Status of Combined Abatement Processes – Liquid Phase NO_x Oxidation

Code No.	Vendor	No. of Plants	Plant size Range MWe	Total MWe	Dates	% Abatement SO_2	NO_x	Ref.
NS31.1	Niro	2	20–96*	116*	1980–81	90–95	20–60	213, 352

*MW thermal

Table 5.8 Status of Combined Abatement Processes – Liquid-Phase NO_x Reduction

Code No.	Vendor	No. of Plants	Plant size Range MWe	Total MWe	Dates	% Abatement SO_2	% Abatement NO_x	Ref.
NS32.1	Asahi	1	0.05*	0.05*	–	99	80–85	419
NS32.2	SHL	1	5*	5*	–	–	–	352
NS32.3	PENSYS	3	1–1.5	3.5	1977–82	Over 95	Over 90	145, 672

* Thousand Nm^3/h
– Indicates information unavailable

Process Code NS32.1 – Sodium Sulphite Scrubbing Process with Chelating Compound (Asahi Chemical Process)

Applicability: The process type is suitable for retrofit and new build applications.

Typical land area requirements: ***

Typical fresh water and water treatment requirements: Published estimates are presented in Table 5.9.

Consumption of reagents: Published estimates are presented in Table 5.9.

End-product or waste disposal requirements: Published estimates are presented in Table 5.9.

Typical power requirements: Published estimates are presented in Table 5.9.

Table 5.9 Published Estimated Requirements

Reference		419
MWe		500
Inlet gas: SO_2 content		
NO_x content		
Cooling water per MWh	tonne	0.9
Steam (11 bar max.)	tonne	0.1
Reagents per MWh:		
Limestone	kg	26
$FeSO_4.7\,H_2O$	kg	0.1
EDTA	kg	0.1
Na_2CO_3	kg	3.4
or NaOH	kg	1.3
Residual oil (cracking)	litre	11
End-product and waste disposal requirement per MWh:		
Gypsum	kg	45
Na_2SO_4	kg	5.3
Waste water for treatment		Small quantity
Electric power:	kWh	54

Typical pressure losses: ***

Typical reductions in combustion plant efficiency: ***

Operating experience [419]: This process has been operated as an integrated system only in a small unit treating 40–60 Nm^3/h of gas from an oil-fired boiler, but the absorption, crystallisation and decomposition parts of the system have been confirmed, and scale-up data obtained, on a bench-scale unit treating 500–600 Nm^3/h of gas. The system has not been tested on flue gas from a coal-fired boiler.

Process development: Development to full scale is needed.

Appraisal:

1.	Information available	1
2.	Process simplicity	0
3.	Operating experience	0
4.	Operating difficulty	0*
5.	Loss of power	1
6.	Reagent requirements	1
7.	Ease of end-product disposal	1
8.	Process applicability	2
	Total	6

*Assumed in absence of information

Process Code NS32.2 – Limestone Slurry Scrubbing with Chelating Compound

Applicability: This process type is suitable for retrofit and new build applications.

Typical land area requirements: ***

Typical fresh water and water treatment requirements: ***

Consumption of reagents: ***

End-product or waste disposal requirements: ***

Typical pressure losses: ***

Typical power requirements: ***

Typical reductions in combustion plant efficiency: ***

Operating experience [352]: The process was reported in 1986 to be under test in a laboratory and on a 5000 m^3/h pilot plant.

Process development: The Pittsburgh Energy Technology Center of the US Department of Energy has conducted development work on the process [685]. Development to full scale needed.

Appraisal:

1. Information available 0
2. Process simplicity 1
3. Operating experience 0
4. Operating difficulty 0*
5. Loss of power 1*
6. Reagent requirements 1
7. Ease of end-product disposal 1
8. Process applicability 2
   Total 6

*Assumed in absence of information

Process Code NS32.3–Sulf-X Process

Applicability: This process type is suitable for retrofit and new build applications. See Section 2.3 for estimates of requirements for application of the process to sulphur capture alone. There appear to be no published estimates of requirements for combined abatement of SO_2 and NO_x.

Typical land area requirements: ***

Typical fresh water and water treatment requirements: ***

Consumption of reagents: ***

End-product or waste disposal requirements: ***

Typical pressure losses: ***

Typical power requirements: ***

Typical reductions in combustion plant efficiency: ***

Operating experience [145, 277]: Experience limited to 1.5 MWt scale, with some stages of the regeneration system omitted.

Process development: Integrated operation, and development to full scale, needed.

Appraisal:

1. Information available 0
2. Process simplicity 0
3. Operating experience 0
4. Operating difficulty 0*
5. Loss of power 1*
6. Reagent requirements 2
7. Ease of end-product disposal 2
8. Process applicability 2
   Total 7

*Assumed in absence of information

CATEGORY NS41

The status of Category NS41 processes is indicated in Table 5.10, which presents information on the size of plants built, year of completion, and percentage abatement of SO_2 and NO_x.

Process Code NS41.1–Oxidation plus Ammonia Scrubbing (Walther Process)

Applicability: This process type is suitable for retrofit and new build applications.

Typical land area requirements: ***

Typical fresh water and water treatment requirements: Published estimates are presented in Table 5.11.

Consumption of reagents: Published estimates are presented in Table 5.11.

End-product or waste disposal requirements: Published estimates are presented in Table 5.11.

Typical power requirements: Published estimates are presented in Table 5.11.

Typical pressure losses: ***

Typical reductions in combustion plant efficiency: ***

Operating experience: Pilot plant operation only.

Table 5.10 Status of Combined Abatement Processes–Liquid-Phase NO_x Oxidation

Code No.	Vendor	No. of Plants	Plant size Range MWe	Total MWe	Dates	% Abatement SO_2	NO_x	Ref.
NS41.1	Walther	–	–	–	–	91	88	**
NS41.2	Kawasaki	2	0.1–5*	5.1*	1974	95	91	419

* Thousand Nm³/h
** Vendor's promotional literature
– Indicates information unavailable

Table 5.11 Published Estimated Requirements

Reference		Walther*
Furnace type		Slag tap
MWe		400
Gas flow	'000 Nm³/h	1300
Inlet gas: SO_2 content	mg/Nm³	2200
NO_x content	mg/Nm³	1640
Abatement: SO_2	%	91
NO_x	%	88
Water and water treatment requirements per MWh:		
Cooling water	tonne	9.7
Water treatment		Nil
Consumption of reagents per MWh:		
Ammonia	kg	5.1
Ozone (see power consumption)	kg	4.0
End-product or waste disposal per MWh:		
$(NH_4)_2SO_4$	kg	13.4
NH_4NO_3	kg	8.2
Ammonium halides, particulates		Site-specific
Power requirements per MWh:		
Ozone production	kWh	42
Other	kWh	20

*Walther promotional literature

Process developments: Needs to be developed to full scale operation.

Appraisal:

1.	Information available	1
2.	Process simplicity	1
3.	Operating experience	0
4.	Operating difficulty	0*
5.	Loss of power	1
6.	Reagent requirements	1
7.	Ease of end-product disposal	2
8.	Process applicability	2
	Total	8

*Assumed in absence of information

Process Code NS41.2 – Kawasaki Process

Applicability: This process type is suitable for retrofit and new build applications.

Typical land area requirements: ***

Typical fresh water and water treatment requirements: ***

Consumption of reagents: ***

End-product or waste disposal requirements: ***

Typical pressure losses: ***

Typical power requirements: ***

Typical reductions in combustion plant efficiency: ***

Operating experience: Limited to operation on a pilot plant handling a flow of 5000 Nm³/h from a coal-fired boiler.

Process developments: Development to full scale is needed.

Appraisal:

1.	Information available	0
2.	Process simplicity	0
3.	Operating experience	0
4.	Operating difficulty	0*
5.	Loss of power	1*
6.	Reagent requirements	1
7.	Ease of end-product disposal	1
8.	Process applicability	2
	Total	5

*Assumed in absence of information

CATEGORY NS42

The status of Category NS42 processes is indicated in Table 5.12, which presents information on the size of plants built, year of completion, and percentage abatement of SO_2 and NO_x.

Process Code NS42.1 – IHI Process

Applicability: This process type is suitable for retrofit and new build applications.

Typical land area requirements: ***

Fresh water and water treatment requirements: Published estimates are presented in Table 5.13.

Consumption of reagents: Published estimates are presented in Table 5.13.

Table 5.12 Status of Combined Abatement Processes – Liquid-Phase NO_x Reduction

Code No.	Vendor	No. of Plants	Plant size Range MWe	Total MWe	Dates	% Abatement SO_2	NO_x	Ref.
NS42.1	IHI	2	5–27*	32*	1975	90	80	419
NS42.2	Fuji Kasui-Sumitomo	1	25*	25*	1976	95	90	419

*Thousand Nm³/h

End-product or waste disposal requirements: ***

Typical pressure losses: ***

Typical power requirements: Published estimates are presented in Table 5.13.

Typical reductions in combustion plant efficiency: ***

Table 5.13 Published Estimated Requirements

Reference		419
Gas flow	'000 Nm³/h	100
Approximate equivalent MWe		40
Boiler fuel		Oil
Inlet gas: SO₂ content	ppmv	1500
NO₂ content	ppmv	180
Particulates content	mg/Nm³	100
Abatement: SO₂ emission	%	90
NOₓ emission	%	80
Water	tonne/MWh	1.4
Steam	tonne/MWh	0.11
Consumption of reagents, kg/MWh:		
Limestone		79
Sulphuric acid		4.5
Slaked lime		4.0
Additives		1.7
Fuel oil		53
Power requirements:	kWh	210

Operating experience: The process was tested on flue gas from an oil-fired boiler in a 5000 Nm³/h pilot plant in a 3000 hour test run. A 27,000 Nm³/h prototype unit was reported in 1979 [419] to be under test.

Process development: Development to full scale needed.

Appraisal:

1. Information available   0
2. Process simplicity   0
3. Operating experience   0
4. Operating difficulty   0*
5. Loss of power   0
6. Reagent requirements   1
7. Ease of end-product disposal 1
8. Process applicability   2
   Total   4

*Assumed in absence of information

Process Code NS42.2–Moretana Calcium Process

Applicability: This process type is suitable for retrofit and new build applications.

Space requirements: ***

Typical land area requirements: ***

Typical fresh water and water treatment requirements: ***

Consumption of reagents: ***

End-product or waste disposal requirements: ***

Typical pressure losses: ***

Typical power requirements: ***

Typical reductions in combustion plant efficiency: ***

Operating experience: Process has been tested on a pilot plant scale, treating 25,000 Nm³/h gas from a sintering furnace, and on the bench scale with a synthetic coal-fired flue gas.

Process developments: Development to full scale needed.

Appraisal:

1. Information available   0
2. Process simplicity   0
3. Operating experience   0
4. Operating difficulty   0*
5. Loss of power   0
6. Reagent requirements   1*
7. Ease of end-product disposal 1
8. Process applicability   2
   Total   4

*Assumed in absence of information

5.4 Process for Detailed Study

The selection of processes for detailed study in this Manual has been based upon their suitability for application in the UK for the two datum combustion systems (Section 1.3) considered:

– Large (450 tonne steam/h) industrial boiler (Datum system 1)

– Small (25 tonne steam/h) factory boiler (Datum system 2)

In principle, all of the processes listed in Section 5.1 can be applied to all combustion plant, but the attraction of many processes diminishes with factors such as decrease in plant operating scale, and increases in Combined Abatement process complexity, reagent costs, and end-product disposal difficulty.

Appraisal of processes

To evaluate some of these factors, a rough appraisal of each process type has been made in Section 5.3 by assigning 'merit points' for a number of features; merit points have been awarded according to the scale:

0   Below average merit

1   Average merit

2   Above average merit

The features to which these points have been assigned are described in Section 1.5: they are briefly:

1. Information available

2. Process simplicity

3. Operating experience–extent and difficulties encountered

4. Operating difficulty–availability, reliability

5. Loss of power sent out–by installation of the FGD process

6. Reagent requirements–quantities

7. Ease of end-product disposal

8. Process applicability–e.g. for retrofit

All of the processes listed in Section 5.1 and outlined in Section 5.2 are appraised in Section 5.3. The merit points assigned to the process types for each of the above features are summarised in Table 5.14. It should be noted that the number of points in the merit point system adopted in other Sections of the Manual are not strictly comparable with those considered here.

Processes suitable for the UK

The principal purpose of assigning merit points to each of the processes was to aid in the selection of processes that could be considered suitable for application in the UK. It was arbitrarily assumed that suitable combined abatement processes would be those having more than 10 merit points.

Selection of processes for detailed study

All of the combined abatement process types are shown in Table 5.15 with an indication of which (if any) of the Datum Combustion systems are considered, from the above criteria, to be suitable for application of the process in the UK

Table 5.14 Summary of Combined Abatement Process Appraisals

Code No. NS	Name	Merit Points for Feature No: 1 2 3 4 5 6 7 8	Total Points
11.1	Active carbon/SCR	1 2 0 1 2 2 2 2	12
11.2	Copper oxide	1 1 1 0 1 2 1 1	8
12.1	Ralph M. Parsons	0 2 0 0 2 2 2 1	9
12.2	Cat Redn./oxidation	0 1 0 0 2 1 2 1	7
21.1	LIMB	0 2 0 0 2 1 1 2	8
22.1	Electron beam	0 2 0 0 1 2 2 2	9
23.1	NOXSO	0 0 0 0 1 1 2 2	6
31.1	Spray dryer	1 2 0 0 2 1 1 2	9
32.1	Asahi	1 0 0 0 1 1 1 2	6
32.2	Limestone/Ferrous EDTA	0 1 0 0 1 1 1 2	6
32.3	Sulf-X	0 0 0 0 1 2 2 2	7
41.1	Walther	1 1 0 0 1 1 2 2	8
41.2	Kawasaki	0 0 0 0 1 1 1 2	5
42.1	IHI	0 0 0 0 0 1 1 2	4
42.2	Moretana	0 0 0 0 0 1 1 2	4

Features:
1. Information available
2. Process simplicity
3. Operating experience
4. Operating difficulty
5. Loss of power
6. Reagent requirements
7. Ease of end-product disposal
8. Process applicability

Table 5.15 Applications of Combined Abatement Processes Considered in Detail

Code No.	Name	Application to Datum System 1	2	Section
NS11.1	Active carbon/SCR	Yes*	–	5.5
NS11.2	Copper oxide	–	–	–
NS12.1	Ralph M. Parsons	–	–	–
NS12.2	Cat. Redn./oxidation	–	–	–
NS21.1	LIMB	–	–	–
NS22.1	Electron beam	–	–	–
NS23.1	NOXSO	–	–	–
NS31.1	Spray dryer	Yes**	Yes**	5.5
NS32.1	Asahi	–	–	–
NS32.2	Limestone/Ferrous EDTA	–	–	–
NS32.3	Sulf-X	–	–	–
NS41.1	Walther	–	–	–
NS41.2	Kawasaki	–	–	–
NS42.1	IHI	–	–	–
NS42.2	Moretana	–	–	–

* In general, only if a centralised reagent reprocessing plant were available

** Tentative evaluation: promise for future application

None of the basic process types is considered to be suitable for immediate application to the smallest operating scale dealt with in this Manual (Datum System 2), as the only contender (Process Code NS11.1–Active Carbon/SCR), although rated as being a simple process, would be too complex for a small-scale boiler.

The otherwise attractive Spray Dryer Process (Process Code NS31.1–Merit Rating 9 points) fails to achieve the criterion of 10 Merit Points because of lack of information available and of operating experience. However, this process is considered to have promise for future application to all operating scales because of its simplicity, and is therefore tentatively evaluated in Section 5.5.

For large industrial boilers, the simplest process for consideration in the UK is: Active carbon/SCR

163

(Process Code NS11.1–merit rating 12 points), evaluated in Section 5.5. This process, though of above average simplicity, would be attractive only where reagent and by-product processing facilities were available.

Owing to the small number of plants built, and the very limited operating experience of combined abatement processes, and in particular of the two processes selected for evaluation, there are insufficient data available to enable reliable estimates of capital and operating costs to be made. The Active Carbon/SCR process (Code NS11.1) is an elaboration of the similar flue gas desulphurisation (FGD) process (S51.1; see Section 2) for which also there were inadequate data for cost estimates. The Lime Spray Dryer Absorber process (Code NS31.1) is an elaboration of the similar FGD process (Code S22.1) for which cost estimates have been made for application to Datum System 1 only. It can be assumed that costs for the combined abatement version of this process would be higher than those for the FGD version.

5.5 Evaluations of Selected Combined Abatement Processes

Evaluation of BF-Mitsui Active Carbon Combined Abatement Process (Process Code NS11.1)

See Section 5.2 for: outline of the basic process; its chemistry; block diagram.

See Section 5.3 for general appraisal of the basic process; name of manufacturers offering this type of equipment.

See Section 5.4 for the reason for choosing the BF-Mitsui (Code NS11.1) basic process.

This basic process type is considered (Section 5.4) to be suitable for application in the UK only to large boilers having access to reagent reprocessing facilities, and hence it is evaluated here only for Datum Combustion System 1. The process, developed by Bergbau-Forschung GmbH (Essen, FRG) and Mitsui Mining Co. Ltd. (Tokyo, Japan), is offered by Uhde GmbH (Dortmund, FRG). See Appendix 2 for details of this manufacturer.

Process Description

Figure 5.17 shows a simplified flow diagram for application of the process to an oil fired boiler. It is assumed that the sulphur dioxide produced in the process is treated to recover elemental sulphur.

Dust and sulphur oxides removal: Gas at a temperature of 120–150°C from the boiler plant, boosted by a fan, is cooled to 120°C if necessary by injection of water. The cooled gas enters the 1st Stage Adsorber containing activated carbon granules (prepared from coal in the form of extruded granules about 5 mm in length) from the 2nd Stage Adsorber. The carbon, which is contained in louvred channels, with the gas flowing across the bed, adsorbs sulphur dioxide, oxygen and water vapour to form adsorbed

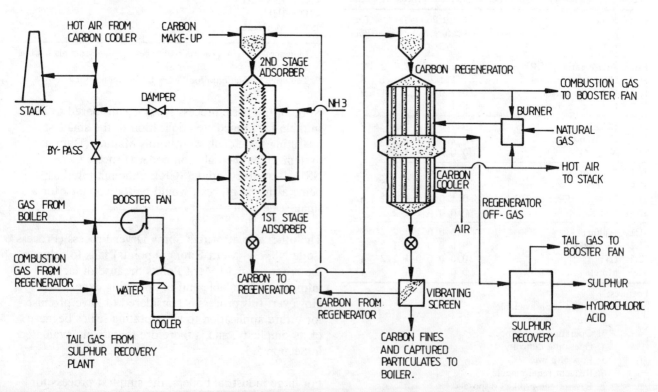

Figure 5.17 BF-Mitsui Active Carbon Adsorption Combined Abatement Process

sulphuric acid. Nitrogen dioxide (forming 5–10% of the total nitrogen oxides content of the gas) is also adsorbed, and the carbon bed filters out much of the particulates content of the gas. Ammonia is then added to the gas before it enters the 2nd Stage Adsorber, which is fed with fresh and regenerated carbon; the carbon, which acts as a Selective Catalytic Reduction catalyst, is again contained within louvred channels with the gas flowing across, allowing the ammonia and nitric oxide to react forming water vapour and elemental nitrogen. Ammonia also reacts with residual sulphur dioxide to form ammonium sulphate and bisulphate which are adsorbed on the surface of the carbon. Carbon passes from the 2nd to the 1st Stage Adsorber. The gas temperature rises by 15–20°C across the Adsorber/SRC system; the cleaned gas passes via a Damper to Stack.

Adsorbent regeneration: The active carbon is removed continuously from the base of the 1st Stage Adsorber and conveyed to the Regeneration section, which can be either on- or off-site. The carbon flows slowly through vertical tubes, which are heated indirectly by hot combustion gas at 600°C derived from combustion of natural gas. The carbon temperature is raised to 400–450°C. The hot combustion gas leaves the heating section at 300°C. The adsorbed gases, sulphuric acid and nitrogen dioxide, react with the carbon, releasing sulphur dioxide, elemental nitrogen and carbon dioxide, and consuming some of the carbon. The ammonium sulphate and bisulphate decompose, giving elemental nitrogen, sulphur dioxide and water vapour. The heated carbon resides in a vessel to complete the regeneration process, and it then passes through vertical tubes in the cooling section, where it is indirectly cooled to 100°C by air; part of the air leaving the cooling section at 250°C supplies combustion air for the natural gas combustion, and the remainder is exhausted to stack. Carbon fines and particulates trapped from the gas are removed by a Vibrating Screen and burned in the Boiler. The oversize material is recycled to the 2nd Stage Adsorber; carbon losses are made up by adding fresh carbon.

Sulphur recovery: The off-gas from the Adsorber, containing SO_2, CO_2 and water vapour then passes via a reduction reactor to a Claus unit to recover sulphur, which is condensed out; tail gas from the Claus unit is returned to the Adsorber.

Status and Operating Experience

The status of the BS-Mitsui process is indicated in Section 5.3, where it is seen that four installations have been (or are being) erected, in the size range 30,000–1,110,000 Nm³/h, totalling 1,505,000 Nm³/h. Three units of a similar process (Sumitomo) have been erected, totalling 312,000 Nm³/h.

It is seen in Section 5.3 that the process has been operated on one pilot plant for 16,000 hours, and on a small commercial unit for a similar period, with very high reliability, especially if maintenance work is carried out during scheduled boiler shut-down periods.

Variations and Development Potential

Design and operating variations can include:

- Adaptation of the process to either adsorption of SO_2 or removal of NO_x alone.

- Direct heating of carbon in the Regenerator with hot sand.

- Alternatives for sulphur recovery: as elemental sulphur by a variety of processes (Alliance process, Claus process, Foster Wheeler 'Resox' process); as sulphuric acid (by conventional catalytic oxidation of SO_2); or as liquid SO_2.

Although complex, the process can achieve high sulphur capture (up to 99%) and high NO_x abatement (up to 90%) without excessive reagent make-up requirements or production of waste products. The process also removes acid halides and particulates.

The sulphur produced can be marketed as sulphuric acid, liquid SO_2 or as elemental sulphur; if it is produced as elemental sulphur, it can be disposed of safely. High operating flexibility is obtainable by installing large surge capacity for regenerated and unregenerated active carbon.

Development and use of lower-cost active carbon is the most likely development.

Process Requirements for Each Application Considered

These are shown in Table 5.16 for the application considered: Datum Combustion System 1.

It is assumed that the pollutant contents of the gas is to be reduced to the following levels:

SO_2: 250 mg/Nm³ (dry), equivalent to 92.0% capture of SO_2
NO_x: 200 mg/Nm³ (dry), equivalent to 51.4% abatement of NO_x
HCl: unaffected
Particulates: 15 mg/Nm³ (dry), equivalent to 61% capture of particulates

It is further assumed that:

Table 5.16 Process Requirements – Datum Combustion System 1

Inlet Gas at full load		
Volume flow	'000 Nm³/h	456
Dry gas	'000 Nm³/h	408
Water Vapour	'000 Nm³/h	48
Actual volume flow	'000 m³/h	708
Temperature	°C	150
Particulates content	mg/Nm³ (dry)	39
SO_2 content	mg/Nm³ (dry)	5010
NO_x content	mg/Nm³ (dry)	1025
HCl content	mg/Nm³ (dry)	3
Exit gas at full load		
Volume flow	'000 Nm³/h	464
Dry gas	'000 Nm³/h	407
Water vapour	'000 Nm³/h	57
Actual volume flow	'000 m³/h	693
Temperature	°C	135
Particulates content	mg/Nm³ (dry)	15 (17)
SO_2 content	mg/Nm³ (dry)	400 (861)
NO_x content	mg/Nm³ (dry)	500 (553)
HCl content	mg/Nm³ (dry)	3 (3)
Reaction temperature	°C	135 (135)
Particulates removal		Simult.
Reagent		Active Carbon
Requirements at full load		
Water	tonne/h	7.3
Active carbon	tonne/h	0.21
Ammonia	tonne/h	−2.28
Natural gas	GJt/h	10.1
Air	tonne/h	24
Electric power	MWe	0.7
Manpower	men/shift	***
Average load factor	%	***

Figures in parentheses are annual average emissions for 90% combined abatement plant availability

- The 1st Stage Adsorber captures 80%, and the 2nd Stage Adsorber 20%, of the total sulphur captured; the sulphur is captured in the 1st Stage as sulphuric acid, and in the 2nd Stage as ammonium sulphate.

- The sulphur is recovered as elemental sulphur in a Claus unit.

- The NO_x content is reduced by 8% by adsorption of NO_2 in the 1st Stage Adsorber; the remaining NO_x abatement occurs in the 2nd Stage Adsorber by SCR with ammonia.

- The mechanical breakdown of carbon leads to a carbon make-up rate 50% higher than the theoretical.

- The requirements for Regenerator off-gas treatment and sulphur recovery are not included.

By-products and Effluents

The by-product from the system is elemental sulphur, and hydrochloric acid, and there is a carbon waste in addition to the captured particulates. The rates of production of the material produced are summarised in Table 5.17 for the Datum Combustion System considered.

Table 5.17 Estimated Rates of Output Solids Production – Oil-Fired Combustion System 1

Rate of production:		
Sulphur (99%)	tonne/h	0.95
Active carbon fines	tonne/h	0.07
Particulates	tonne/h	0.01

*Total from four 500 MWe boiler plants

Table 5.18 Efficiency and Emission Factors – Oil-Fired Combustion System 1

Oil heat input (gross)	MWt	464
Oil fired	tonne/h	39.4
Abatement plant power	MWe	0.7
Equivalent oil input	tonne/h	0.2*
Natural gas for Regenerator	GJt/h	10.1
Equivalent oil input	tonne/h	0.2
Useful energy from system	GJt/h	1468
Total equivalent oil input	tonne/h	39.8 (a)
Efficiency factor	GJt/tonne	36.9 (a)
Emissions		
Sulphur in SO_2	kg/h	82 (176)
Nitrogen in NO_x	kg/h	62 (69)
Chlorine in HCl	kg/h	1.2 (1.2)
Particulates	kg/h	6.1 (7.1)
Emission factors (per tonne oil)		(a)
Sulphur	kg/tonne	2.06 (4.42)
Nitrogen	kg/tonne	1.56 (1.73)
Chlorine	kg/tonne	0.03 (0.03)
Particulates	kg/tonne	0.15 (0.18*)

*Calculated assuming overall power generation efficiency = 0.33
(a) Based on oil fired to boiler plus oil equivalent to electric power and natural gas consumed
Figures in parentheses are annual average emissions for 90% combined abatement plant availability

Efficiency and Emission Factors

The efficiency and emission factors for the process are summarised in Table 5.18 for the application considered: Datum Combustion System 1. In calculating the efficiency factor, it is assumed that at full load, the combined abatement unit consumes:

- 10.1 GJt/h of natural gas for the Regenerator, equivalent to an increase of 0.24 tonne/h of oil;

- and 0.7 MWe of electrical power, equivalent to the combustion of a further 0.2 tonne/h of oil at a power station, assuming an overall power generation efficiency of 33%. This additional oil is arbitrarily assumed to be included with the coal burned in the boiler for calculating the efficiency factor.

For illustrating purposes, the annual average emissions and emission factors for combined abatement plant availabilities of 100% and 90% are given in Tables 5.16 and 5.18. Further details are given in Section 1.6.

Effect of load variations: ***

Effect of design variations: ***

Limitations: ***

Costs

Capital Costs: ***

Annual running costs and cost factors: ***

Effects of design variations on costs: ***

Effects of annual load patterns on annual running costs: ***

Process Advantages and Drawbacks

The advantages are that the process can be retrofitted; process readily adaptable to either NO_x or SO_2 abatement alone; process also removes particulates and other air pollutants; no gas reheat needed; high sulphur capture efficiency (over 90% capture) and NO_x abatement efficiency (50–90%) can be attained; operating flexibility; no requirement for large quantities of reagent; no large quantities of waste products; by products potentially marketable or (if elemental sulphur is produced) can be easily and safely disposed of.

The disadvantages are the complexity of the process; need for experienced labour.

Evaluation of Lime Spray Dryer Combined Abatement Process (Process Code NS31.1)

See Section 5.2 for: outline of the basic process; its chemistry; block diagram.

See Section 5.3 for a list of manufacturers offering this type of equipment.

See Section 5.3 for general appraisal of the basic process.

See Section 5.4 for the reason for choosing the Code S31.1 basic process.

This basic process type is considered (Section 5.4) to be potentially suitable for future application in the UK to large boilers and small factory boilers, and hence it is tentatively evaluated here for Datum Combustion Systems 1 and 2. The process presented here is of the type using lime slurry, with an additive (sodium hydroxide) in a Spray Dryer Absorber, producing a dry end-product. The process is offered in Europe by Fläkt Industri AB (Vaxjo, Sweden) in collaboration with A/S Niro Atomizer (Copenhagen, Denmark), and in the USA by Joy-Niro. See Appendix 2 for details of these manufacturers.

Process Description

Figure 5.18 shows a simplified flow diagram for application of the process to an oil fired boiler. It is assumed that for all applications, the solid end-product will be disposed of to landfill, but because the end-product contains soluble components, the precise details of its disposal will be site-specific.

Gas from the boiler plant of the electrostatic precipitator or baghouse enters a Spray Dryer Absorber (SDA) fed with a spray of calcium hydroxide slurry containing sodium hydroxide in solution. Operating conditions are chosen to give a gas temperature leaving the SDA of about 100°C, at which temperature the maximum extent of combined abatement occurs. The residence time of the slurry droplets in the SDA is about 10 seconds. The gas then passes to an Electrostatic Precipitator (ESP), or more usually a Baghouse, to remove the gas-borne reaction products which have not been collected in the hopper base of the SDA. A significant proportion (10–20%) of the total abatement occurs in the ESP or Baghouse. The collection hoppers of the SDA and of the ESP or Baghouse are electrically heated to prevent solids build-up resulting from condensation. The cleaned gas then passes, via a Fan, to the Stack; reheating is not required. Part of the collected solids is returned via a Recycle Product Silo to the Feed Tank, where it is mixed with fresh lime/caustic soda slurry. The fresh lime is slaked with water in the Slaker, and the calcium hydroxide is slurried with water and sodium hydroxide solution in the Slurry Tank before being pumped to the Feed Tank. The dry waste end-product contains captured particulates, sodium and calcium sulphite, sulphate, nitrite, nitrate, and unreacted hydroxide.

Status and Operating Experience

The status of the process is indicated in Section 5.3, where it is seen that two units (70 MWt and 96 MWt) have been installed and operated as combined abatement plant.

Experience is limited to short-term; information is

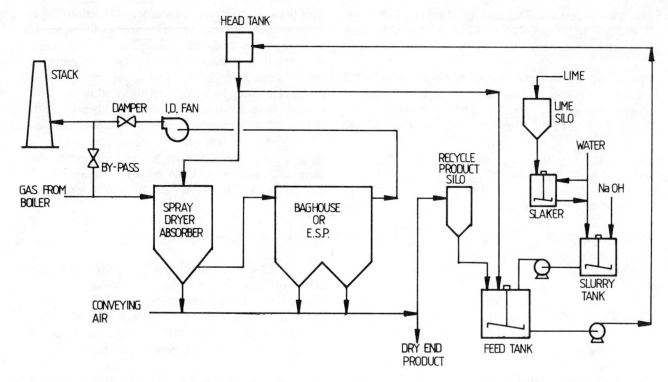

Figure 5.18 Lime Spray Dryer Combined Abatement Process

scanty. Experiences of operating lime spray dryers for FGD alone are mentioned in Section 2.5.

Variations and Development Potential

Design and operating variations can include:

- Number, type and location of slurry atomisers in the SDA.

- Alternatives of Electrostatic Precipitator and Baghouse for removal of gas-borne end-product.

The simplicity of the process is its chief characteristic, making it suitable for application on even small-scale plant (e.g. System 2).

More operating experience is needed to establish operating data.

Process Requirements for Each Application Considered

Estimated requirements are shown in Table 5.19 for the applications considered: Datum Combustion Systems 1 and 2.

It is assumed that the pollutant contents of the gas are to be reduced to the following:

- System 1:
SO_2 1000 mg/Nm³ (dry), equivalent to 80.1% capture of SO_2
NO_x 500 mg/Nm₃ (dry), equivalent to 51.4% abatement of NO_x
Particulates 15 mg/Nm³ (dry), equivalent to 61.2% capture of particulates

- System 2:
SO_2 1000 mg/Nm³ (dry), equivalent to 78.6% capture of SO_2
NO_x 400 mg/Nm³ (dry), equivalent to 35.2% abatement of NO_x
Particulates 15 mg/Nm³ (dry), equivalent to 58.4% capture of particulates

It is further assumed that:

- The molar ratio of (fresh lime)/(captured SO_2 + NO_x) is 1.5 mol Ca per mol (SO_2 + NO_x absorbed).

- The molar ratio of (fresh Na)/(fresh Ca) is 0.5.

For System 2 it is assumed that the lime would be supplied as calcium hydroxide.

By-products and Effluents

It is arbitrarily assumed that, in addition to captured particulates and to impurities present in the lime, the composition of the end-product is made up as follows:

- All of the input NaOH is converted to sodium sulphate decahydrate.

- All of the captured NO_x appears in the end-product as calcium nitrate.

Table 5.19 Process Requirements

Oil-Fired Combustion System		1	2
Inlet Gas at full load			
Volume flow	'000 Nm³/h	456	23.22
Dry gas	'000 Nm³/h	408	20.92
Water vapour	'000 Nm³/h	48	2.30
Actual volume flow	'000 m³/h	707	42.78
Temperature	°C	150	230
Particulates content	mg/Nm³ (dry)	39	36
SO_2 content	mg/Nm³ (dry)	5010	4675
NO_x content	mg/Nm³ (dry)	1025	615
HCl content	mg/Nm³ (dry)	3	3
Exit Gas at full load			
Volume flow	'000 Nm³/h	477	25.50
Dry gas	'000 Nm³/h	407	20.89
Water vapour	'000 Nm³/h	70	4.62
Actual volume flow	'000 m³/h	652	34.85
Temperature	°C	100	100
Particulates content	mg/Nm³ (dry)	15 (17)	15 (17)
SO_2 content	mg/Nm³ (dry)	1000 (1401)	1000 (1368)
NO_x content	mg/Nm³ (dry)	500 (526)	400 (422)
HCl content	mg/Nm³ (dry)	3 (3)	3 (3)
Reaction temperature	°C	90	90
Particulates removal		Simult.	Simult.
Lime supplied as:		CaO	$Ca(OH)_2$
Slurry composition			
$Ca(OH)_2$	wt. %	15.7	6.8
NaOH	wt. %	4.2	1.8
Molar ratio			
Ca/(Captured SO_2 + NO_x)		1.5	1.5
Na/(Captured SO_2 + NO_x)		0.75	0.75
Requirements at full load			
Quicklime (94.8%)	tonne/h	2.68	–
Slaked lime (96%)	tonne/h	–	0.15
Caustic soda (47%) liquor	tonne/h	0.91	0.08
Water (slaking)	tonne/h	0.82	–
Water (slurrying)	tonne/h	16	1.94
Electric power	MWe	3.3	0.2
Manpower	men/shift	***	***
Average load factor	%	***	***

Figures in parentheses are annual average emissions for 90% combined abatement plant availability ***data not available

- The captured SO_2 not appearing in the end-product as sodium sulphate appears in equimolar proportions as calcium sulphate dihydrate (gypsum) and as calcium sulphite hemihydrate.

- Calcium hydroxide that has not combined with SO_2 and NO_x appears in the end-product as the unchanged hydroxide.

The end-product has to be disposed of, e.g. to a landfill, but with precautions taken to deal with the soluble components – sodium salts, and calcium nitrate and chloride – and the excess alkalinity.

The estimated rates of production and composition of the material produced are summarised in Table 5.20 for the two Datum Combustion Systems.

Table 5.20 Estimated Properties of Waste Solids Produced

Oil-Fired Combustion System		1	2
Rate of production:	tonne/h	8.6	0.37
Composition of wet product:			
Moisture	wt. %	1.8	1.8
Particulates	wt. %	0.1	0.1
Calcium hydroxide	wt. %	24.8	24.0
Calcium sulphate	wt. % (a)	14.2	14.8
Calcium sulphite	wt. % (b)	10.7	12.6
Calcium nitrate	wt. %	4.5	2.2
Sodium sulphate	wt. % (c)	42.4	42.9
Other impurities	wt. %	1.5	1.6

(a) As gypsum
(b) As hemihydrate
(c) As decahydrate

Efficiency and Emission Factors

The efficiency and emission factors for the process are summarised in Table 5.21 for the two applications considered: Datum Combustion Systems 1 and 2.

– For Datum System 1, the combined abatement unit consumes 3.3 MWe of electric power, equivalent to the combustion of a further 0.8 tonne/h of oil at a power station, assuming an overall power generation efficiency of 33%. This additional oil is arbitrarily assumed to be included with the oil burned in the boiler for calculating the efficiency factor.

– For Datum System 2, the combined abatement unit consumes 0.2 MWe of electric power, equivalent to the combustion of a further 0.05 tonne/h of oil at a power station, assuming an overall power generation efficiency of 33%. This additional oil is arbitrarily assumed to be included with the oil burned in the boiler for calculating the efficiency factor.

For illustration purposes, the annual average emissions and emission factors for combined abatement plant availabilities of 100% and 90% are given in Tables 5.18 and 5.20. Further details are given in Section 1.6.

Effect of load variations: ***

Effect of design variations: ***

Limitations: ***

Costs

Capital costs: ***

Annual running costs and cost factors: ***

Effects of design variations on costs: ***

Effects of annual load patterns on annual running costs: ***

Process Advantages and Drawbacks

The advantages are that the basic process is well-established in USA and Europe for FGD; potentially simple process allowing application to small scale without high labour demands; land area required is small; fine grinding not needed as caustic soda solution and lime are used as absorbent (calcium hydroxide is a fine powder); simple chemistry; fairly high sulphur capture efficiency (over 90% capture) and NO_x abatement (up to 60% abatement) has been demonstrated; waste product can be easily and safely disposed of if account is taken of its soluble salts content and excess alkalinity.

The disadvantages are uncertain, but (based on FGD experience) likely to include tendency for deposits to form on SDA walls.

Table 5.21 Efficiency and Emission Factors

Oil-Fired Combustion System		1	2
Oil heat input (gross)	MWt	464	22.2
Oil fired	tonne/h	39.4	1.88
Abatement plant power	MWe	3.3	0.2
Equivalent oil input	tonne/h	0.8*	0.05*
Useful energy from system	GJt/h	468	67.8
Total equivalent oil input	tonne/h	40.2 (a)	1.93 (a)
Efficiency factor	GJt/tonne	36.5 (a)	35.1 (a)
Emissions			
Sulphur in SO_2	kg/h	204 (286)	10.5 (14.3)
Nitrogen in NO_x	kg/h	63 (69)	2.5 (2.6)
Chlorine in HCl	kg/h	1.2 (1.2)	0.06 (0.06)
Particulates	kg/h	6.1 (7.1)	0.31 (0.36)
Emission factors (per tonne coal)		(a)	(b)
Sulphur	kg/tonne	5.07 (7.11)	5.44 (0.19)
Nitrogen	kg/tonne	1.54 (1.72)	1.30 (1.35)
Chlorine	kg/tonne	0.03 (0.03)	0.03 (0.03)
Particulates	kg/tonne	0.15 (0.18)	0.16 (0.19)

*Calculated assuming overall power generation efficiency = 0.33
(a) Based on oil fired to boiler plus oil equivalent to electric power consumed
Figures in parentheses are annual average emissions for 90% combined abatement plant availability

Appendices

Appendix 1: Bibliography

Appendix 2: Vendor Information

Appendix 3: Cost Estimates and Procedures

Appendix 4: Subject Index

Appendix 1

Bibliography

GENERAL PAPERS ON ACID EMISSIONS ABATEMENT

General Approach to Acidic Emissions Abatement

1 Farrell R.J. and Ziegler E.N. 'Processes for reducing emissions.'

2 Mori A. (Hitachi Ltd.), Yamada H. (Hitachi P.E. and C. Co. Ltd.) and Kuroda H. (Babcock-Hitachi K.K.) 'Flue gas treatment system for coal-fired power plants'. *Hitachi Review*, 1980, **29** (6) 303–306.

3 Kyte W.S., Bettelheim J. and Cooper J.R.P. (CEGB). 'Possible fossil fuel developments within the electric power generation industry and their impact on other industries'. Symposium Series 78, 1983.

4 Van der Brugghen F.W. (Kema). 'Technological measures to reduce air pollution during coal firing'. *Resource and Conservation*, 1981, **7**, 133–143.

5 Meagher J.F., Stockburger L., Bananno R.J., Bailey E.M. and Luria M. (Tennessee Valley Authority). 'Atmospheric oxidation of flue gases from coal fired power plants–a comparison between conventional and scrubbed plumes'. *Atmospheric Environment*, 1981, **15** (5), 749–762.

6 Jahnig C.E. and Shaw H. (Exxon). 'A comparative assessment of flue gas treatment processes, Part 1–Status and design basis.' *JAPCA*, 1981 (Apr), **31** (4), 421–428.

7 Jahnig C.E. and Shaw H. (Exxon). 'A comparative assessment of flue gas treatment processes, Part 2–Environmental and cost comparison'. *JAPCA*, 1981 (May), **31** (5), 596–604.

8 Kyte W.S. (CEGB) 'Some implications of possible emission control technologies in the electric power generation industry'. Paper to I. Chem. E/SCI Symposium, Control of Acid Emissions in the UK, November 1985.

9 Catalano L., Elliott T.C. and Makansi J. 'Control technologies mature as policy debate lulls'. *Power*, 1985 (May), 13–20.

10 Environmental Resources Ltd. 'Acid rain–a review of the phenomenon in the EEC and Europe', Report for the Commission of the European Communities. Graham and Trotman Ltd., 1983.

11 Szabo M., Shah Y. and Abraham J. (PEDCO) 'Acid rain: control strategies for coal fired utility boilers. Volume 1'. Report No. DOE/METC-82-42 (Vol. 1) to US Department of Energy, May 1982.

12 Institution of Chemical Engineers/Society of Chemical Industry. 'Control of acid emissions in the UK' London, November 1985.

13 Dudley N., Barret M. and Baldock D. (Earth Resources Research). 'The acid rain controversy', 1985.

14 Anonymous. 'Acid rain. Its causes, effects and abatement'. *Sulphur* 1983, (Nov–Dec), (169), 32–37.

15 Cope D.R. (IEA Coal Research). 'Acid rain: The available control technology'. Paper to Acid Rain Inquiry, Edinburgh, 29 Sept, 1984.

16 Cope D.R. (IEA Coal Research). 'Control of acidic emissions from static plant.' Conference, 1985.

17 Wallin S.C. (Warren Spring Laboratory). 'Abatement systems for SO, NO particles–Technical options'. Paper to Institution of Environmental Sciences Seminar, An Update on Acid Rain, London, November 1984.

18 Elliott T.C. and Schwieger R.G. (editors) 'The acid rain source book'. New York: McGraw-Hill Inc, 1984.

19 Schwieger R.G. and Elliott T.C. (editors). 'Acid rain engineering solutions, regulatory aspects'. New York: McGraw-Hill Publications Co., 1985.

20 Goklany I.M. and Hoffnagle, G.F. (TRC Environmental Consultants). 'Trends in emissions of PM, SO_x and NO_x and VOC: NO_x ratios and their implications for trends in pH near industrialised areas'. *JAPCA*., 1984, 844–846.

21 Beer J.M. 'Clean combustion of coal, research and applications. An overview of recent developments in the USA.' *J. Inst. Energy*, 1986 (March), **59** (438), 3–19.

22 Truchot A. et al. 'Contribution of petroleum refineries to emissions of nitrogen oxides.' Report No. 9/84 to CONCAWE, May 1984.

23 De Meulemeester A. et al. 'Sulphur dioxide emissions from oil refineries and combustion of oil products in Western Europe in 1979 and 1982.' Report No. 10/84 to CONCAWE, May 1985.

24 Nakabayashi Y. (EPDC). 'A future forecast of the research and evaluation on the integrated flue gas treatment system'. Paper to EPDC/IIP/VDI, NO_x Symposium, Karlsruhe, West Germany, February 1985. Paper I.

25 Hovey H.H., Davis E., Sistla G., Glavin P., Twaddell R. and Rao S.T. (New York State Department of Environmental Conservation). 'Evaluation, selection and economic assessment of control strategies for acidic deposition.' Paper to APCA, 78th Annual Meeting, Detroit, USA, June 1985. Paper 85–1B.3.

26 Miller M.J. (EPRI). 'Retrofit SO_2 and NO_x control technologies for coal-fired power plants'. Paper to APCA, 78th Annual Meeting, Detroit, USA, June 1985. Paper 85-1A.2.

27 Rubin E.S. (Carnegie-Mellon University). 'Air pollution constraints on increased coal use by industry–an international perspective.' *JAPCA*, 1981 (Apr), **31** (4), 349–360.

28 Sedman C.B. (US EPA) and Ellison W. (Ellison Consultants). 'German FGD/deNO$_x$ experience'. Paper to Third Annual Pittsburg Coal Conference, Pittsburgh, USA, September 1986.

29 Koch H. (VEBA Kraftwerke Ruhr). 'Effects of environmental protection measures on planning, operation and costs of conventional thermal power stations.' *VGB Kraftwerkstechnik* 84, 1984 (Dec), (12), 935–941.

30 Anonymous. 'CEGB claims credit for acid rain move as international pressure intensifies.' *ENDS Report* 140, 1986 (Sept), 3–5.

31 Anonymous. 'Acid emissions: an opportunity for British industry?' *ENDS Report* 140, 1986 (Sept), 13–15.

32 Kimura T., Nakabayashi T. and Mouri K. (Electric Power Development Co.). 'Overall flue gas treatment technology from coal-fired power plants in Japan.' Paper to 5th International Conference on Coal Research, Düsseldorf, FRG, September 1980, Paper D-5, 313–333.

33 Barsin J.A. (Babcock & Wilcox). 'Options for reducing NO_x and SO emissions during combustion'. Paper to Power Magazine, First International Conference on Acid Rain, Washington, USA, March 1984.

34 Ando J. (Chuo University). 'Recent developments in SO_2 and NO_x abatement technology in Japan.' Paper to EPA/EPRI, Symposium on Flue Gas Desulphurisation, Cincinnati, USA, June 1985.

35 National Coal Board. 'Study into the emission of air pollutants coming from the use of coal within the United Kingdom.' Report No. EUR 6853 EN to Commission of the European Communities, Part 1, p. 14, 1980.

36 National Coal Board (British Coal), Coal Research Establishment. CRE Technology Review No. 2, 1982 (December).

37 Chem Systems International Ltd. 'Reducing pollution from selected energy transformation sources'. A study for the Commission of the European Communities, Environment and Consumer Protection Service. London: Graham & Trotman Ltd., 1976.

38 Buckley-Golder D.H. (ETSU). 'Acidity in the environment'. ETSU Report No. R.23, June 1984.

39 Derwent R.G. (ETSU). 'The nitrogen budget for the UK and NW Europe'. ETSU Report No. R.37, April 1986.

40 Department of the Environment. 'Digest of environmental protection and water statistics: 1986, No.9. HMSO.

41 Walker D.S., Galbraith R. and Galbraith J.M.

(Warren Spring Laboratory). 'Survey of nitrogen oxides, carbon monoxide and hydrocarbon emissions from industrial and commercial boilers in Scotland'. WSL Report No. LR524 (AP)M.

Emission Limitation Regulations

42 Evans P. (Department of the Environment, UK). 'The EC Directive on smoke and SO_2: The future for smoke control'. Paper to National Society For Clean Air, 47th Annual Conference, Bournemouth, September 1980.

43 Parkinson G. 'SO_2–Removal techniques ready for tighter curbs.' *Chemical Engineering*, 1983 (25 July), 17–20.

44 Catalano L. and Makansi J. 'Acid rain: New SO_2 controls inevitable.' *Power*, 1983 (Sept), 25–33.

45 Remirez R., Hoppe R., McQueen S. and Smith J. 'Acid rain: Europe, Canada, Act; the US Dithers'. *Chemical Engineering*, 1983 (11 July), 29–31.

46 Anonymous. 'An administration backed acid rain bill may be forthcoming'. *Chemical Engineering*, 1983 (22 August), 20.

47 Anonymous. 'An acid rain study is slated in Europe; action is expected in US as well'. *Chemical Engineering*, 1983 (19 Sept), 10.

48 Siegfriedt W.E. and Ludwig M. (Fluor). 'Desulphurisation processes in West Germany–an overview'. Power conference? 1984.

49 Select Committee on the European Communities. 'Air pollution–22nd Report, Session 1983–84'. Report No. HL 265 to House of Lords, 26 June 1984.

50 Department of the Environment. 'Acid rain: The Government's reply to the fourth report from the Environment Committee, Session 1983–84'. Report No. HC446-1 to Parliament, December 1984.

51 Environment Committee. 'Acid rain. Vol. 1. Report together with the proceedings of the committee relating to the report.' Fourth report to the House of Commons, Session 1983–84 (HMSO 446-1).

52 Westaway M.T. and McKay J. (British Petroleum). 'The impact of legislative requirements and legislative change on industry.' National Society for Clean Air, Oxford, March 1984.

53 Brady G.L. (President's Council on Environmental Quality, Washington) and Conway G. (Imperial College). 'Market approaches for sulphur dioxide management: A comparative analysis of Great Britain and the United States.' Proposal to Economic and Social Research Council, September 1984.

54 Clarke A.J. 'European legislative position'. Watt Committee Acid Rain Working Group.

55 Means C.S. (Associated Electric Cooperative Inc.) and Landwehr J.B. (Burns and McDonnell). 'Estimated economic impact of proposed acid rain control regulations on Associated Electrics Cooperative Inc., Springfield, Missouri'. Paper to Coal Technology '83, 6th International Coal and Lignite Utilisation Conference, Houston, USA, November 1983, **3**, 269–280.

56 Trisko E.M. (Stern Brothers). 'Potential impacts of acid rain control legislation.' Paper to Coal Technology '83, 6th International Coal and Lignite Utilisation Conference, Houston, USA, November 1983, **3**, 233–257.

57 Commission of the European Communities. 'Proposal for a Council Directive on limitation of emissions of pollutants into the air from large combustion plants'. Report COM (83) 704 Final, Brussels, 15 December 1983. Amended Proposal, COM (85) 47 Final, 18 February 1985.

58 Dacey P.W. (IEA Coal Research). 'An overview of international NO control regulations.'

59 Organisation for Economic Co-operation and Development. 'Emission standards for major air pollutants from energy facilities in OECD member countries'. Paris, 1984.

60 Aniansson B. 'A firm commitment'. *Acid Magazine 3*, Autumn 1985.

61 Haigh N. 'EEC environmental policy and Britain'. London: Environmental Data Services Ltd, 1984.

62 Short H., Herd J., McQueen S., and Smith J. 'EC producers grapple with a plan that limits air emissions'. *Chemical Engineering*, 1984 (14 May), 20E–20H.

63 Catalano L., 'Acid rain controls defeated in surprising sub-committee vote.' *Power*, 1984 (June), 9–10.

64 Catalano L., 'Environmentalists sue EPA to

issue industrial NSPS'. *Power*, 1984 (May), 9–10.

65 Catalano L., 'SO_2 standard not included in EPA's proposed NSPS for industrial boilers'. *Power*, 1982 (June), 9–10.

66 Clean Air (Emission of grit and dust from furnaces) Regulations 1971, (Statutory Instrument 1971 No. 162).

General Papers on Acidic Emissions Abatement Costs

67 Leggett A., Rubin E. and Torrens M. (OECD). 'Comparing the costs of flue gas treatment systems internationally'. Paper to ECE, Fourth Seminar on the Control of Sulphur and Nitrogen Oxides from Stationary Sources, Graz, Austria, May 1986.

68 Eriksson S., Forrester R., Johnston R. and Teper M. (IEA Coal Research). 'Economic and technical criteria for coal utilisation plant. Part 1: Economic and financial conventions'. Report No. A1/77, December 1977.

69 Torrens I.M. (OECD). 'Coal pollution abatement costs'. Paper to Coal Technology '83, 6th International Coal and Lignite Utilisation Conference, Houston, USA, November 1983, **3**, 257–267.

70 Maxwell J.D., Humphries L.R. (Tennessee Valley Authority), and Mobley J.D. (US Environmental Protection Agency). 'Economics of NO_x, SO_2 and ash control systems for coal-fired utility power plants.' Joint Symposium on Stationary Combustion NO_x Control, Boston, USA, May 1985. Paper 8a–7.

71 Remirez R. 'Looking into cheaper ways to deal with acid rain'. *Chemical Engineering*, 1986 (7 July), 17–19.

72 Ireland P.A. and Keeth R.J. (Stearns-Roger). 'Economic comparison of wet vs dry FGD'. Paper to National Lime Association Conference, Effective Use of Lime for Flue Gas Desulphurisation, Denver, USA, September 1983.

73 Organisation for Economic Co-operation and Development, Environment Directorate. 'Understanding pollution abatement cost estimates'. Report No. W.0067, March 1986.

74 Rubin E.S. and Torrens I.M. (Editors). 'Cost of coal pollution abatement: results of an international symposium'. Paris, OECD, 1983.

75 Ponder T.C., Yerino L.V., Katari V., Shah Y. and Devitt T.W. (PEDCO). 'Simplified procedures for estimating FGD system costs'. Report No. EPA-600/2-76-150 to US EPA, June 1976.

76 Kaplan N., Lachapelle D.G. and Chappell J. (US Environmental Protection Agency). 'Control cost modelling for sensitivity economic comparison'. Paper ID 5A.

77 Reisdorf J.B., Keeth R.J., Miranda J.E., Scheck R.W. (Stearns-Roger) and Morasky T.M. (Electric Power Research Inst.). 'Economic evaluation of FGD systems'. Paper to EPA/EPRI Flue Gas Desulphurisation Symposium, New Orleans, USA, November 1983.

78 Bobman M.H., Weber G.F. (University of North Dakota) and Dorchak T.P. (US Department of Energy). 'Comparative costs of flue gas desulphurisation: advantages of pressure hydrated lime injection'. Paper to EPA/EPRI Joint Symposium on Dry SO_2 and Simultaneous SO_2/NO_x Control Technologies, Raleigh, USA, June 1986. Report DOE/FE/60181-177.

79 Mora R.R., Ireland P.A. (Stearns-Roger) and Morasky T. (Electrical Power Research Institute). 'Estimating procedure for retrofit FGD costs'. Paper to EPA/EPRI Flue Gas Desulphurisation Symposium, New Orleans, USA, November, 1983. Paper 2B.

80 Damon J.E., Scheck R.W. (Stearns-Roger Engineering Co.) and Cichanowicz J.E. (Electric Power Research Institute). 'Economics of SCR post combustion NO_x control processes'. Paper to Joint Symposium, Stationary Combustion NO_x Control, 1982.

81 Ireland P.A., Brown G.D., Sebesta J.J. (Stearns Catalytic Corp.) and McElroy M.W. (Electric Power Research Institute). 'Economics of furnace sorbent injection for SO_2 Emission control'. Paper to EPA/EPRI Symposium on Dry SO_2 and Simultaneous SO_2/NO_x Control Technologies, Raleigh, USA, June 1986. Paper 5B.

82 McMahon T.C. and Rigsby L.S. (Ashland Coal). 'The high road to sulphur dioxide reduction'. Paper to Coal Technology Europe, 3rd European Coal Utilisation Conference, Amsterdam, The Netherlands, October 1983, **4**, 137–147.

83 Lachapelle D.G., Kaplan N. and Chappell J. (US Environmental Protection Agency).

'EPA's LIMB cost model: development and corporative care studies'. Paper 6H.

84 Storm J. (Niro Atomizer). 'Economics of dry FGD and byproducts handling'. Paper to OECD International Symposium, The Economic Aspects of Coal Pollution Abatement Technologies, Petten, The Netherlands, May 1982.

85 Naulty D.J. (Stearns-Roger), Muzio L.J. (KVB Corp.) and Hooper R. (Electric Power Research Institute). 'Economics of dry FGD by sorbent injection'. Paper to Coal Technology '83, 6th International Coal and Lignite Utilisation Conference, Houston, USA, November 1983, **3**, 209–229.

86 Krieg J.P., Mortimer G.W. and Weiss L.H. (Cogit Consulting Group). 'Economics of sulphur removal using "dry"-carbonated trona ore FGD sorbent'. Paper No. 5D.

87 Musgrove J.G. and Donnelly J.R. (Bechtel Power Corp.) 'Dual-alkali flue gas desulphurisation system cost versus operating availability'. Paper to Coal Technology '83, 6th International Coal and Lignite Utilisation Conference, Houston, USA, November 1983, **3**, 139–153.

88 Hollinden G.A., Stephenson G.A. (Tennessee Valley Authority) and Stensland J.G. (FMC Corp). 'An economic evaluation of limestone double-alkali flue gas desulphurisation systems'. Paper to Coal Technology '83, 6th International Coal and Lignite Utilisation Conference, Houston, USA, November 1983, **3**, 155–187.

89 Scharer B. and Haug N. 'The cost of flue gas desulphurisation and denitrification in the Federal Republic of Germany'. Paper to Fourth Seminar, Control of Sulphur and Nitrogen Oxides from Stationary Sources, Austria, May 1986.

90 Torrens I.M. (OECD). 'Coal pollution abatement costs'. Paper to Coal Technology Europe, Third European Coal Utilisation Conference, Amsterdam, October 1984, **4**, 213–223.

91 Sutherland H. et al. 'Cost of control of sulphur dioxide, nitrogen oxides and particulates emissions from large combustion plants in oil refineries'. Report No. 7/84 to CONCAWE, September 1984.

92 Samish N.C. (Shell Development Company). 'The cost of FGD'. Paper to 14th Annual Meeting of the Battelle Stack Gas Assessment and Technologies Survey Programme, London, September 1986.

93 Bakke E. (Peabody Process Systems). 'Economical retrofit of wet scrubbers to coal fired boilers'. 1986.

94 Morasky T.M., Dalton S.M. and Preston G.T. (EPRI). 'Economic assessment and operating and maintenance costs of FGD systems'. Paper to Seminar on Flue Gas Desulphurisation, Ottawa, Canada, September 1983.

95 Keeth R.J., Miranda J.E., Reisdorf J.B. and Scheck R.W. (Stearns-Roger). 'Economic evaluation of FGD systems. Volumes 1–3.' Report No. CS-3342 to EPRI, December 1983.

96 Shattuck D.M. et al. (Stearns Catalytic Corp.). 'Retrofit FGD cost-estimating guidelines'. Report No. CS-3696 to EPRI, October 1984.

97 Rubin E.S. 'Contribution of pollution costs to total annual revenue requirements, US, Germany and Japan (1982 US $)'. *Env. Sci. Technol.*, 1983, **17** (8), 366A–377A.

98 Highton N.H. and Webb M.G. 'Pollution abatement costs in the electricity supply industry in England and Wales'. *Journal of Industrial Economics*, 1981 (September), **3–** (1), 49–65.

99 Anonymous. 'Will pollution control costs hinder the return of coal?' (Summary of OECD Symposium). *ENDS Report*, 1983 (October), (105), 10–13.

100 Scharer B. and Haug N. 'On the economics of flue gas desulphurisation–measures, costs and effectiveness'. Paper to OECD, International Symposium on the Economic Aspects of Coal Pollution Abatement Technologies, Petten, The Netherlands, May 1982. OECD Proceedings, 1983. (Editors: Rubin E.S. and Torrens I.M.).

101 Wijdeveid H.W.J. (ESTS). 'Recent developments in reduction of cost of FGD systems in the Netherlands'. Paper to OECD/ENEA Enclair '86 Symposium, Taormina, Italy, October 1986.

102 Leggett J.A. (OECD). 'Costs and cost-effectiveness of techniques to reduce emission of NO_x and VOC: development of the OECD compendium'. Paper to OECD/ENEA Enclair '86, Taormina, Italy, October 1986.

103 Smith T.F. (CEGB). 'Factors affecting costs of power station emission controls with particular reference to the UK' Paper to OECD/ENEA Enclair '86, Taormina, Italy, October 1986.

Effects of Acidic Gases on Equipment and Materials of Construction

104 Verhoff F.H. and Choi M.K. (West Virginia University). 'Effects of sulphuric acid condensation on stack gas equipment'. *Journal of Institute of Energy*, 1980 (Jun), 92–99.

105 Berger D.M., Trewella R.J., and Wummer C.J. (Gilbert/Commonwealth). 'Evaluating linings for power plant SO$_2$ scrubbers'. *Power Engineering*, 1980 (Nov), 71–75.

106 Ellis P.F., Anliker D.M., Jones G.D. (Radian Corporation), and Steward D.A. (EPRI). 'FGD system failure analyses of metallic components'. *Materials Performance*, 1986 (Mar), 15–23.

107 Beavers J.A. and Koch G.H. 'Review of corrosion related failures in FGD systems'. *Materials Performance*, 1982 (Oct), 13–25.

108 Ellison W. (consultant) and Lefton S.A. (Aptech). 'FGD's reliability: What's being done to achieve it?' *Power*, 1982 (May), 71–76.

109 Kyte W.S. (CEGB). 'Corrosion in FGD plant'. Chapter 6, 'Dewpoint corrosion'. (Holmes, Editor). Chichester: Ellis Horwood Ltd., June 1985.

110 Javetski J. 'Solving corrosion problems in air-pollution control equipment, Part I'. *Power*, 1978 (May), 72–77.

111 Javetski J. 'Solving corrosion problems in air-pollution control equipment, Part II'. *Power*, 1978 (June), 80–97.

112 Dene C.E., Syrett B.C., Koch G.M. and Beavers J.A. 'Alloys and coatings for SO$_2$ scrubbers'. Paper to American Power Conference, Chicago, USA, April 1982.

113 Forsythe R.C., Hirt F.K. and Richards W.E. 'Chimney liner experience at the Bruce Mansfield plant'. Paper to American Power Conference, Chicago, USA, April, 1982.

114 Lee T.S. (LaQue Centre for Corrosion Technology) and Lewis R.O. (Montano State University). 'Evaluation of corrosion behaviour of materials in a model SO$_2$ scrubber system'. *Materials Performance*, 1985 (May), 25–32.

115 Pitt W.G. and Andersen T.N. (Kennecott Minerals Co.). 'Corrosion of alloys in simulated smelter FGD scrubber solutions'. *Materials Performance*, 1982 (May), 26–29.

116 Dille E.R., Froelich D.A. and Weilert C.V. (Burns & McDonnell) 'Tame the latest FGD-system corrosion pest: fluorides'. *Power*, 1983 (Aug) 41.42.

117 Rosenberg H.S., Koch G.H. (Battelle Columbus), Meadows M.L. (Black & Veatch) and Steward D.A. (EPRI). 'Materials for outlet ducts in wet FGD systems'. *Materials Performance*, 1986 (Feb), 41–55.

118 Deleted

119 Johnson C.A. (Peabody Process Systems). 'Evaluation of materials of construction for Alabama Electric Co-operative's limestone FGD system'. Paper to National Association of Corrosion Engineers, Denver, USA, August 1981.

120 Anonymous. 'Why put a fan on the wet side of a scrubber'. *Power*, 1986 (Sept.), 151–152.

PAPERS ON FLUE GAS DESULPHURISATION

General Approach to Flue Gas Desulphurisation and Recovery Processes

121 Makansi J. 'Optimizing today's processes for utility and industrial power plants. A special report'. *Power*, 1982 (Oct), S.1–S.24.

122 Kyte W.S. (CEGB). 'Some chemical and chemical engineering aspects of FGD'. *Trans. I. Chem. E.*, 1981, **59**, 219–228.

123 United Nations, Economic Commission for Europe. 'Air pollution studies 1. Air-borne sulphur pollution: effects and control'. Report No. GE.84–40823, 1984.

124 Moser R.E. (Brown and Root). 'FGD options offer environmental trade-offs'. *Hydrocarbon Processing*, 1981 (Oct), 88–92.

125 Melia M.T., McKibben R.S. and Pelsor B.W. (PEDCo). 'Project summary. Utility FGD survey: October 1983–September 1984'. Report No. EPA–340/1-85-014 to US Environmental Protection Agency, October 1984.

126 Rittenhouse R.C. 'Equipment retrofits add conformance to emissions control'. *Power Engineering*, 1986 (Sept), **90** (9), 18–24.

127 Kyte W.S., Bettelheim J. and Cooper J.R.P. (CEGB). 'Sulphur oxides control options in the UK electric power generation industry'. *I.*

Chem. E. Symposium Series, (77), Loughborough, April 1983.

128 Marx J.A. and Nagaraja M.L. (M.W. Kellogg). 'Structural engineering vital to FGD design'. *Electrical World*, 1983 (July), 93–95.

129 Martel G. (Northeast Utilities Service) and Veratti T. (Nalco). 'Reduce impact of acid emissions from your oil-fired boiler'. *Power*, 1983 (Oct), 105–106.

130 Wall J.D. 'Control FCC SO_x emissions'. *Hydrocarbon Processing*, 1984 (Oct), 45–46.

131 Sheppard S.V. (Ceilcote). 'Tailor air pollution control equipment to applications and requirements'. *Power Engineering*, 1986 (Feb), 32–35.

132 Ellison W. (Ellison), Leimkuhler J. (GEA) and Makansi J. 'West Germany meets strict emission codes by advancing FGD'. *Power*, 1986 (Feb), 29–33.

133 Mobley J.D. (US EPA) and Dickerman J.C. (Radian). 'Commercial utility FGD systems'. *Mechanical Engineering*, 1984 (July), 62–71.

134 Beals J. (Pennwalt), Cannell L. and Hengel J. (Black and Veatch). 'How FGD reagent quality affects system performance'. *Power*, 1984 (Mar), 27–30.

135 Anonymous. 'Product guide: flue gas desulphurisation'. *Modern Power Systems*, 1986 (June), 99–101.

136 Schwieger R., and Hayes A. 'Reliability concerns, regulations lead to virtual standardisation of air pollution control systems'. *Power*, 1985 (April), **129** (4), 81–93.

137 Morrison G.F. (IEA Coal Research). 'Control of sulphur oxides from coal combustion'. Report No. ICTIS/TR21, IEA Coal Research, London, November 1982.

138 CONCAWE. 'SO_2 emission trends and control options in Western Europe'. Report No. 1/82, 1982 (Jan).

139 The Watt Committee on Energy. 'Acid rain'. Report No. 14, August 1984.

140 Meyer C.E. 'Flue gas heat exchanger increases efficiency, reduces pollution'. *Power Engineering*, 1986 (Mar), 30–31.

141 Gillette J.L. and Chiu S.Y. (Argonne National Laboratory). 'FGD: Review of selected commercial and advanced technologies'. Report No. ANL/FE-81-51 for US Department of Energy, February 1981.

142 NATO Committee on the Challenges of Modern Society. 'FGD. First follow-up report: Control of air pollution from coal combustion.' Report No. 138, Ottawa, Canada, June 1982.

143 NATO Committee on the Challenges of Modern Society. 'FGD. Second follow-up report: Control of air pollution from coal combustion–Focus on NO_x and Limestone injection multistage burner NO_x/SO_x control technology'.

144 United Kingdom contribution to NATO-CCMS FGD Study Group. 'Control of air pollution from coal combustion'. Vienna, Austria, May 1986.

145 Lunt R.R. and Mackenzie J.S. (United Engineers & Constructors). 'Longer-term options for reducing SO_2 emissions'. Paper to Power Magazine, First International Conference on Acid Rain, Washington, USA, March 1984.

146 Pruce L.M. 'Why so few regenerative scrubbers'. *Power*, 1981 (June), 73–76.

147 VEW Special Steels (UK) Ltd. 'Materials for FGD Plant'. February 1986.

148 Steiner P. (Foster Wheeler), Dalton S.M. (EPRI) and Knoblauch K. (Bergbau Forschung). 'Capture and conversion of SO_2 RESOX prototype demonstration in Germany'. *Combustion*, 1980 (Jan), 28–31.

149 Gutterman C. and Steiner P. (Foster Wheeler). 'Continuous testing of the RESOX process–Final Report'. Report No. FWC/FWDC/TR-84/28 to Electric Power Research Institute, August 1984.

150 Gutterman C., Steiner P., Aiello M. and Violante D. (Foster Wheeler). 'RESOX process for urban FGD system–Final Report'. Report No. FWC/FWDC/TR-85/9 to Empire State Electric Energy Research Corporation, June 1985.

151 Ellis R.J. et al. 'Sulphur emissions from combustion of residual fuel oil based on EEC energy demand and supply, 1980–2000'. Report No. 5/86 to CONCAWE, July 1986.

152 Rittenhouse R.C. 'Additives: a lower cost alternative to hardware retrofits'. *Power Engineering*, 1986 (June), **90** (6), 18–24.

153 Anonymous. 'Growing FGD usage brings its own problems'. *Process Engineering*, 1985 (May), 41–42.

154 Neukam H. 'Flue gas desulphurisation with Ljungstrom heat exchanger'. *Chem. Tech.*, 1983, **12** (1), 18–20.

155 Rosenberg H.S. and Choi P.S.K. (Battelle). 'Energy aspects of FGD and stack gas reheat'. *AIChE Symp. Ser.*, 1980, **76** (196), 28–37.

156 Bettelheim J., Kyte W.S. and Littler A. (CEGB). 'Fifty years' experience of FGD at power stations in the UK'. *The Chemical Engineer*, 1981 (June), 275–278.

157 Kohl H. and Riesenfeld F. 'Gas Purification'. Gulf Publishing Co., Third Edition, 1979.

158 McIlvaine Scrubber Manual, Vol. IV, 1979.

Non-Regenerable Solution-Based Wet Processes (Category S10)

159 Reason J., Baur P. and Makansi J. 'Finch, Pruyn cleans air and water while increasing steam production'. *Power*, 1981 (Nov), 73–83.

160 Brady J.D. (Andersen 2000). 'Particulate and SO_2 removal with wet scrubbers'. *CEP*, 1982 (June), 73–77.

161 Anonymous. 'NaOH scrubbing process most attractive for small-scale SO_2 pollution abatement'. *TI Chem.E.* 1971 (Jan), **16** (1), 7–9.

162 Ponder W.H. (US EPA), Fischer W.H. (Gilbert Associates) and Zaharchuk R. (Firestone). 'Environmental assessment of the dual alkali FGD system applied to an industrial boiler firing coal and oil'. *A.I.Ch.E. Symposium Series*, 1980, **76** (201), 80–95.

163 Henry J.R., Wrobel B.A. (Northern Indiana Public Service Company), Ellefson D.W., Katzberger S.M., Predick P.R. (Sargent and Lundy). 'Lime handling and preparation for two double-alkali FGD systems'. Paper to Coal Technology '83, 6th International Coal and Lignite Utilisation Conference, Houston, USA, November 1983, **2**, 37–48.

164 Kirchgessner D.A., Gullett B.K. (US Environmental Protection Agency) and Lorrain J.M. (Acurex). 'Physical parameters governing the reactivity of $Ca(OH)_2$ with SO_2'. Paper to EPA/EPRI Joint Symposium on Dry SO_2 and Simultaneous SO_2/NO_x Control Technologies, Raleigh, USA, June 1986. Paper 2D.

165 Deleted

166 Anonymous. 'FMC announces pre-engineered double alkali scrubber system'. 1981 (8 Jul), 4.

167 Brady J.D. (Andersen 2000). 'Sulphur dioxide removal from exhaust gases'. *CEP*, 1984 (Sept), 59–62.

168 Anonymous. 'Dry scrubber overcomes scale-up problems at Coyote'. *Power*, 1983 (Apr), 114–115.

169 Lewis M.F. (Montana-Dakota Utilities), and Gehri D.C. (Rockwell). 'Atomisation–The key to dry scrubbing at the Coyote Station'. Paper to EPA/EPRI Symposium, Flue Gas Desulphurisation, May 1982.

170 Stern J.L. (Joy). 'Dry scrubbing for FGD'. *CEP*, 1981 (Apr), 37–42.

171 Francis D.V. (Arco Chemical Co.), Biolchini R.J. and Coons J.D. (FMC). 'Operating experience with high sulphur coal in an industrial double alkali FGD system'. Paper to Coal Technology '81, 4th International Coal Utilisation Conference, Houston, USA, November 1981, **3**, 215–229.

172 Tamaki A. (Chiyoda Chemical Engineering & Construction Co. Ltd). 'Commercial application of dilute sulphuric and/gypsum (the Chiyoda Thoroughbred 101) FGD process for large power plant boilers'. Paper to 66th Annual AIChE Meeting, Washington, USA, December 1974.

Non-Regenerable Slurry-Based Wet Processes (Category S20)

173 Sugita Y., Oguri H. and Sakamoto Y. (IHI). 'State of the art in desulphurisation system for preventing environmental pollution caused by flue gas from coal fired power stations.' *IHI Engineering Review*, 1984 (Oct), **14** (4), 29–35.

174 Esche M. (Saarberg-Hölter). 'Stack gas desulphurisation without reheating'. Paper to Coal Technology Europe, 3rd European Coal Utilisation Conference, Amsterdam, Netherlands, October 1983, **4**, 93–102.

175 Johnson C.A. (Peabody Process Systems Inc.). 'Flyash alkali technology–Low cost FGD'. Paper to Coal Technology '80, Houston, USA, November 1980, 569–588.

176 Kunzweiler V.L., Landwehr J.B., Collier C.W. and Froelich D.A. (Burns and McDonnell). 'Start-up experience of five FHD systems'. Paper to American Power Conference, Chicago, USA, April 1980.

177 Van Ness R.P. (Louisville Gas and Electric Co), Kingston W.H. and Borsare D.C. (Combustion Engineering). 'Operation of C-E FGD system for high sulphur coal at Louisville Gas and Electric Co., Cane Run 5'. *Combustion*, 1980 (Feb), 10–16; and Vol. 41 of Proceedings of the American Power Conf. 1979, pp. 656–664.

178 Makansi J. 'Wet venturi doubles as an SO_2 scrubber'. *Power*, 1981 (Dec), 71–72.

179 Yeager K. (EPRI). 'Advanced SO_2 control'. *EPRI Journal*, 1981 (Mar), 38–39.

180 Johnson C.A. (Peabody Process Systems). 'Minnesota Power's operating experience with integrated particulate and SO_2 scrubbing'. *Journal of the Air Pollution Control Association*, 1981 (Jun), **31** (6), 701–705.

181 Hoffman D.C. (Dravo Lime). 'Thiosorbic lime for FGD processes'. *Mining Engineering*, 1981 (Nov), 1628–1631.

182 Catalano L. (compiler). 'FGD improves with adipic acid'. *Power*, 1982 (Jul), 84–85.

183 Nesbit W. 'Scrubbers: The technology nobody wanted'. *EPRI Journal*, 1982 (Oct), 8–15.

184 Ellison W. (consultant) and Kutemeyer P.M. (Bischoff). 'New developments advance forced-oxidation FGD'. *Power*, 1983 (Feb), 43–45.

185 Anonymous. 'Research boosts thiosulphate for FGD systems'. *Chemical Engineering*, 1983 (8 Aug), 11–12.

186 Chang J.C.S. (Acurex) and Mobley J.D. (US EPA). 'Testing and commercialisation of by-product dibasic acids as buffer additives for limestone FGD systems'. *JAPCA*, 1983 (Oct), (10), 955–956.

187 Esche M. and Igelbuscher, H. (Saarberg-Hölter). 'Technical solutions for the new SO_2 legislation in West Germany–FGD without reheating'. Paper to Air Pollution Control Association, 77th Conference, San Francisco, USA, June 1984. Paper No. 84–971.

188 Makansi J. 'A limestone FGD system'. *Power*, 1985 (Sept), 107–109.

189 Kojima T., Shikishima S., Kanamori A. and Torii M. (IHI). 'Operating results of FGD system for unit No.3 (700MW) at Takehara thermal power station of the Electric Power Development Co. Ltd.' *IHI Engineering Review*, **17** (2), 1–5.

190 Murphy K.R., Shilling N.Z. (General Electric Environmental Services Inc) and Pennline B.H. (US Department of Energy). 'Low cost in-duct scrubbing system will be tested at Muskingum River'. *Modern Power Systems*, 1986 (Jun), 79–83.

191 Kirchner R.W. (Cabot Corp.). 'Materials of construction for flue gas desulphurization systems'. *Chemical Engineering*, 1986 (19 Sept), 81–86.

192 Rock K.L. (D.M. International), Glamser J.H. (Davy McKee) and Esche M. (Saarberg-Hoelter). 'Commercial operating history and latest developments to the Davy S-H process'. Paper to Coal Technology '81, 4th International Coal Utilisation Conference, Houston, USA, November 1981, **3**, 261–280.

193 Rader P.C., Hansen R.W. and Borsare D.C. (Combustion Engineering). 'Design of lime/limestone flue gas desulphurisation systems for high chlorides'. Paper to Coal Technology 1981, 4th International Coal Utilisation Conference, Houston, USA, November 1981, **3**, 231–261.

194 Anonymous. 'High sulphur coal tests demonstrate the successful operation of dry scrubbing'. *Chemical Engineering*, 1982 (22 Mar), 18.

195 Karlsson H.T., Klingspor J., Linne M. and Bjerle I. (Chemical Centre). 'Activated wet-dry scrubbing of SO_2'. *JAPCA*, 1983 (Jan), **33** (1), 23–28.

196 Blythe G.M., Burke J.M., Kelly M.E., Rohlack L.A. (Radian) and R.G. Rhudy (EPRI). 'EPRI spray drying pilot plant status and results'. Paper to EPA/EPRI Symposium.

197 Mudgett J.S. (Strathmore paper), Sadowski R.S. (Riley Stoker), West W.W. (Mikropul) and Mutsakis M. (Koch). 'Dry SO_2 scrubbing achieved with spray dryer and fabric filter'. Reprint from the 'Eighth Annual Industrial Plant–Energy Systems Guidebook', McGraw-Hill, 1982.

198 Rainauer, T.V. (Mikropul), Monat J.P. (Abcor) and Mutsakis M. (Koch). 'Dry FGD on an industrial boiler'. *CEP*, 1983 (Mar), 74–81.

199 Emerson R.D. (Sunflower Electric). 'Dry FGC system: Start-up, performance and acceptance tests'. *Power Engineering*, 1984 (Oct), 50–52.

200 Kaplan S.M. and Felsvang K. (Niro Atomizer). 'Spray dryer absorption of SO_2 from industrial boiler flue gas'. *A.I.Chem. E. Symposium Series*, 1980, **76** (201), 23–30.

201 Meyler J.A. (Joy). 'Case history of a dry scrubber application at Northern States Power Company'. Paper to Coal Technology '80, 1980, 589–596.

202 Felsvang K. (Niro Atomizer). 'Results from operation of Riverside dry scrubber'. Paper to Riverside Dry FGD Symposium, Minneapolis, USA, June 1981.

203 Gude K.E. 'The spray dryer absorber concept for FGD'. Paper to Symposium on Danish Know-How and Technology on Energy and Pollution Control, Peking, November 1983.

204 Hansen S.K., Felsvang K.S. (Niro Atomizer), Morford R.M. and Spencer H.W. (Joy). 'Status of the Joy/Niro Atomizer dry FGD system and its future application for the removal of high sulphur, high chloride and NO_x from flue gases'. Paper to ASME Joint Power Generation Conference, September 1983.

205 Thousig J.T., Jorgensen C. and Fallenkamp B. (Niro Atomizer). 'Dry scrubbing of toxic incinerator flue gas by spray absorption'. Paper to ENVITEC 83, Düsseldorf, February 1983.

206 Schwartzback C. (Niro Atomizer). 'The science and art of spray dryer design for FGD'. Paper to Coal Technology '82, 5th International Coal Utilisation Conference, Houston, USA, December 1982.

207 Felsvang K. (Niro Atomizer). 'Desulphurization of low rank, high sulphur coal by dry flue gas desulphurisation'. Paper to 8th International Congress of Chemical Engineering, Chemical Equipment Design and Automation, Praha, Czechoslovakia, September 1984.

208 Jacobson P. (Fläkt Industri AB) and Madhuk R. (Niro Atomizer). 'Flakt dry FGD Systems'. Paper to Flue Gas Desulphurisation Seminar, Bombay and New Delhi, Jan/Feb 1986.

209 Horn R.J. (Ecolair Environmental Co.). 'Installation and operation of a retrofit dry flue gas desulphurisation system'. Paper to Coal Technology '83, 6th International Coal and Lignite Utilisation Conference, Houston, USA, November 1983, **3**, 115–139.

210 Downs W., Sanders W.J. and Miller C.E. (Babcock and Wilcox). 'Control of SO_2 emissions by dry scrubbing', 262–271.

211 Burnett T.A., Threet G.E., Humpries L.R., Robards R.F. and Runyan R.A. (Tennessee Valley Authority). 'Spray dryer/baghouse flue gas desulphurisation – evaluation for high-sulphur utility applications'. Paper to American Institute of Chemical Engineers, Winter Annual Meeting, Chicago, USA, November 1985.

212 Robards R.F., DeGuzman J.S., Runyan R.A. and Flora H.B. (Tennessee Valley Authority). 'Spray Dryer/ESP testing for utility retrofit applications on high-sulphur coal'. Paper to American Power Conference, Chicago, USA, April 1986.

213 Livengood C.D. and Farber P.S. (Argonne National Laboratory). 'Performance and economics of a spray-dryer FGD system used with high-sulphur coal'.

214 Hammer P.R.R. (Niro Atomizer) 'Desulphurisation of flue gases from coal burning by spray absorption'. Paper to Coal Technology Europe, 2nd International Coal Utilisation Conference, Copenhagen, Denmark, September 1982, **3**, 297–311.

215 Colley J.D. (Radian Corporation), Donaldson T. (Central Illinois Light Co) and Stewart D. (Electric Power Research Institute). 'Process troubleshooting at a utility limestone F.G.D. system'. Paper to Coal Technology '83, 6th International Coal and Lignite Utilisation Conference, Houston, USA, November 1983, **3**, 187–207.

216 Hargrove O.W., Colley J.D. (Radian Corp.) and Mobley J.D. (US Environmental Protection Agency). 'Adipic acid-enhanced limestone flue gas desulphurisation system commercial demonstration'. Paper to Coal Technology '81, 4th International Coal Utilisation Conference, Houston, USA, November 1981, **3**, 201–213.

217 Felsvang K. (Niro Atomizer). 'Acid rain control through dry scrubbing'. Paper to Power Magazine, First International Conference on Acid Rain, Washington, USA, March 1984.

218 Makansi J. 'New processes enhance the in-duct emissions-control option'. *Power*, 1986 (July), **139** (7) 27–29.

219 Buschmann J.C., D'Ambrosi F.D. and Mezner M. (Fläkt). 'Start-up and operating experience of the University of Minnesota dry FGD system'. Paper to APCA, 78th Annual Meeting, Detroit, USA, June 1985, Paper 85-58.3.

220 Widico M.J. and Dhargalkar P.H. (Research-Cottrell). 'Dry FGD process for various coals'. Paper to APCA, 78th Annual Meeting, Detroit, USA, June 1985. Paper 85-58.4.

221 Davidson L.N., Goffredi R.A. and Wedig C.P. (Stone and Webster). 'The importance of maintenance for lime FGD systems'. Paper to APCA, 78th Annual Meeting, Detroit, USA, June 1985. Paper 85-58.6.

222 Cannall A.L. and Meadows M.L. (Black and Veatch). 'Effects of recent operating experience on the design of spray dryer FGD systems'. Paper to APCA, 78th Annual Meeting, Detroit, USA, June 1985. Paper 85-58.8, and *JAPCA*, 1985, **35** (7), 782-788.

223 Burnett G.F. and Basel B.E. (Burns and McDonnell). 'The status of dry scrubbing in the United States'. Paper to APCA, 78th Annual Meeting, Detroit, USA, June 1985. Paper 85-58.1.

224 Mobley J.D. (US EPA), Cassidy M. and Dickerman J. (Radian). 'Organic acids can enhance wet limestone flue gas scrubbing'. *Power Engineering*, 1986 (May) 32-35.

225 Ashley M. (Lodge-Cottrell). 'Spray dry desulphurisation plant requires lower capital investment'. *Modern Power Systems*, 1985 (May).

226 Murphy K.R., Shilling N.Z. (GEESI) and Pennline H. (US Department of Energy). 'In-duct scrubbing pilot study'. *JAPCA*., 1986 (Aug), **36** (8), 953-958.

227 Ashley M.J. (Lodge-Cottrell). 'The Lodge-Cottrell spray dry desulphurisation system'. Paper to 14th Annual Meeting of the Battelle Stack Gas Assessment & Technology Survey Programme, London, September 1986.

228 Dhargalkar P.H. (Research-Cottrell), and Ford P.G. (Davy McKee). 'Performance of two established FGD processes'. Paper to Coal Tech '85, 5th International Conference on Coal Utilisation and Trade, London, December 1985, **2**, 307-334.

229 Richman M. (Research-Cottrell Inc.). 'Advanced FGD technology for V.Y. Dallman Station'. Paper to Coal Technology '80, 1980, 597-607.

230 Yeargan R.D. (TVA). 'Paradise Fossil Plant: Units 1 & 2 scrubber operating experience'. Paper to EPRI, FGD Users Conference, Farmington, USA, June 1986.

231 Wallenwein E.H. (Bischoff). 'Desulphurisation plant developed by West Germany utility'. *Modern Power Systems*, 1985 (May), 33-37.

232 Hargrove O.W., Colley J.D. (Radian Corp.), Wadlington M. (Texas Utilities Generating Co.) and Stewart D.A. (EPRI). 'FGD system and water balance improvements at Texas Utilities Generating Company's Martin Lake station'. Paper to Symposium on Advances in Fossil Power Plant Water Management, Orlando, Florida, February 1986.

233 Mori T., Matsuda S., Nakajima F., Nishimura T. and Arikawa Y. (Hitachi). 'Effect of Al^{3+} and F^- on desulphurisation re-action in the limestone slurry scrubbing process'. *Ind. Eng. Chem. Process Des. Dev.*, 1981, **20**, 144-147.

234 Anonymous. 'Forced-O_2 FGD system achieves 99.8% availability in first year'. *Power*, 1986 (April), 19-20.

235 Martin J.R., Ferguson W.B. and Frabotta D. 'C-E dry scrubber systems: application to Western coals'. *Combustion*, 1981 (February), 12-20.

236 Crowe R.B. (Celanese Fibres), Lane J.F. (Rockwell International) and Petti V.J. (Wheelabrator-Frye). 'Early operation of the Celanese Fibres Company coal-fired boiler using the dry flue gas cleaning system'. *Combustion*, 1981 (February), 34-37.

237 Borgwardt R.H. (US EPA). 'Combined flue gas desulphurisation and water treatment in coal-fired power plants'. *Env. Sci. Technol.*, 1980 (March), **14** (2), 294-298.

238 Shattuck D.M., Stenby E.W., Lacey J.N. and Layton K.F. 'Utah Power and Light's experience with wet scrubbing of SO_2 at the Huntington and Hunter plants'. Paper to 42nd American Power Conference, Chicago, USA, April 1980.

239 Massey C.L., Moore N.D., Munson G.T., Runyan R.A. and Wells W.L. 'Forced oxidation of limestone scrubber sludge at TVA's Widows Creek Unit 8 steam plant'. Paper to US EPA, 6th Symposium on Flue Gas Desulphurisation, Houston, USA, October 1980.

240 Chan P.K. and Rochelle G.T. 'Limestone dissolution: effects of pH, CO_2 and buffers modelled by mass transfer'. Paper to ACS National Meeting, Atlanta, USA, March 1981.

241 Burke J.M., Metcalfe R.P., Cmiel R. and Mobley J.D. 'Technical and economic evaluation of organic acid addition to the San Miguel FGD system'. Paper to EPA's Industry Briefing on the Organic Acid Enhanced Limestone FGD Process, San Antonio, USA, July 1984.

242 Benson L.B. 'The role of magnesium in increasing SO_2 removal and improving the reliability in magnesium-enhanced FGD systems'. Paper to 2nd Annual Coal Conference, Pittsburgh, USA, September 1985.

243 Mobley J.D. (US EPA) and Chang J.C.S. (Accurex Corp.). 'The adipic acid enhanced limestone flue gas desulphurisation process: an assessment. *JAPCA*, **31** (12), 1249–1253.

244 Chang J.C.S., Kaplan N. and Brna T.G. (Accurex Corp.). 'Effects of Mg^{++} and Cl^- ions on limestone dual alkali system performance'. *ACS Div. Fuel Chem.*, 1985, **30** (2), 1145–161.

245 Smith T. (Consultant), Colley D. (Radian Corp.) and Steward D. (Electric Power Research Institute). 'Apply process-chemistry know-how to your FGD system'. *Power*, 1985 (Sept), 35–37.

246 Dharmarajan (Central & South West Services Inc.). 'Stirred mill proves its worth for FGD lime-slaking duties'. *Power*, 1985 (Oct), 61–63.

247 Friedlander G.D. 'Huge scrubber retrofitted at four corners'. *Electrical World*, 1984 (Mar), 71–72.

248 Beals J. (Pennwalt), Cannell L. and Hengel J. (Black & Veatch). 'How FGD reagent quality affects system performance'. *Power*, 1984 (Mar), 27–30.

249 Ellison W. (Consultant) and Egan R. (Munters Corp.) 'Incorporate the latest FGD trends into mist-eliminator design'. *Power*, 1984 (Mar) 35–37.

250 Makansi J. 'Particulate and SO_2 scrubbers that require no wetted-surface internals'. *Power*, 1983 (Mar) 119.

251 Anonymous. 'SO_2 scrubber makes saleable gypsum'. *Oil & Gas Journal*, 1979 (5 Mar) 180.

252 Stowe D.H., Henzel D.S. and Hoffman D.C. (Dravo Lime Co.) 'The FGD reagent dilemma: lime, limestone or thiosorbic lime'. Report No. EPA-600/7-79-16TB to US EPA, July 1979.

253 Anonymous. 'Dry scrubber overcomes scale-up problems at Coyote'. *Power*, 1983 (Apr), 114–115.

254 Anonymous. 'Bechtel offers partial desulphurisation to reduce costs'. *Process Engineering*, 1986 (Oct), 15.

255 Laslo D. and Bakke E. (Peabody Process Systems). 'State-of-the-art design applications on a closed-loop FGD system'. Paper to EPA/EPRI, FGD Symposium, Cincinnati, USA, June 1985.

256 Laslo D. (Peabody Process Systems), Chang J.C.S. (Acurex) and Mobley J.D. (US EPA). 'Pilot plant tests on the effects of dissolved salts on lime/limestone FGD chemistry'. Paper to EPA/EPRI, Symposium on FGD, New Orleans, USA, November 1983.

257 Laslo D. and Bakke E. (Peabody Process Systems). 'The effect of dissolved solids on limestone FGD scrubbing chemistry'. Paper to ASME, 1983 Joint Power Generation Conference, Indianapolis, USA, September 1983.

258 Bakke E. (Peabody Process Systems). 'Cost effective wet FGD systems on medium to high sulphur coals'. Paper to 1985 Joint Power Generation Conference, Milwaukee, USA, October 1985.

259 Anonymous. 'Catenany-grid scrubber'. *Chemical Engineering*, 1986 (13 Oct), 39.

260 Fahlenkamp H. (Deutsche Babcock Anlagen). 'Recent developments in West Germany's limestone-based FGD technology'. Paper to Joint ASME/IEEE, Power Generation Conference, Portland, USA, October 1986.

261 Chang J.C.S. (Acurex Corp.) and Laslo D. (Peabody Process Systems). 'Chloride ion effects on limestone FGD system performance'. Paper to EPA/EPRI, FGD Symposium, Hollywood, Florida, USA, May 1982.

262 Tearney J.F., Froelich D.A. and Graves G.M. (Burns & McDonnell). 'SO_2 control of non-regenerable wet FGD systems'. Paper to Power Magazine, First International Conference on Acid Rain, Washington, USA, March 1984.

263 Wataya K., Hon A. (Toyama Hyoda), Hashimoto N., Koshizuka H. (Chiyoda) and Clasen D.D. (Chiyoda International). 'Operating results of Toyama Kyoda Electric Powers' Chiyoda Thoroughbred 121 FGD system'. Paper EPA/EPRI, 9th FGD Symposium, Cincinnati, USA, June 1985.

264 Wiitala W.W. (Marguette Board of Light & Power), Arello J. (Lutz, Daily & Brain), Martinelli R. and Lapp D. (GEESI). 'Spray dry scrubbers at Marquette's Shiras power plant'. Paper to American Public Power Association, 28th Annual Engineering & Operations Workshop, Toronto, Canada, March 1984.

Regenerable Solution-Based Wet Reagent Processes (Category S30)

265 Bettelheim J., Cooper J.R.P., Kyte W.S. and Rowlands D.T.H. (CEGB). 'The integration of a regenerable FGD plant on to a 2000 MW coal fired power boiler station site in the UK'. *I. Chem. E. Symposium Series*, 1981, (72).

266 Dhargalkar P.H. (Research-Cottrell) and Ford P.G. (Davy McKee). 'Performance of two established FGD processes'. Paper to Coal Tech '85, 5th International Conference on Coal Utilisation and Trade, London, December 1985, **2**, 307–334.

267 Madenburg R.S. and Seesee T.A. (Morrison-Knudsen Co.). 'H_2S reduces SO_2 to desulphurise flue gas'. *Chemical Engineering*, 1980 (14 Jul), 88–89.

268 Farrington J. and Bengtsson S. (Fläkt Inc.) 'Citrate solution absorbs SO_2'. *Chemical Engineering*, 1980 (16 Jun), 88–89.

269 Makansi J. (compiler). 'Regenerative FGD: progress is slow but steady'. *Power*, 1983 (Aug), 36–37.

270 Walker R.J., Wildman D.J. and Gasior S.J. (US Department of Energy). 'Evaluation of some regenerable SO_2 absorbents for FGD'. *JAPCA*, 1983 (Nov), **33** (11), 1061–1067.

271 Munson R.A., Fitch W.N., and Nissen W.I. '50 MW power plant demonstration of the removal of sulphur oxides from stack gases using the Bureau of Mines citrate process'. 95–98.

272 Langenkamp H. and Van Velzen D. (CEC). 'FGD by the Mark 13A process'.

273 Van Velzen D., Langenkamp H. and Ferrari, A. 'The Mark 13A Process for FGD'. Excerpts from Programme Progress Reports, Hydrogen Production Energy Storage and Transport, Jan 1983–Jun 1984.

274 Anonymous. 'The European Commission to co-finance a project against acid deposition'. *Biomass News International*, 1986 (Jun), (18), 7.

275 US Bureau of Mines, Report of Investigations 8638 (1982). 'FGD: evaluation of the modified citrate process draws important conclusions'. *Sulphur*, 1983 (Jan/Feb), (164), 43–45.

276 Neumann U., Vangala R. and Giovanetti A. (Davy McKee). 'Wellman-Lord SO_2 recovery operating experience serving coal fired boilers'. Paper to Coal Technology Europe '82, 2nd International Coal Utilisation Conference, Copenhagen, Denmark, September 1982, **3**, 281–296.

Regenerable Slurry-Based Wet Reagent Processes (Category S40)

277 Makansi J. 'New regenerative FGD system demonstrated at state hospital'. *Power*, 1983 (Oct), 131–132.

278 Anonymous. 'Regenerable scrubber meets EPA limits, utility requirements'. *Power*, 1984 (Apr), 33–34.

279 Marawczyk C., MacKenzie J.S. (United Engineers and Constructors), and Bitsko R. (Philadelphia Electric). 'The outlook for regenerative magnesium oxide FGD' *CEP*, 1984 (Sep), 62–68.

280 Makansi J. 'MgO scrubber links utility to chemical firms'. *Power*, 1981 (Dec).

281 PETC Quarterly Progress Report, 30 September 1985.

Regenerable Dry Reagent Processes (Category S50)

282 Ploeg J.E.G. (Shell Internationale), Akagi E. (Showa) and Kishi K. (Japan Shell). 'How Shell's FGD unit has worked in Japan'. *Petroleum International*, 1974 (Jul) **14** (7), 50–58.

283 Steiner P. (Foster Wheeler Energy Corp.). 'Pollution control system and method for the removal of sulfur oxides'. UK Pat. Appl. GB 2,009,117A, 13 June 1979 (filed 7 September 1978).

284 Townley D. and Winnick J. 'Flue gas desulphurisation using an electrochemical sulfur oxide concentrator'. *Ind. Eng. Chem. Process Des. Dev.*, 1981, **20**, 435–440.

285 Bee, R., Reale R. and Walls A. (Mitre Corp.). 'Demonstration/evaluation of the Cat-Ox flue gas desulphurisation system–final report'. Report No. EPA-600/2-78-063 to US Environmental Protection Agency, March 1978.

286 Steiner P. (Foster Wheeler Development Corp.), Dalton S.M. (EPRI) and Knoblauch K. (Bergbau Forschung). 'Capture and conversion of sulphur dioxide at the ReSOx prototype demonstration in Germany'. Proceedings of the American Power Conference, 1979, **41**, 719–723.

Non-Regenerable Dry Reagent Applied to Flue Gas (Category S60)

287 Yeager K. (EPRI). 'SO$_2$ control by dry sorbent injection'. *EPRI Journal*, 1983 (Mar), 36–37.

288 Samuel E.A., Furlong D.A. (Envirotech), Brna T.G. (US EPA) and Ostop R.L. (City of Colorado Springs). 'SO$_2$ removal using dry sodium compounds'. *AIChE Symposium Series*, 1981, **77** (211), 54–60.

289 Anonymous. 'Dry capture of SO$_2$'. *EPRI Journal*, 1984 (Mar), 14–21.

290 Hamala S. (Tampella Ltd.). 'LIFAC cuts SO$_x$ in Finland'. *Modern Power Systems*, 1986 (June), 87–91.

291 Forsythe R.C. (Dravo Lime Co.) and Kaiser R.A. (Ohio Edison Co.). 'Hydrate addition at low temperature: SO$_2$ removal in conjunction with a baghouse'. Paper to 2nd Annual Pittsburgh Coal Conference, Pittsburgh, USA, September 1985.

292 Yoon H., Ring P.A. and Burke F.P. (Conoco). 'Coolside SO$_2$ abatement technology–1 MW field tests'. Paper to Coal Technology '85, Pittsburgh, USA, November 1985.

293 Graf R., 'Lurgi dry FGD processes based on the circulating fluid bed principle and the spray absorber system'. Paper to Technical Academy, Wuppertal, 1983. (In German.)

Non-Regenerable Dry Reagent Applied in Furnace (Category S70)

294 Maulbetsch J. (EPRI). 'Status of furnace sorbent injection technology'. Paper to 47th American Power Conference, Chicago, USA, April 1985.

295 Chughtai M.Y. and Michelfelder S. 'Direct desulphurisation through additive injection in the vicinity of the flame'. Paper to EPA/EPRI Flue Gas Desulphurisation Symposium, New Orleans, USA, November 1983. Paper 4C.

296 Parkinson G. and McQueen S. 'A shot of limestone may cure SO$_2$-removal woes'. *Chemical Engineering*, 1984 (20 Feb), 30–35.

297 Bortz S, and Flament P. (International Flame Research Foundation). 'Recent IFRF fundamental and pilot scale studies on the direct sorbent injection process'. Papers of First Joint Symposium, Dry SO$_2$ and Simultaneous SO$_2$/NO$_x$ Control Technologies, San Diego, USA, November 1984, and Symposium, Schone Verbranding van Steenkool, Noordwijkerhout, Netherlands, January 1985.

298 Case P.L., Ho L., Clark W.D., Kau E., Pershing D.W., Payne R. and Heap M.P. (Energy and Environmental Research Corporation). 'Testing of wall-fired furnaces to reduce emissions of NO$_x$ and SO$_x$. Volume 1. Final report'. Report No. EPA/600/7-85/026a for US EPA, June 1985.

299 Doyle J.B. and Jankura B.J. (Babcock and Wilcox). 'Furnace limestone injection with dry scrubbing of exhaust gases'. Paper to 1982 Spring Technical Meeting of the Central States Section of the Combustion Institute, Columbus, USA, March 1982.

300 Gallaspy D.T. (Southern Company Services, Inc.). 'Dry sorbent emission control prototype conceptual design and cost study'. Paper 6G.

301 Kokkinos A., Lewis R.D., Borio D.C., Plumley A.L. (Combustion Engineering) and McElroy M.W. (Electric Power Research Institute). 'Feasibility of furnace injection of limestone for SO$_2$ Control'. Paper to Joint Symposium on Stationary Combustion NO$_x$ Control 1982. EPRI, Proceedings, CS 3182, July 1983.

302 Burdett N.A., Cooper J.R.P. (CEGB), Dearnley S. (UK Department of Energy), Kyte W.S. (CEGB) and Tunnicliffe M.F. (Health & Safety Executive). 'The application of direct limestone injection to UK power stations'. *J. Inst. Energy*, 1985 (June), **58** (435), 64–69.

303 Ness H., Dorchak T.P. (US Dept. of Energy) and Reese J.R. (Energy & Environmental Research Corp.). 'Experience with furnace injection of pressure hydrated lime at the 50 MW Hoot Lake Station'. *Inside R&D*, 1985 (20 Mar.).

304 Anonymous. 'SO$_2$ scrubbing: more work for sodium'. *Chemical Week*, 1984 (18 July), 34–35.

Waste Product Disposal

305 Haynes L.H. (Central Illinois Light Co.), Ansari A.H., and Owen J.E. (Gilbert/Commonwealth). 'Ash/FGD waste disposal options: A comparative study for CILCO Duck Creek site'. *Combustion*, 1980 (Jan), 21–27.

306 Kyte W.S. and Cooper J.R.P. (CEGB). 'The disposal of products from FGD processes'. Paper to Second International Conference, Ash Technology and Marketing, London, September 1984.

307 Kyte W.S. and Cooper J.R.P. (CEGB). 'The disposal of products and wastes from FGD processes'. *I. Chem. E. Symposium Series*, (96), 1986, 233–247.

308 Donnelly J.R., Jons E. (Niro Atomizer) and Webster W.C. (Webster and Associates). 'Synthetic gravel from dry FGD end-products'. Paper to 6th International Ash Utilisation Symposium, Reno, USA, March 1982.

309 Donnelly J.R. (Niro Atomizer), Webster W.C. (Webster and Associates), Duedall I.W., Hsu J., Parker J.H. and Woodhead P.M.J. (NY State University). 'Ocean disposal of consolidated spray dryer FGD wastes'. Paper to International Conference, Coal-fired Power Plants and The Aquatic Environment, Copenhagen, Denmark, August 1982.

310 Donnelly J.R. (Niro Atomizer). 'Disposal and utilisation of spray dryer FGD end-products'. Paper to Canadian Electrical Association Seminar, SO_2 Removal by Dry Process, Ottawa, Canada, October 1982.

311 Weis J.G., Hendry D.W. (Burns and McDonnell) and Baumgardner D. (Plains Electric Generation and Transmission Co-operative). 'Centrifuging FGD sludge can eliminate thickening step'. *Power*, 1985 (Oct), 67–69.

312 Johnson C.A. (Peabody Press Systems). 'Alternative methods of handling waste from flue gas desulphurisation systems'. Paper to Coal Technology Europe, 3rd European Coal Utilisation Conference, Amsterdam, The Netherlands, October 1983, **4**, 81–92.

313 Smith C.L. and Rau E. (IU Conversion Systems). 'Stabilised FGD sludge goes to work'. Paper to Coal Technology '81, 4th International Coal Utilisation Conference, Houston, USA, November 1981, **2**, 247–258.

314 Adams D.F. and Farwell S.O. (University of Idaho). 'Sulphur gas emissions from stored flue gas desulphurisation sludges'. *JAPCA*, 1981 (May), **31** (5), 557–564.

315 Goodwin R.W. (General Electric Environmental Services). 'Effect of auto-oxidation on treatment and disposal properties of lignite derived flue gas desulphurisation sludge'. Paper to Coal Technology '83, 6th International Coal Utilisation Conference, Houston, USA, November 1983, **6**, 263–284.

316 Johnson C.A. (Peabody Process Systems). 'FGD sludge stabilisation and fixation: an alternative disposal technique to produce a commercial gypsum'. Paper to Coal Tech '85, 5th International Conference on Coal Utilisation and Trade, London, December 1985, **2**, 349–369.

317 Ellison W. (Ellison Consultants). 'F.G.D. Gypsum: Utilisation vs disposal'. Background paper for 8th FGD Symposium, New Orleans, USA, November 1983.

318 Cope D.R. and Dacey P.W. (IEA Coal Research). 'Solid residues from coal use–disposal and utilisation'. Report No. ICEAS/B3, IEA Coal Research, London, July 1984.

319 Ellison W. (Consultant) and Luckevich L.M. (Ontario Research Foundation). 'FGD waste: Long-term liability or short-term asset?' *Power*, 1984 (June), 79–82.

320 Bengtsson S., Ahman S., Lillestolen T. (Fläkt), and Koudijs G. (Dorr-Oliver). 'Thermal oxidation of spray dryer FGD waste product'. Paper for EPA/EPRI Symposium, Flue Gas Desulphurisation, Cincinnati, USA, June 1985.

321 Goodwin R.W. (Chemico). 'Waste treatment and disposal aspects: combustion and air pollution control processes'. *JAPCA*, 1981 (July), **31** (7), 744–747.

322 Anonymous. 'Environmental impacts of a flue gas desulphurisation programme'. ENDS Report No. 117, October 1984, 9–11.

323 Goodwin R.W. (GEESI). 'Resource recovery from flue gas de-sulphurisation systems'. *JAPCA*, 1982 (September), **32** (9), 986–989.

324 Mzyk D. (Texas Utilities Generating Co.) and Zmuda J. (Research Cottrell). 'By-product gypsum production at a 2300 MW power plant'. Paper to US EPA, 9th Symposium on Flue Gas Desulphurisation, Cincinnati, USA, June 1985.

325 Rosenstiel T.L. and Debus A.A.G. (US Gypsum Co.). 'Process for preparing wastes for non-pollutant disposal'. UK Patent GB 2,097,990B, 2 January 1986. (Appl. 8204421 filed 15 February 1982).

326 Boldt K.R., Tusa W. and Streets D. (Fred C. Hart Associates and Argonne Nat. Lab.). 'Analysis of industrial boiler solid waste impacts'. *JAPCA*, 1981 (July), **31** (7), 753–760.

327 Weeter D.W. 'Utilisation of dry calcium based flue gas desulphurisation waste as a hazardous waste fixation agent'. *JAPCA*, 1981 (July), **31** (7), 751–753.

328 Wirsching F., Poch W., Huller R. and Hamm

H. 'Environmentally safe disposal of coal-fired power station waste'. Eur. Pat. Appl. EP 139,953 (Cl. A62D3/00), 8th May 1985. DE Appl. 3,329,972, 19 August 1983.

329 Krueger B. and Kraus M. (Bischoff). 'Converting the residue from FGD installations to alpha-calcium sulphate hemihydrate crystals'. Ger. Pat. DE 3,331,838, 21 March 1985. Appl. 3 September 1983.

330 Hoelter H., Ingelbuescher H., Gresch H. and Dewert H. 'Making residues from coal-fired power plants environmentally favourable'. Ger. Pat. DE 3,322,539, 17 January 1985. Appl. 23 June 1983.

331 Mitsubishi Heavy Industries Ltd. 'Treatment of FGD wastewater'. Japan Pat. 60 60,886, 5 January 1985. Appl. 83/106,768, 16 June 1983.

332 Jons E. (Niro Atomizer) 'Properties of stabilised desulphurisation products from spray-dry process'. Niro Atomizer A/S, Soeberg, Denmark. Report No. NP-5750388 (Order No. T185750388), 1984. (In Danish.)

333 Aggour M.S. and Stanbro W.D. (Univ. of Maryland). 'Field ageing of fixed sulphur dioxide scrubber waste'. *J. Energy Eng.*, 1985, **111** (1), 62–73.

334 Sayre W.G. 'Selenium: a water pollutant from FGD'. *JAPCA*, 1980 (October), **30** (10), 1134.

335 Thompson C.M. (Radian Corp.). 'Chemical and physical characterisation of Western low-rank coal waste materials. Part 1: By-products from sodium-based dry scrubbing systems, final report'. Report No. DOE/FC/10200-T2 (DE83001167) to US Department of Energy, August 1982.

336 Rittenhouse R.C. 'Additives: the answer to freezing, dust and sludge instability'. *Power Engineering*, 1986 (July), **90** (7), 38–41.

337 Jons E. (Niro Atomizer), 'The use of spray drying absorption FGD products in building materials'. Paper to FGD Symposium, Leningrad, USSR, July 1986.

338 Jons E. (Niro Atomizer). 'SDA-ash as the only residue from flue gas cleaning'. Paper to ACI/RILEM Joint Seminar, Monterrey, Mexico, March 1985.

PAPERS ON NITROGEN OXIDES ABATEMENT

General Approach to Nitrogen Oxides Emissions Abatement

339 Ogunsola O.I. and Reuther J.J. (Pennsylvania State University). 'Relationship between fuel-nitrogen-to-NO_x conversion efficiency and boiling range for coal-derived liquid fuel combustion'. Report No. PSU-FCL-C-80-78.

340 Artem'ev Y.P., Verbovetskii E.K. and Kozhanov D.S. (VTI). 'The effect of air pre-heat temperature on the formation of nitrogen oxides'. *Thermal Engineering*, 1980, **27** (9), 527–528.

341 Parkinson G. 'NO_x controls: Many new systems undergo trials'. *Chemical Engineering*, 1981 (9th March), 39–43.

342 McCartney M.S. and Cohen M.B. (Combustion Engineering). 'Techniques for reducing NO_x emissions from coal fired steam generators'. Paper to Power Magazine, First International Conference on Acid Rain-Regulatory Aspects and Engineering Solutions, Washington, USA, March 1984.

343 Morrison G.F. (IEA Coal Research). 'Nitrogen oxides from coal combustion–abatement and control'. Report No. ICTIS/TR 11, IEA Coal Research, London, November 1980.

344 Dacey P. (IEA Coal Research). 'Developments in NO_x control for coal-fired boilers'. Working paper 67, IEA Coal Research, London, November 1984.

345 Siddiqi A.A. and Tenini J.W. (ARCO). 'NO_x controls in review'. *Hydrocarbon Processing*, 1981 (Oct), 115–124.

346 Parker L.B. and Trumbule R.E. 'Opportunities for increased control of nitrogen oxides emissions from stationary sources: Implications for mitigating acid rain'. Report No. 82–217 ENR to Congressional Research Service, December 1982.

347 Yanai M. (Kawasaki Heavy Industries Ltd). 'Kawasaki's technology on NO_x abatement'. Paper to EPDC/IIP/VDI, NO_x Symposium, Karlsruhe, West Germany, February 1985. Paper R.

348 Ishimoto R. and Miyamae S. (IHI). 'NO_x abatement technologies in IHI'. Paper to EPDC/IIP/VDI, NO_x Symposium, Karlsruhe, West Germany, February 1985. Paper N.

349 Kuroda H. and Masai T. (Babcock Hitachi). 'Babcock Hitachi NO_x abatement technology'. Paper to EPDC/IIP/VDI/NO_x Symposium, Karlsruhe, West Germany, February 1985. Paper L.

350 Ando J. (Chuo University). 'Review of Japanese NO_x abatement technology for stationary sources'. Paper to EPDC/IIP/VDI, NO_x Symposium, Karlsruhe, West Germany, February 1985. Paper A.

351 Jones G.D. (Radian) and Mobley J.D. (US EPA). 'Review of US NO_x abatement technology'. Paper to APCA, 78th Annual Meeting, Detroit, USA, June 1985. Paper 85-55.2.

352 ECE NO_x Task Force. 'Technologies for controlling NO_x emissions from stationary sources'. Report No. IIP4/1986, April 1986.

353 Davids P., Oels H.J. and Rosenbusch K. (Umweltbundesamt). 'Technical consequences of NO_x emission limits in West Germany'. *Gaswarme Int.*, 1986 (May–June), **35** (4), 178–786.

354 Kircher U. 'NO_x emissions and reduction measures in the glass industry'. *Gaswarme Int.*, 1986 (May–June), **35** (4), 207–212.

355 Bergsma F. (TNO) 'Abatement of NO_x from coal combustion. Chemical background and present state of technical development'. *Ind. Eng. Chem. Process Des. Dev.*, 1985, 24 (1), 1–7.

356 Moore T. (EPRI). 'The retrofit challenge in NO_x control'. *EPRI Journal*, 1984 (Nov), 26–33.

357 Mason H.B. *et al*. 'Environmental assessment of stationary source NO_x control technologies'. Third Stationary Source Combustion Symposium, Vol. IV. US Environmental Protection Agency, EPA-600/7-79-050d, February 1979.

358 Ferrari L.M. *et al*. 'Nitrogen oxides emissions and emission factors for stationary sources in New South Wales'. Proceedings, International Clean Air Conference, Brisbane, Australia, May 1978. (Ann Arbor Science, 1978.)

359 US Environmental Protection Agency. 'Compilation of air pollutant emission factors, 2nd edition'. AP-42, US EPA, Research Triangle Park, N.C., 1975.

360 MacCurley W.R., Moscowitz C.M., Ochsner J.C. and Reznik R.B. 'Source assessment; dry bottom industrial boilers firing pulverised bituminous coal'. Report No. EPA-600/2-79019e to US Environmental Agency, June 1979.

Abatement by Combustion Modifications

361 Sekinguchi Y., Okigami N., Taninaka I. and Sakai S. (Hitachi Zosen). 'Development of new NO_x combustion control method'. Report Number UDC 661.5: 662.9, 67–77.

362 Kawamura T. (Mitsubishi) and Frey D.J. (Combustion Engineering). 'Current developments in low NO_x firing systems'. Paper to EPA/EPRI Joint Symposium, Stationary Combustion NO_x Control, Denver, USA, October 1980.

363 Wheeler W.H. (Urquhart). 'Chemical and engineering aspects of low NO_x concentration'. *The Chemical Engineer*, 1980 (Nov), 693–699.

364 Coe, W.W. (CEA). 'How burners influence combustion'. *Hydrocarbon Processing*, 1981 (May), 179–184.

365 Parkinson G. 'Catalytic burning tries for NO_x control jobs'. *Chemical Engineering*, 1981 (15 June), 51–55.

366 Bell C.T. and Warren S. (Airoil-Flaregas). 'Experience with burner NO_x reduction'. *Hydrocarbon Processing*, 1983 (Sept), 145–147.

367 Ando J. (Chuo University) and Mobley J.D. (US EPA). 'Low NO_x burners for pulverised-coal-fired boilers in Japan'. Paper to FGD Pilot Study Group of NATO Committee on the Challenges of Modern Society, York, May/June 1984.

368 Vatsky J. (Foster Wheeler). 'Industrial and utility boiler NO_x control'. EPA/EPRI Symposium.

369 Vatsky J. (Foster Wheeler). 'High capacity low NO_x coal burner for retrofit and new units'. *Power Engineering*, 1982 (Jan).

370 Vatsky J. (Foster Wheeler). 'Modern combustion systems for coal-fired steam generators'. Paper to Pacific Coast Electric Association Conference, San Francisco, USA, March 1980.

371 Pruce L. 'Reducing NO_x emissions at the burner, in the furnace, and after combustion'. *Power* 1981 (Jan), **125** (1), 33–40.

372 Phelan W.J. (International Flame Research Foundation). 'The effect of pulverised coal type

and burner parameters when staging air combustion for NO_x reduction'. Paper to Coal Technology Europe, 3rd European Coal Utilisation Conference, Amsterdam, Netherlands, October 1983, **1**, 85–110.

373 Bancel P.L. and Massoudi M.S. (Kaiser Engineers Inc). 'Gas turbine NO_x controlled with steam and water injection'. *Power Engineering*, 1986 (June), **90** (6), 34–37.

374 Moore T. 'The retrofit challenge in NO_x control'. *EPRI Journal*, 1984 (Nov), **9** (9), 26–33.

375 Campbenedetto E.J. (Babcock & Wilcox) and Schuster H. (Deutsche Babcock). 'Development of low-NO_x pulverised coal firing system'. Paper to Coal Technology Europe '81, Cologne, West Germany, June 1981.

376 Takahashi Y., Tokuda K., Sengoku T., Nakashima F. and Kaneko S. (Mitsubishi). 'Evaluation of Tangential fired low NO_x burners'. Paper to EPA/EPRI Joint Symposium on Stationary Combustion NO_x Control, Dallas, USA, November 1982.

377 Phelan W.J. (IVO), 'The influence of P.F. burner design parameters on the NO_x-emission and char burnout when staging the combustion air'. Paper to Symposium, Noordwijkerhout, January 1985.

378 Masai T., Morita S., Akiyama I. (Babcock-Hitachi) and Ohtsuka K. (Hitachi). 'Low NO_x combustion technology for pulverised coal fuel'. *Hitachi Review*, 1985, **34** (5), 207–212.

379 Mason H.B. (Acurex). 'Survey of control techniques for nitrogen oxide emissions from stationary sources'. *AIChE Symposium Series*, 1979, **75** (188), 1–13.

380 Hunter, S.C. and Carter W.A. (KVB Inc). 'Application of combustion technology for NO_x emissions reduction on petroleum process heaters'. *AIChE Symposium Series*, 1979, **75** (188), 14–26.

381 Sakai M., Fujima Y., Namiki T. and Okada M. (Mitsubishi). 'Development on low NO_x combustion technology'. Paper to EPDC/-IIP/VDI, NO_x Symposium, Karlsruhe, West Germany, February 1985. Paper P.

382 Mahjoob A.L., Singh S.N. and Yokosh S.M. (Aqua-Chem Inc). 'An experimental investigation of the effects of flue gas recirculation on NO_x formation'. Paper to APCA, 78th Annual Meeting, Detroit, USA, June 1985. Paper 85-55.3.

383 Lisauskas R.A., Snodgrass R.J. (Riley Stoker), Johnson S.A. (Physical Sciences Inc.) and Eskinazi D. (EPRI). 'Experimental investigation of retrofit low-NO_x combustion systems'. Paper to EPA/EPRI, 1985 Symposium on Stationary Combustion NO_x Control, Boston, USA, May 1985. EPRI, Proceedings, CS-4360, January 1986.

384 Folsom B., Abele A. and Reese J. (Energy & Environmental Research Corp.). 'Field evaluation of the distributed mixing burner'. Paper to 1985 Symposium on Stationary Combustion NO_x Control, Boston, USA, May 1985. EPRI, Proceedings, CS4360.

385 Lisauskas R.A., Itse D.C. (Riley Stoker) and Masser C.C. (EPA). 'Extrapolation of burner performance from single burner tests to field operation'. Paper to EPA/EPRI, 1985 Symposium on Stationary Combustion NO_x Control, Boston, USA, May 1985. EPRI, Proceedings. CS-4360, January 1986.

386 Mulholland J.A. and Hall R.E. (US EPA). 'The effect of fuel nitrogen in reburning application to a firetube package boiler'. Paper to EPA/EPRI, 1985 Symposium on Stationary Combustion NO_x Control, Boston, USA, May 1985. EPRI, Proceedings, CS-4360, January 1986.

387 Yang R.J., Garacia F.J. and Hunter S.C. (KVB Inc). 'Screening and optimisation of in-furnace NO_x-reduction processes for refinery process heater applications'. Paper to EPA/EPRI, 1985 Symposium on Stationary Combustion NO_x Control, Boston, USA, May 1985. EPRI, Proceedings, CS-4360, January 1986.

388 England G., Kwan Y. and Payne R. (Energy & Environmental Research Corporation). 'Development and field-demonstration of a low-NO_x burner for TEOR steamers'. Paper to EPA/EPRI, 1985 Symposium on Stationary Combustion NO_x Control, Boston, USA, May 1985. EPRI, Proceedings, CS-4360, January 1986.

389 Hunter S.C. and Benson R.C. (KVB Inc). 'Reduction of nitric oxide emissions on a full-scale cement kiln using primary air vitiation'. Paper to EPA/EPRI, 1985 Symposium on Stationary Combustion NO_x Control, Boston, USA, May 1985. EPRI, Proceedings, CS-4360, January 1986.

390 Fleming D.K. (Institute of Gas Technology) and Kurzynske F.R. (Gas Research Institute). 'NO_x control for glass-melting tanks'. Paper to EPA/EPRI, 1985 Symposium on Stationary Combustion NO_x Control, Boston, USA, May

1985. EPRI, Proceedings, CS-4360, January 1986.

391 Suzuki T., Morimoto K., Ohtani K., Odawara R., Kohno T., Matsuda Y. and Suyari M. (Kobe Steel). 'Development of Low-NO$_x$ combustion for industrial applications'. Paper to EPA/EPRI, 1985 Symposium on Stationary Combustion NO$_x$ Control, Boston, USA, May 1985. EPRI, Proceedings, CS-4360, January 1986.

392 Kesselring J.P. and Krill W.V. (Alzeta Corp.). 'A low-NO$_x$ burner for gas-fired firetube boilers'. Paper to EPA/EPRI, 1985 Symposium on Stationary Combustion NO$_x$ Control, Boston, USA, May 1985. EPRI, Proceedings, CS-4360, January 1986.

393 Wendt J.O.L. (University of Arizona). 'Fundamental coal combustion mechanisms and pollutant formation in furnaces'. *Prog. Energy Combust. Sci.*, 1980, **6**, 201–222.

394 Waibel R. and Nickeson D. (John Zink Co.) 'Staged fuel burners for NO$_x$ control'. Paper to International Flame Research Foundation, 8th Members Conference, Noordwijkerhout, The Netherlands, May 1986.

395 Takahashi Y., Sakai M., Junimoto T., Haneda H., Hawamura T. and Kaneko S. (Mitsubishi). 'Development of MACT: In-furnace NO$_x$ removal process for utility steam generators'. Paper to American Power Conference, Chicago, USA, April 1982. Proceedings 1982, **44**, 402–412.

396 Penterson C.A. (Riley Stoker Corp.). 'Development of an economical low-NO$_x$ firing system for coal fired steam generators'. ASME Paper 82-JPGC-Pwr-43, 1982.

397 Hunter S.C. (KVB Inc.). 'Refinery process heater NO$_x$ control by staged combustion air lances'. Paper to 38th Petroleum Mechanical Engineering Workshop Conference, Philadelphia, USA, 1982.

398 Schaedel S.V. (GRI). 'Pyrocore–radiant burner with a bright future'. *Gas Research Institute Digest*, 1984 (July/Aug), **7** (4), 4–9.

399 Whitehead D.M. and Butcher R.W. (British Petroleum). 'Forced draft burners compared'. *Hydrocarbon Processing*, 1984 (July), 51–55.

400 Anonymous. 'Oil power–how they're making it more acceptable'. *Achievement*, 1986 (June), 19–20.

401 Makansi J. (compiler). 'Low-NO$_x$ burners can play key role in retrofits, upgrades'. *Power*, 1986 (Sept), 61–62.

402 Lim K.J. et al. 'Technology assessment report for industrial boiler applications: NO$_x$ combustion modifications'. Report No. EPA-600/7-79-178f to US Environmental Protection Agency, December 1979.

403 Gabrielson J.E., Langsjoen P.L. and Kosvic T.C. 'Field tests of industrial stoker coal-fired boilers for emission control and efficiency improvement'. Report No. EPA-600/7-79-130a, May 1979.

404 British Gas–Private communication.

405 Fenumore

406 Martenay

407 Clark A.G.

Abatement by Flue Gas Treatment

408 Makansi J. 'Controlling NO$_x$ emissions from utility power plants'. *Power*, 1985 (Sept), 107–109.

409 Makansi J. (Compiler). 'Meeting future NO$_x$ caps goes beyond furnace modifications'. *Power*, 1985 (Sept), 45–46.

410 Anonymous. 'Flue gas treatment aims for process simplicity, NO$_x$ control'. *Power*, 1985 (May), **129** (5), 31–32.

411 Karlsson H.T. and Rosenberg H.S. (Battelle). 'Flue gas denitrification. Selective catalytic oxidation of NO to NO$_2$'. *Ind. Eng. Chem. Process Des. Dev.*, 1984, **23** (4), 808–814.

412 Sengoku T., Miyake J., Suzuki T., Seto T., Nishimoto Y., Lida K., Sera T. and Mitsuoka S. (Mitsubishi). 'A consideration on NO$_x$ reduction catalysts for coal-fired boilers'. *Mitsubishi Heavy Industries Technical Review*, 1983 (Feb), **20** (1), 1–7.

413 Iwata K., Nishimoto Y. and Muraishi K. (Mitsubishi). 'Selective catalytic reduction'. *Modern Power Systems*, 1985 (Dec), 33–51.

414 Hyrst B.E. (Exxon Research and Engineering Co.). 'Thermal denox technology update'. 1985 Joint Symposium on Stationary Combustion NO$_x$ Control, Boston, USA, May 1985.

415 Nagai K. and Tanaka S. (Hitachi Zosen). 'NO$_x$

abatement systems developed by Hitachi Zosen'. Paper to EPDC/IIP/VDI, NO$_x$ Symposium, Karlsruhe, West Germany, February 1985, Paper M.

416 Hurst B.E. (Exxon Research & Engineering Co.). 'Thermal deNO$_x$: the practical approach to deep NO$_x$ reduction'. Paper to 32nd Canadian Chemical Engineering Conference, Vancouver, Canada, October 1982.

417 Hurst B.E. (Exxon Research & Engineering Co.) 'Exxon thermal deNO$_x$ process for stationary combustion sources'. Paper to US–Dutch International Symposium on Air Pollution by Nitrogen Oxides, Maastricht, The Netherlands, May 1982.

418 Kerry H.A. and Weir A. 'Catalytical DeNO$_x$ demonstration system at Huntington Beach Generating Station Unit 2'. Paper to Joint Symposium on Stationary Combustion NO$_x$ Control, Dallas, USA, November 1982.

419 Faucett H.L., Maxwell J.D. and Burnett T.A. (Tennessee Valley Authority). 'Technical assessment of NO$_x$ removal processes for utility application'. Report No. AF-568 to EPRI, March 1978.

PAPERS ON COMBINED FGD AND/OR NO$_x$ ABATEMENT AND/OR HALIDES ABATEMENT

Combined Flue Gas Desulphurisation and Nitrogen Oxides Abatement

420 Knoblauch K., Richter E. and Juntgen H. (Bergbau-Forschung). 'Application of active coke in processes of SO$_2$- and NO$_x$-removal from flue gases'. *Fuel*, 1981 (Sept), **60**, 832–838.

421 Vatsky J. and Schindler E.S. (Foster Wheeler). 'Limestone injection with an internally-staged low-NO$_x$ burner'. Paper to EPA/EPRI, 1st Joint Symposium on Dry SO$_2$ and Simultaneous SO$_2$/NO$_x$ Control Technologies, San Diego, USA, November 1984.

422 Felsvang K., Morsing P. and Veltman P. (Niro Atomizer). 'Acid rain prevention through new SO$_x$/NO$_x$ dry scrubbing process'. Paper to Eighth Symposium, Flue Gas Desulphurisation, New Orleans, USA, November 1983.

423 Takenouchi S., Takahashi K., Atsumi T. and Tanaka H. (Sumitomo). 'Simultaneous NO$_x$/SO$_x$ removal from sinter waste gas by dry process'. *Transactions ISIJ*, 1983, **23**, 1076–1084.

424 Hoffmann V. (Uhde). 'Activated coke will reduce emissions in Arzberg'. Flue Gas Desulphurisation, 1986 (June), 71–77.

425 Rosenberg H.S. (Battelle). 'Combined NO$_x$/SO$_2$ removal for flue gases'. *CME*, 1985 (Jan), 48.

426 Drehmel D.C., Martin G.B. and Abbott J.H. (US EPA). 'Results from EPA's development of limestone injection into a low NO$_x$ furnace'.

427 Dykema O.W. (Rockwell). 'SO$_x$ and NO$_x$ control in combustion'. Paper to Coal Technology '83, 6th International Coal and Lignite Utilisation Conference, Houston, USA, November 1983, **3**, 321–343.

428 Richter E. and Knoblauch K. (Bergbau-Forschung). 'BF-Process for SO$_2$- and NO$_x$-removal from flue gases'. Paper to Coal Tech. '85, 5th International Conference on Coal Utilisation and Trade, London, December 1985, **2**, 335–348.

429 Marshall A.R., Goldsack J.S. and Gray J.S. (Babcock Power). 'Reduction of sulphur and nitrogen oxide emissions from utility and industrial boilers'. Paper to Coal Technology Europe '84, 4th European Coal Utilisation Conference, Messe Essen, FRG, September 1984, **2**, 87–125.

430 Furusawa T., Koyama M. and Tsujimura M. (University of Tokyo). 'Nitric oxide reduction by carbon monoxide over calcined limestone enhanced by simultaneous sulphur retention'. *Fuel*, 1985 (March), **64**, 413–415.

431 Drehmel D.C., Martin G.B., Milliken J.O. and Abbott J.H. (US EPA). 'Low NO$_x$ combustion systems with SO$_2$ control using limestone'. Paper to APCA Annual Meeting, Atlanta, USA, June 1983. Paper No. 83-38.7.

432 Drummond C.J., Markussen J.M., Plantz A.R. and Yeh J.T. (US Department of Energy). 'Advanced environmental control technologies for the simultaneous removal of sulphur dioxide and nitrogen oxides from the flue gas'.

433 Ito Y., Fujimoto T. and Nagaoka, O. (Mitsui). 'Mitsui-BF simultaneous SO$_x$ and NO$_x$ removal system'. Paper No. 8C.

434 Haslbeck J.L., Neal L.G. and Wang C.J. (NOXSO). 'The NOXSO process: a dry simultaneous SO$_2$/NO$_x$ control technology'. Paper to EPA/EPRI, First Joint Symposium, Dry SO$_2$ and Simultaneous SO$_2$/NO$_x$ Control Technologies, San Diego, USA, November 1984.

435 Richter E. and Knoblauch K. (Bergbau-Forschung). 'BF Process for SO_2- and NO_x-removal from flue gases. Paper to Coal Tech. '85, 5th International Conference on Coal Utilisation and Trade, London, December 1985, **2**, 335–348.

436 Barnes H.L. and Shapiro E. (Pittsburgh Environmental & Energy Systems). 'Process for removing sulphur and/or nitrogen oxide or oxides from other gases containing such oxide or oxides'. UK Pat. Appl. No. GB 2,003,126A, 7 March 1979 (filed 21 August 1978).

437 Flament G. 'The simultaneous reduction of NO_x and SO_2 in coal flames by direct injection of sorbents in a staged mixing burner'. International Flame Research Foundation, Document No. G19/a/10, September 1981.

438 Chang S.G., Littlejohn D. and Lyon S. 'Effects of metal chelates on wet flue gas scrubbing chemistry'. *Env. Sci. Technol.*, 1983, 17 (11), 649–653.

439 Ploeg J.E.G. (Shell International Research). 'A process for the simultaneous removal of nitrogen oxides and sulphur oxides from a gas stream'. Eur. Pat. Appl. No. 80200733.6, 18 February 1981 (filed 31 July 1980).

440 Staudinger G. and Schrofelbauer H. 'Laboratory tests, field trials and application of furnace limestone injection in Austria'. Paper to EPRI/EPA, 1st Joint Symposium on dry SO_2 and simultaneous SO_2/NO_x Control Technologies, San Diego, USA, November 1984.

441 Dalton S.M. 'Current status of dry NO_x–SO_x emission control process'. Paper to Joint Symposium on Stationary Combustion NO_x control, Dallas, USA, November 1982.

442 Anonymous. 'Process scrubs both sulphur and nitrogen oxides'. *Chemical Engineering*, 1983 (31 Oct), 21–22.

443 Anonymous. 'Wet-type simultaneous SO_x, NO_x removing process developed'. *IHI Bulletin*, 1976 (Nov), 1.

444 Gleason R.J. and Helfritch D.J. (Cottrell Environmental Sciences). 'Alternative electron beam SO_x and NO_x control systems.' Paper to AIChE Spring National Meeting, Houston, USA, March 1985.

Combined Flue Gas Desulphurisation and Acid Halides Abatement

445 Anonymous. 'Gas cleaning: Combined removal of sulphur, dust and fluorine'. *Sulphur*, 1983 (March/April), (165), 42–43.

446 Uchida S. and Tsuchiya K. (Shizuoka University). 'Simulation of spray drying absorber for removal of HCl in flue gas from incinerators'. *Ind. Eng. Chem. Process Des. Dev.*, 1984, **23** (2), 300–307.

447 Kyte W.S., Bettelheim J. (CEGB), Nicholson N.E. and Scarlett J. (Davy McKee). 'Selective absorption of hydrogen chloride from flue gases in the presence of sulphur dioxide'. *Environmental Progress*, 1984 (Aug), **3** (3), 183–187.

448 Nippon Kokan K.K. 'HCl removal from flue gas'. Jap. Pat. JP 60 38,024 (85 38,024), 27 February 1985. Appl. 83/146,978, 10 August 1983.

449 Deguchi A., Kochiyama Y., Hosoda H., Miura M., Hirama T., Nishizaki H. and Horio M. 'The search for an absorbent for HCl and SO_2 removal at high temperature'. *Nenryo Gakkaishi*, 1982, **61** (668), 1105–1108.

450 Mitsubishi Heavy Industries. 'HCl removal'. JP 60 90,028 (85 90,028), 21 May 1985 (Appl. 83/196,128, 21 October 1983).

BATTELLE BI-MONTHLY REPORTS

451 Battelle Bi-Monthly Report No. 37, 10 March 1980

452 Battelle Bi-Monthly Report No. 38, 10 May 1980

453 Battelle Bi-Monthly Report No. 39, 10 July 1980

454 Battelle Bi-Monthly Report No. 40, 10 September 1980

455 Battelle Bi-Monthly Report No. 41, 10 November 1980

456 Battelle Bi-Monthly Report No. 42, 10 January 1981

457 Battelle Bi-Monthly Report No. 43, 10 March 1981

458 Battelle Bi-Monthly Report No. 44, 10 May 1981

459 Battelle Bi-Monthly Report No. 45, 10 July 1981

460 Battelle Bi-Monthly Report No. 46, 10 September 1981

461 Battelle Bi-Monthly Report No. 47, 10 November 1981

462 Battelle Bi-Monthly Report No. 48, 10 January 1982

463 Battelle Bi-Monthly Report No. 49, 10 March 1982

464 Battelle Bi-Monthly Report No. 50, 10 May 1982

465 Battelle Bi-Monthly Report No. 51, 10 July 1982

466 Battelle Bi-Monthly Report No. 52, 10 September 1982

467 Battelle Bi-Monthly Report No. 53, 10 November 1982

468 Battelle Bi-Monthly Report No. 54, 10 January 1983

469 Battelle Bi-Monthly Report No. 55, 10 March 1983

470 Battelle Bi-Monthly Report No. 56, 10 May 1983

471 Battelle Bi-Monthly Report No. 57, 10 July 1983

472 Battelle Bi-Monthly Report No. 58, 10 September 1983

473 Battelle Bi-Monthly Report No. 59, 10 November 1983

474 Battelle Bi-Monthly Report No. 60, 10 January 1983

475 Battelle Bi-Monthly Report No. 61, 10 March 1983

476 Battelle Bi-Monthly Report No. 62, 10 May 1984

477 Battelle Bi-Monthly Report No. 63, 10 July 1984

478 Battelle Bi-Monthly Report No. 64, 10 September 1984

479 Battelle Bi-Monthly Report No. 65, 10 November 1984

480 Battelle Bi-Monthly Report No. 66, 10 January 1985

481 Battelle Bi-Monthly Report No. 67, 10 March 1985

482 Battelle Bi-Monthly Report No. 68, 10 May 1985

483 Battelle Bi-Monthly Report No. 69, 10 July 1985

484 Battelle Bi-Monthly Report No. 70, 20 September 1985

485 Battelle Bi-Monthly Report No. 71, 10 November 1985

486 Battelle Bi-Monthly Report No. 72, 10 January 1986

487 Battelle Bi-Monthly Report No. 73, 10 March 1986

Appendix 2

Vendor Information

(a) FGD AND SCR PROCESSES

AIR-FROHLICH AG
Romanshornerstrasse 100
CH-9320 Arbon
Switzerland

Tel: 010-41-71-465525
Tlx: 71400 AIRAG CH
Mr. R. Allemann

ANDERSEN 2000 INC.
306 Dividend Drive
Peachtree City
Georgia 30269
USA

Tel: 010-1-404-997-2000
Tlx: 542858 ANDERSEN PECH
Mr. J.D. Brady, President

BABCOCK-HITACHI K.K.
6, 2 Chome Ote Mashi
Chiyoda Ku
Tokyo
Japan

Tel:
Tlx: 3822467 BHKYW J

BABCOCK POWER LTD
165 Great Dover Street
London SE1 4YB

Tel: 01-407-8383
Tlx: 884151/2/3
Mr. R.G.J. Baker

BERGBAU-FORSCHUNG GmbH
Franz-Fischer-Weg 61
D-4300 Essen 13
West Germany

Tel: 010-49-201-1059456
Tlx:

BIONEER OY
PO Box 537
SF-13111 Hämeenlinna
Finland

Tel: 010-358 17 23371
Tlx: 2314 PERA SF
Mr. K. Harsunen
Managing Director

G. BISHOFF GmbH & CO. KG
Postf. 10 05 33,
Gartnerstrasse 44
D-4300 Essen
West Germany

Tel: 010-49-201-8112-0
Tlx: 857779 GASBI D
T Niess DV/AG

CHIYODA CORPORATION
PO Box 10
Tsurumi
Yokahama
Japan

Tel: 010-81-45-521-1231
Tlx: CHIYO J47726
Mr. K. Kohya
General Manager
Licensing Group

DAVY McKEE
15, Portland Place
London W1A 4DD

Tel: 01-637 2821
Tlx: 22604
Mr. R.C. Akroyd
Chief Executive
Environmental Projects

DB GAS CLEANING CORP.
14 Orinda Way
PO Box 944
Orinda CA 94563
USA

Tel: 010-1-415-254-4164
Tlx: 171256
Mr. P.B. Slakey,
Vice President

DEUTSCHE BABCOCK ANLAGEN AG
Postf. 4 & 6
Krefeld 11
West Germany

Tel: 010-49-2151-448571
Tlx: 853824 BSHK D D-4150
Dr. C. Hemmer, Sales Director

DOWA MINING CO.
8–2 Marounouchi 1-chome
J26298 Chiyoda-Ku
Tokyo
Japan

Tel:
Tlx: DOWAMICO

EBARA CORPORATION
Haneda Asahicho
EBARAC J
Ota-Ku
Tokyo 144
Japan

Tel:
Tlx: 2466091
Mr. Keita Kawamura

ESTS BV
PO Box 1000
1970 CA Ijmuiden
The Netherlands

Tel: 010-31-2510-9922
Tlx: 35211 HOVS NL
Mr. H Daalder, Sales Manager

FLÄKT INDUSTRI AB
S-35187
Vaxjo
Sweden

Tel: 010-46-470-87000
Tlx: 52132 FLAKTV S
Mr. Per Jacobson

GENERAL ELECTRIC ENVIRONMENTAL
SERVICES
200 North 7th Street
Lebanon–PA. 17042
USA

Tel: 010-1-717-274-7218
Tlx: 842332 GEESI A LEBA
Mr. R. Snaddon, Manager
International Marketing

HALDOR TOPSOE A/S
Nymollevej 55
DK-2800 Lyngby
Denmark

Tel: 010-45-2878100
Tlx: 37444 HTAS DK
Mr. Frands E. Jensen
Area Sales Manager

INSTITUT FRANCAIS DU PETROLE
1 et 4 Ave. de Bois-Preau
92506 Rueil Malmaison Cedex
France

Tel: 010-33-47526000
Tlx: 203050 F
Alphonse Hennico

ISHIKAWAJIMA-HARIMA HEAVY
INDUSTRIES CO.
30–13, 5-Chome
Toyo,
Koto-Ku
Tokyo 135
Japan

Tel:
Tlx: IHIHET J22232
Mr. H. Ikeno, Air Pollution
Control Design Department

KAWASAKI HEAVY INDUSTRIES
2-16-1 Nakamachidori
Ikeda-Ku
Kobe
Japan

Tel:
Tlx:

KRC UMWELTTECHNIK GmbH
Alfred Nobel Strasse 20
D-8700 Würzburg
West Germany

Tel: 010-49-931-90890
Tlx: 9318129 KRCWZB
Mr. W Zabel

LINDE AG
Werksgruppe TVT Muenchen
Dr. Carl-von-Linde-Strasse 6–14
D-8023 Hoellriegelskreuth
West Germany

Tel: 010-49-89-72731
Tlx: 5283270 LI D
Dr. H Becker

LODGE COTTRELL
Division of Dresser UK Ltd
George Street Parade
Birmingham
West Midlands B3 1QQ

Tel: 021-236-3388
Tlx: 338458
Dr. M J Ashley
Deputy Managing Director

LURGI GmbH
Postfach 111231
Gervinusstrasse 17/19
D-6000 Frankfurt
West Germany

Tel: 010-49-69-1571
Tlx: 412360 IG D
Mr. Wenzel Von Jordan,
Hauptbevollmachtigter

MITSUBISHI HEAVY INDUSTRIES LTD
Bow Bells House
Bread Street
London EC4M 9BQ

Tel: 01-248-8821
Tlx: 888994 MHI LN G
Mr. T. Ono,
Senior Manager

MITSUI MIIKE MACHINERY CO. LTD
Mitsui Building
1-1 Nihonbashi Muromachi 2-chome
Chuo-Ku
Tokyo 103
Japan

Tel: 010-81-270-3481
Tlx: MMMCO J24529
Mr. K. Nagamatsu
Deputy General Manager

NEI INTERNATIONAL COMBUSTION LTD
Sinfin Lane
Derby

Tel: 0332 760223
Tlx:
Mr. V D Trimm

A/S NIRO ATOMIZER
Gladsaxevej 0305
DK-2860 Soeborg
Denmark

Tel: 010-45-169-1011
Tlx: 15603 ATOMN DK
Mr. R. Madhok

OTTO H. YORK CO. INC
Box 3100, 42 Intervale Road
Parsippany
New Jersey 07054-0918
USA

Tel: 010-1-201-299-9200
Tlx: 139134 OTTO YORK FFLD
Mr. K. Schifftner,
Scrubber Department

PEABODY PROCESS SYSTEMS
201 Merritt 7, Corporate Park
Box 6037
Norwalk CT 06852
USA

Tel: 010-1-203-846-1600
Tlx: 965870 PEABODYSYS STD
Dr. E. Baake,
Vice President

SAARBERG-HOLTER UMWELTTECHNIK
GmbH
Hafenstrasse 6
D-6600 Saarbrücken 2
West Germany

Tel: 010-49-681-32104/5/6/7
Tlx: 4421124 SHU D
Mr. Michael Esche

L & C STEINMULLER GmbH
Postf. 100855/100865
Fabrikstrasse 1
D-5270 Gummersbach 1
West Germany

Tel: 010-49-2261-852920
Tlx: 884 5510 SG D
Dr. Ing H. Voos,
General Manager, Environmental Protection

SUMITOMO HEAVY INDUSTRIES LTD
2-1 Otemachi, 2-chome
Chiyoda-Ku
Tokyo 100
Japan

Tel: 010-81-3-245-4321
Tlx: SUMIJUKI J22264

TAMPELLA LTD
Boiler Division
PO Box 626
SF-33101 Tampere 10
Finland

Tel: 010-358-31-32400
Tlx: 22666 TABOI SF

TENNESSEE VALLEY AUTHORITY
3N 78A Missionary Ridge Place
Chattanooga
Tennessee 37401
USA

Tel:
Tlx: 9103333745 ZARSKI/CHATTAN
Dr. M. D. High
Director of Energy
Demonstrations & Technology

THYSSEN ENGINEERING GmbH
Postf. 10 38 54
Am Thyssenhaus 1
D-4300 Essen 1
West Germany

Tel: 010-49-201-1061
Tlx: 8579881-0 TI D
Mr. U. Gebhard
Sales Dept.

UHDE GmbH
Technical Dept. V1
Friedrich-Unde-Strasse 15
D-4600 Dortmund 1
West Germany

Tel: 010-49-231-547 2373
Tlx: 822841–26 UD D
Dr. Ulrich Neumann,
Technical Div. V1

WAAGNER-BIRO AG
Waagner-Biro-Strasse 98
Postf. 1004
A-8021 Graz
Austria

Tel: 010-43-316-5010
Tlx: 31316 WABIG A
Mr. Weitzer

WALTHER & CIE AG
Postf. 85 05 61
Waltherstrasse 51
D-5000 Köln 80
West Germany

Tel: 010-49-221-67850
Tlx: 8873341 WAL D
Mr. S E Christeleit,
Export Sales Manager

(b) DENOX BURNER MANUFACTURERS

AIROIL BURNER CO. (GB) LTD
Horton Road
West Drayton
Middlesex UB7 8BG

Tel: 08954-44031
Tlx: 23923 AIROIL G
Mr. D.W. Harckham
Sales/Marketing Director

BABCOCK POWER LTD
165 Great Dover Street
London SE1 4YB

Tel: 01-407-8383
Tlx: 884151/2/3
Mr. J.S. Goldsack
Chief Engineer,
Engineering Services

DEJONG COEN BV
PO Box 5
3100 AA Schiedam
The Netherlands

Tel: 010-376166
Tlx: 24372
Arie W. Spoormaker

DUNPHY OIL & GAS BURNERS LTD
Queensway
Rochdale
Lancashire OL11 2SL

Tel: 0706-49217
Tlx: 635071 A/B DUNPHY G
Mr. M.P. Dunphy
Managing Director

HAMWORTHY COMBUSTION LTD
Fleets Corner
Poole
Dorset BH17 7LA

Tel: 0202-675123
Tlx: 41226
Mr. J B Champion
Engineering & Development Manager

NU-WAY LTD
PO Box 1
Vines Lane
Droitwich
Worcester WR9 8NA

Tel: 0905-772331
Tlx: 338551 NUWAY G
Mr. J W Findlay

PEABODY HOLMES LTD
(Combustion Division)
Brenchley House
123–135 Week Street
Maidstone
Kent ME14 1RF

Tel: 0622-671381
Tlx: 965850 PHMAID G
Mr. J Lisowski
Chief Combustion Engineer

SAACKE LTD
Fitzherbert Road
Farlington
Portsmouth PO6 1RX

Tel: 0705-383111
Tlx: 86212 SAACKE G
Mr. J N Bartlam

STORDY COMBUSTION ENGINEERING LTD
Heath Mill Road
Wombourne
Wolverhampton WV5 8BD

Tel: 0902-89217
Tlx: 338528 STORCO G
Mr. R.A. Freeman
Managing Director

JOHN ZINK CO LTD Tel: 0727-61451
Alban Park Tlx: 265930
Hatfield Road
St Albans
Herts AL4 0JJ

Appendix 3

Cost Estimates and Procedures

1. PROCEDURE FOR ESTIMATING CAPITAL COSTS FOR NEW-BUILD FGD PLANT

(a) *Source*
Full details of the costing procedure adopted are available in EPRI Report Number CS-3342, Volumes 1–3, December 1983 [95].

(b) *Base Plant*

- 1000 MWe size (2×500 MWe)
- Location: Wisconsin, USA, 200 metres above sea-level
- Seismic Zone 1 (US), i.e. minor risk of damage
- Plant design life of 30 years
- Coal analysis: 4.0% Sulphur
       HHV of 23.5 MJ/kg

(c) *Base Capital Costs*
The total constructed cost of on-site FGD and related facilities, including direct and indirect construction costs. Items included are:

- Civils
- Process equipment
- Piping
- Electrics
- Instruments and control
- Insulation and painting
- Direct field costs
- Indirect field costs (taxes, insurance, construction supplies and equipment, temporary facilities, vendor fees)

The estimate is divided according to functional systems, some of which may not be applicable to certain processes:

- Reagent feed
- Sulphur dioxide removal
- Flue gas
- Regeneration
- By-product
- Waste handling
- General support
- Particulate removal

(d) *Process Adjustments*
Adjustment factors are applicable for the following process variations:

- Unit size (100 to 700 MWe)
- Flue gas flowrate (1.1 to 2.5 m^3/s per net MWe)
- Sulphur content of coal (1.0 to 6.0%)

(e) *Location*
Location adjustment factors are available for the following:

- Seismic zone
- Climate
- Soil conditions
- Material and labour cost index

(f) *Escalation Adjustments*
The EPRI cost estimation procedure is based on December 1982 costs. Although adjustments can be made for start-up dates up to 1993, they are based on EPRI's standard escalation rate of 8.5%.

(g) *Project Contingency*
The project contingency covers additional equipment or other costs which would result from a more detailed design. The contingency factors are based on EPRI Class II guidelines (e.g. 15–30%) and are applied on a system basis. Higher factors are applied to equipment items of special design, and lower factors to standard items.

(h) *Process Contingency*
The process contingency applies to new technology to quantify the design uncertainty and cost of the commercial-scale equipment. The contingency factors are based on EPRI guidelines (e.g. 10–50%) and are applied on a system basis.

(i) *General Facilities*
It is assumed that a major paved road will have to be built along with the necessary area drainage. A new laboratory, office building and a warehouse are to be constructed. This equates to 10 percent of the Escalated Total Process Capital in the EPRI procedure.

(j) *Engineering and Office Fees*
The engineering hours spent by the equipment supplier, architect-engineer and utility to place a

total system in operation. A base fee of 10 percent of the Total Process Capital is used, increasing to a maximum of 15 percent depending on the effects of location and retrofit factors.

(k) *Allowance for Funds During Construction*
The duration of the engineering, procurement and construction phases are expected to vary from 1 to 3 years.

(l) *Royalty Allowance*
The royalty allowance as established by EPRI is 0.5% of the process capital.

(m) *Pre-production Costs*
These costs are intended to cover operator training, equipment check-out, major changes in plant equipment, extra maintenance and inefficient use of materials during plant start-up. In addition, it covers fixed and variable operating costs for one month. This equates to 2 percent of the Total Retrofit Investment.

(n) *Inventory Capital*
The inventory capital includes the value of raw material and other consumables on a capitalised basis. A raw material supply of 60 days is assumed for a base-loaded unit at full capacity operation.

2. PROCEDURES FOR ESTIMATING CAPITAL COSTS FOR RETROFIT FGD PLANT

(a) *Source*
Full details of the costing procedure are available in EPRI Report Number CS-3696, October 1984 [96].

(b) *Base Plant*

- 1000 MWe size (2×500 MWe)
- Location: Wisconsin, USA, 200 metres above sea-level
- Seismic Zone 1 (US), i.e. minor risk of damage

(c) *Base Capital Costs*
The total constructed cost of on-site FGD and related facilities, including direct and indirect construction costs. Items included are:

- Civils
- Process equipment
- Piping
- Electrics
- Instruments and control
- Insulation and painting
- Direct field costs
- Indirect field costs (taxes, insurance, construction supplies and equipment, temporary facilities, vendor fees)

The estimate is divided according to functional systems, some of which may not be applicable to certain processes:

- Reagent feed
- Sulphur dioxide removal
- Flue gas
- Regeneration
- By-product
- Waste handling
- General support
- Particulate removal

(d) *Scope Adjustments*
Adjustment factors are available for the following:

- Chimney work
- Boiler reinforcement
- Draft controls
- Demolition and relocation (buildings, ductwork, piping and electrics)

(e) *Process Adjustments*
Adjustment factors are applicable for the following process variations:

- Unit size (100 to 700 MWe)
- Flue gas flowrate (1.1 to 2.5 m^3/s per net MWe)
- Sulphur content of coal (1.0 to 6.0%)

(f) *Location*
Location adjustment factors are available for the following:

- Seismic zone
- Climate
- Soil conditions
- Material and labour cost index

(g) *Retrofit Adjustments*
Difficulties associated with plant retrofit are site specific and the EPRI Report deals with a number of factors including:

- Accessibility and congestion
- Underground obstructions
- Ductwork length and distance from scrubber to tie-ins

(h) *Escalation Adjustments*
The EPRI cost estimation procedure is based on December 1982 costs. Although adjustments can be made for start-up dates up to 1993, they are based on EPRI's standard escalation rate of 8.5%.

(i) *Project Contingency*
The project contingency covers additional equipment or other costs which would result from a more detailed design. Higher factors are applied to equipment items of special design, and lower factors to standard items. The contingency factors are based

on EPRI Class II guidelines (e.g. 15–30%) and are applied on a system basis.

(j) *Process Contingency*
The process contingency applies to new technology to quantify the design uncertainty and cost of the commercial-scale equipment. The contingency factors are based on EPRI guidelines (e.g. 10–50% and are applied on a system basis.

(k) *General Facilities*
It is assumed that a major paved road will have to be built along with the necessary area drainage. A new laboratory, office building and a warehouse are to be constructed. This equates to 10 percent of the Escalated Total Process Capital in the EPRI procedure.

(l) *Engineering and Office Fees*
The engineering hours spent by the equipment supplier, architect-engineer and utility to place a total system in operation. A base fee of 10 percent of the Total Process Capital is used, increasing to a maximum of 15 percent depending on the effects of location and retrofit factors.

(m) *Allowance for Funds During Construction*
The duration of the engineering, procurement and construction phases are expected to vary from 1 to 3 years.

(n) *Royalty Allowance*
The royalty allowance as established by EPRI is 0.5% of the process capital.

(o) *Pre-production Costs*
These costs are intended to cover operator training, equipment check-out, major changes in plant equipment, extra maintenance and inefficient use of materials during plant start-up. In addition, it covers fixed and variable operating costs for one month. This equates to 2 percent of the Total Retrofit Investment.

(p) *Inventory Capital*
The inventory capital includes the value of raw material and other consumables on a capitalised basis. A raw material supply of 60 days is assumed for a base-loaded unit at full capacity operation.

3. PROCEDURE FOR ESTIMATING GENERALISED FIRST-YEAR OPERATING COSTS FOR FGD PLANT

(a) *Fixed O & M Costs*
These costs include:

– Operating Labour
– Maintenance Labour
– Maintenance Material
– Administrative and Support Labour

It is assumed that these costs remain constant and do not vary with capacity factor.

(i) *Operating Labour Costs*
Operating Labour Cost
$= (MW \times LR \times 40 \times 52)/(Plant\ net\ kW)$

Where MW = Number of man-weeks per week. It is based on the number of operators required per 40 hour week, assuming 4.2 shifts are needed per week for 24 hour coverage of a single job.

LR = Average labour rate at FGD system start-up. It is assumed to be $18.30 (£11.00) per hour (EPRI) for January 1983 start-up.

(ii) *Maintenance Labour Costs*
Maintenance Labour Costs $= PC \times MF \times 0.40$

Where PC = Process Capital Cost ($/kW)
The base fixed operating costs quoted by EPRI are based on maximum process capital costs for a specific FGD system.

MF = Maintenance Factor
Based on EPRI guidelines. An average of 3.75% of the total process capital is used, and is based on a weighted average of all the process area costs.

N.B. It is assumed that 40% of maintenance costs are attributable to labour.

(iii) *Maintenance Material Costs*
Maintenance Material Costs $= PC \times MF \times 0.60$

Where PC and MF are defined above, and it is assumed that 60% of maintenance costs are for material.

(iv) *Administrative and Support Labour Costs*
Administrative and support labour costs are assumed to be 30% of the total maintenance and operating labour costs.

(b) *Variable Costs*
These costs include (where applicable):

(i) Reagent consumption
(ii) Reheat steam
(iii) Power usage
(iv) Water usage

(v) Methane usage
(vi) Waste disposal costs

These costs vary directly with unit size, percentage SO_2 removal, percent sulphur in coal and capacity factor. Except for spray drying, all base costs assume 90% SO_2 removal, 4% S in coal, 1.15 reagent stoichiometry and 65% capacity factor. Adjustments are made for deviations from these base cases. For spray drying base costs assume 70% SO_2 removal and 0.5% S in coal.

The following costs (December 1982) have been assumed:

Limestone	$ 14	(£8.40)	per tonne
Lime	$ 60	(£36)	per tonne
Sodium Carbonate	$140	(£84)	per tonne
High pressure steam	$ 8.27	(£5)	per tonne
Low pressure steam	$ 6.66	(£4)	per tonne
Power	$ 0.045	(2.7p)	per kWh
Methane	$ 0.30	(18p)	per m^3

Sludge disposal costs are based on 10 years remaining unit life (in the case of Retrofit plant) and 10 miles to disposal site.

(c) *Escalation Factor*
Fixed and variable operating costs are escalated at EPRI's standard escalation rate (8.5% per annum) over the period of engineering and construction (1–3 years).

4. CAPITAL COST ESTIMATES

4.1a Limestone Gypsum Process–S21.1 (Newbuild)

	£/kWe
1) Base capital cost (see Table 2.65)	51.29
2) Scope adjustments–none	
Total process Newbuild capital	51.29

3) Process adjustments
   (a) Unit size × 1.00
   (b) Gas flow/MWe × 0.81
   (c) Percentage S in Coal × 0.87

4) Location factors
   (a) Cost index × 1.043
   (b) Soils × 1.05

5) Retrofit factors–none

6) Escalation adjustments × 1.158 Total escalation factor = 0.893 Escalated total process capital (excluding contingencies)	45.82
7) Project contingencies (10%)	4.58
8) Process contingency (5%)	2.29
Escalated total process capital (ETPC)	52.70
9) General facilities (5% of ETPC)	2.63
10) Engineering and office fees (12% of ETPC)	6.32
Total plant Newbuild costs (TPC)	61.65
11) Allowance for funds during construction (2.2% of TPC)	1.36
Total plant Newbuild investment (TPI)	63.01
12) Royalties (5% of TPI)	3.15
13) Pre-production costs (2% of TPI)	1.26
14) Inventory capital	0.71
	68.13

Total capital Newbuild cost for 2000 MWe
£136 million

4.1b Limestone Gypsum Process–S21.1 (Retrofit)

	£/kWe
1) Base capital cost (see Table 2.66)	51.29
2) Scope adjustments (4%)	2.05
Total process Retrofit capital	53.34

3) Process adjustments
   (a) Unit size × 1.00
   (b) Gas flow/MWe × 0.81
   (c) Percentage S in Coal × 0.87

4) Location factors
   (a) Cost index × 1.043
   (b) Soils × 1.05

5) Retrofit factors
   (a) Site accessibility and congestion × 1.25
   (b) Underground obstruction × 1.02
   (c) Ductwork length and scrubber location × 1.12

6) Escalation adjustments × 1.158 Total escalation factor = 1.276 Escalated total process capital (excluding contingencies)	68.05
7) Project contingencies (10%)	6.81

8) Process contingency (5%)	3.40
Escalated total process capital (ETPC)	78.26
9) General facilities (5% of ETPC)	3.91
10) Engineering office fees (12% of ETPC)	9.39
Total plant Retrofit costs (TPC)	91.56
11) Allowance for funds during construction (2.2% of TPC)	2.01
Total plant Retrofit investment (TPI)	93.58
12) Royalties (5%)	4.68
13) Pre-production costs (2% of TPI)	1.87
14) Inventory capital	0.71
	100.84

Total capital Retrofit cost for 2000 MWe £202 million

4.2a Chiyoda CT121 Process–S21.1 (Newbuild)

	£/kWe
1) Base capital cost (see Table 2.65)	47.35
2) Scope adjustments–none	
Total process Newbuild capital	47.35
3) Process adjustments	
(a) Unit size × 1.00	
(b) Gas flow/MWe × 0.81	
(c) Percentage S in Coal × 0.87	
4) Location factors	
(a) Cost index × 1.043	
(b) Soils × 1.05	
5) Retrofit factors–none	
6) Escalation adjustments × 1.158	
Total escalation factor = 0.893	
Escalated total process capital (excluding contingencies)	42.30
7) Project contingencies (10%)	4.23
8) Process contingency (5%)	2.12
Escalated total process capital	48.65
9) General facilities (5%)	2.43
10) Engineering and office fees (12%)	5.84
Total plant Newbuild costs	56.92

11) Allowance for funds during construction (2.2%)	1.25
Total plant Newbuild investment	58.17
12) Royalties (5%)	2.91
13) Pre-production costs (2%)	1.16
14) Inventory capital	0.63
	62.87

Total capital Newbuild cost for 2000 MWe £126 million

4.2b Chiyoda CT121 Process–S21.1 (Retrofit)

	£/kWe
1) Base capital cost (see Table 2.66)	47.35
2) Scope adjustments (4%)	1.89
Total process Retrofit capital	49.24
3) Process adjustments	
(a) Unit size × 1.00	
(b) Gas flow/MWe × 0.81	
(c) Percentage S in Coal × 0.87	
4) Location factors	
(a) Cost index × 1.043	
(b) Soils × 1.05	
5) Retrofit factors	
(a) Site accessibility and congestion × 1.25	
(b) Underground obstruction × 1.02	
(c) Ductwork length and scrubber location × 1.12	
6) Escalation adjustments × 1.158	
Total escalation factor = 1.276	
Escalated total process capital (excluding contingencies)	62.82
7) Project contingencies (10%)	6.28
8) Process contingency (5%)	3.14
Escalated total process capital	72.25
9) General facilities (5%)	3.61
10) Engineering office fees (12%)	8.67
Total plant Retrofit costs	84.53
11) Allowance for funds during construction (2.2%)	1.86
Total plant Retrofit investment	86.39

12) Royalties (5%)	4.32
13) Pre-production costs (2%)	1.73
14) Inventory capital	0.63
	93.07

Total capital Retrofit cost for 2000 MWe £186 million

4.3a Saarberg-Hölter Process—S21.1 (Newbuild)

	£/kWe
1) Base capital cost (see Table 2.65)	46.54
2) Scope adjustments—none	
Total process Newbuild capital	46.54
3) Process adjustments	
(a) Unit size × 1.00	
(b) Gas flow/MWe × 0.81	
(c) Percentage S in Coal × 0.87	
4) Location factors	
(a) Cost index × 1.043	
(b) Soils × 1.05	
5) Retrofit factors—none	
6) Escalation adjustments × 1.158	
Total escalation factor = 0.893	
Escalated total process capital (excluding contingencies)	41.58
7) Project contingencies (10%)	4.16
8) Process contingency (5%)	2.08
Escalated total process capital	47.82
9) General facilities (5%)	2.39
10) Engineering and office fees (12%)	5.74
Total plant Newbuild costs	55.94
11) Allowance for funds during construction (2.2%)	1.23
Total plant Newbuild investment	57.18
12) Royalties (5%)	2.86
13) Pre-production costs (2%)	1.14
14) Inventory capital	0.71
	61.89

Total capital Newbuild cost for 2000 MWe £124 million

4.3b Saarberg-Hölter Process—S21.1 (Retrofit)

	£/kWe
1) Base capital cost (see Table 2.66)	46.54
2) Scope adjustments (4%)	1.86
Total process Retrofit capital	48.40
3) Process adjustments	
(a) Unit size × 1.00	
(b) Gas flow/MWe × 0.81	
(c) Percentage S in Coal × 0.87	
4) Location factors	
(a) Cost index × 1.043	
(b) Soils × 1.05	
5) Retrofit factors	
(a) Site accessibility and congestion × 1.25	
(b) Underground obstruction × 1.02	
(c) Ductwork length and scrubber location × 1.12	
6) Escalation adjustments × 1.158	
Total escalation factor = 1.276	
Escalated total process capital (excluding contingencies)	61.75
7) Project contingencies (10%)	6.17
8) Process contingency (5%)	3.09
Escalated total process capital	71.01
9) General facilities (5%)	3.55
10) Engineering office fees (12%)	8.52
Total plant Retrofit costs	83.08
11) Allowance for funds during construction (2.2%)	1.83
Total plant Retrofit investment	84.91
12) Royalties (5%)	4.25
13) Pre-production costs (2%)	1.70
14) Inventory capital	0.71
	91.57

Total capital Retrofit cost for 2000 MWe £183 million

4.4a Lime-Spray Drying Process – S22.1 (Newbuild)

	£/kWe
1) Base capital cost (see Table 2.65)	40.65
2) Scope adjustments – none	
Total process Newbuild capital	40.65

3) Process adjustments
 (a) Unit size × 1.00
 (b) Gas flow/MWe × 0.81
 (c) Percentage S in Coal × 1.08

4) Location factors
 (a) Cost index × 1.043
 (b) Soils × 1.05

5) Retrofit factors – none

6) Escalation adjustments × 1.158 Total escalation factor = 1.109 Escalated total process capital (excluding contingencies)	45.08
7) Project contingencies (10%)	4.51
8) Process contingency (5%)	2.25
Escalated total process capital	51.85
9) General facilities (5%)	2.59
10) Engineering and office fees (12%)	6.22
Total plant Newbuild costs	60.66
11) Allowance for funds during construction (2.2%)	1.33
Total plant Newbuild investment	61.99
12) Royalties (5%)	3.10
13) Pre-production costs (2%)	1.24
14) Inventory capital	0.24
	66.57

Total capital Newbuild cost for 2000 MWe £133 million

4.4b Lime-Spray Drying Process – S22.1 (Retrofit)

	£/kWe
1) Base capital cost (see Table 2.66)	40.65
2) Scope adjustments (4%)	1.63
Total process Retrofit capital	42.28

3) Process adjustments
 (a) Unit size × 1.00
 (b) Gas flow/MWe × 0.81
 (c) Percentage S in Coal × 1.08

4) Location factors
 (a) Cost index × 1.043
 (b) Soils × 1.05

5) Retrofit factors
 (a) Site accessibility and congestion × 1.25
 (b) Underground obstruction × 1.02
 (c) Ductwork length and scrubber location × 1.12

6) Escalation adjustments × 1.158 Total escalation factor = 1.584 Escalated total process capital (excluding contingencies)	66.95
7) Project contingencies (10%)	6.70
8) Process contingency (5%)	3.85
Escalated total process capital	77.00
9) General facilities (5%)	3.85
10) Engineering office fees (12%)	9.24
Total plant Retrofit costs	90.09
11) Allowance for funds during construction (2.2%)	1.98
Total plant Retrofit investment	92.07
12) Royalties (5%)	4.60
13) Pre-production costs (2%)	1.84
14) Inventory capital	0.24
	98.75

Total capital Retrofit cost for 2000 MWe £198 million

4.5a Wellman-Lord Process – S31.1 (Newbuild)

	£/kWe
1) Base capital cost (see Table 2.65)	76.54
2) Scope adjustments – none	
Total process Newbuild capital	76.54

3) Process adjustments
 (a) Unit size × 1.00
 (b) Gas flow/MWe × 0.81
 (c) Percentage S in Coal × 0.695

4) Location factors
   (a) Cost index × 1.043
   (b) Soils × 1.05

5) Retrofit factors – none

6) Escalation adjustments × 1.158
   Total escalation factor = 0.714
   Escalated total process capital
   (excluding contingencies)   54.63

7) Project contingencies (10%)   5.46

8) Process contingency (5%)   2.73

   Escalated total process capital   62.82

9) General facilities (5%)   3.14

10) Engineering and office fees (12%)   7.54

    Total plant Newbuild costs   73.50

11) Allowance for funds during construction (2.2%)   1.62
    Total plant Newbuild investment   75.12

12) Royalties (5%)   3.76

13) Pre-production costs (2%)   1.50

14) Inventory capital   1.81

    82.18

Total capital Newbuild cost for 2000 MWe
    £164 million

4.5b Wellman-Lord Process – S31.1 (Retrofit)

    £/kWe

1) Base capital cost (see Table 2.66)   76.54

2) Scope adjustments (4%)   3.06

   Total process Retrofit capital   79.60

3) Process adjustments
   (a) Unit size × 1.00
   (b) Gas flow/MWe × 0.81
   (c) Percentage S in Coal × 0.695

4) Location factors
   (a) Cost index × 1.043
   (b) Soils × 1.05

5) Retrofit factors
   (a) Site accessibility and congestion × 1.25

   (b) Underground obstruction × 1.02
   (c) Ductwork length and scrubber location × 1.12

6) Escalation adjustments × 1.158
   Total escalation factor = 1.019
   Escalated total process capital
   (excluding contingencies)   81.13

7) Project contingencies (10%)   8.11

8) Process contingency (5%)   4.06

   Escalated total process capital   93.30

9) General facilities (5%)   4.66

10) Engineering office fees (12%)   11.20

    Total plant Retrofit costs   109.16

11) Allowance for funds during construction (2.2%)   2.40
    Total plant Retrofit investment   111.56

12) Royalties (5%)   5.58

13) Pre-production costs (2%)   2.23

14) Inventory capital   1.81

    121.18

Total capital Retrofit cost for 2000 MWe   £242 million

4.6a Magnesia (MgO) Scrubbing Process – S41.1 (Newbuild)

    £/kWe

1) Base capital cost (see Table 2.65)   87.48

2) Scope adjustments – none
   Total process Newbuild capital   87.48

3) Process adjustments
   (a) Unit size × 1.00
   (b) Gas flow/MWe × 0.81
   (c) Percentage S in Coal × 0.72

4) Location factors
   (a) Cost index × 1.043
   (b) Soils × 1.05

5) Retrofit factors – none

6) Escalation adjustments × 1.158
   Total escalation factor = 0.739
   Escalated total process capital
   (excluding contingencies)   64.68

7) Project contingencies (10%)	6.47	8) Process contingency (5%)	4.80
8) Process contingency (5%)	3.23	Escalated total process capital	110.47
Escalated total process capital	74.38	9) General facilities (5%)	5.52
9) General facilities (5%)	3.72	10) Engineering office fees (12%)	13.26
10) Engineering and office fees (12%)	8.93	Total plant Retrofit costs	129.24
Total plant Newbuild costs	87.03	11) Allowance for funds during construction (2.2%)	2.84
11) Allowance for funds during construction (2.2%)	1.91	Total plant Retrofit investment	132.09
Total plant Newbuild investment	88.94	12) Royalties (5%)	6.60
12) Royalties (5%)	4.45	13) Pre-production costs (2%)	2.64
13) Pre-production costs (2%)	1.78	14) Inventory capital	1.81
14) Inventory capital	1.81		143.14
	96.98		

Total capital Newbuild cost for 2000 MWe   £194 million

Total capital Retrofit cost for 2000 MWe   £286 million

4.6b Magnesia (MgO) Scrubbing Process—S41.1 (Retrofit)

	£/kWe
1) Base capital cost (see Table 2.66)	87.48
2) Scope adjustments (4%)	3.50
Total process Retrofit capital	90.98
3) Process adjustments (a) Unit size × 1.00 (b) Gas flow/MWe × 0.81 (c) Percentage S in Coal × 0.72	
4) Location factors (a) Cost index × 1.043 (b) Soils × 1.05	
5) Retrofit factors (a) Site accessibility and congestion × 1.25 (b) Underground obstruction × 1.02 (c) Ductwork length and scrubber location × 1.12	
6) Escalation adjustments × 1.158 Total escalation factor = 1.056 Escalated total process capital (excluding contingencies)	96.06
7) Project contingencies (10%)	9.61

5. OPERATING COST ESTIMATES

5.1 Limestone Gypsum Process—S21.1

	$/kW-year
1) Total fixed cost	9.7
2) Limestone	2.16
3) High-pressure steam	4.8
4) Power	7.9
5) Waste disposal	1.84
Total Operating Costs (Dec '82)	26.4

Total (Dec '86) Costs = £20.75/kW-year

Escalated Total Operating Costs = £24.02/kW-year

5.2 Chiyoda CT-121 Process—S21.1

	$/kW-year
1) Total fixed cost	9.6
2) Limestone	1.92
3) High-pressure steam	4.9
4) Power	7.4

5) By-product credit	(1.6)
6) Waste disposal	1.84
Total Operating Costs (Dec '82)	24.05

Total (Dec '86) Costs = £18.90/kW-year

Escalated Total Operating Costs = £21.88/kW-year

5.3 Saarberg-Hölter Process–S21.1

	$/kW-year
1) Total fixed cost	6.9
2) Lime	4.8
3) Formic acid	0.12
4) High-pressure steam	4.9
5) Power	6.6
6) Waste disposal	1.84
Total Operating Costs (Dec '82)	25.16

Total (Dec '86) Costs = £19.78/kW-year

Escalated Total Operating Costs = £22.90/kW-year

5.4 Lime Spray Drying Process–S22.1

	$/kW-year
1) Total fixed cost	7.0
2) Lime	7.41
3) Power	3.9
4) Waste disposal	13.37
Total Operating Costs (Dec '82)	31.68

Total (Dec '86) Costs = £24.90/kW-year

Escalated Total Operating Costs = £28.82/kW-year

5.5 Wellman-Lord Process–S31.1

	$/kW-year
1) Total fixed cost	10.6
2) Sodium carbonate	0.52
3) Cooling water	1.9
4) High-pressure steam	5.1
5) Low-pressure steam	3.28
6) Power	9.8
7) Fuel	5.12
8) By-product credit	(2.8)
9) Waste disposal	1.95
Total Operating Costs (Dec '82)	35.47

Total (Dec '86) Costs = £27.88/kW-year

Escalated Total Operating Costs = £32.27/kW-year

5.6 Magnesia Scrubbing Process–S41.1

	$/kW-year
1) Total fixed cost	11.7
2) Magnesia	2.88
3) Fuel oil	4.76
4) Cooling water	2.7
5) High-pressure steam	4.7
6) Power	6.3
7) Waste disposal	1.84
8) By-product credit	(6.32)
Total Operating Costs (Dec '82)	28.56

Total (Dec '86) Costs = £22.34/kW-year

Escalated Total Operating Costs = £25.86/kW-year

6. EQUIPMENT COST ESTIMATE SHEETS

Appendix B of EPRI Report No. CS-3342 [95] includes detailed lists (and 1982 costs) of equipment for several FGD processes. These lists were used, with little modification, for estimating UK-supplied equipment costs for six selected processes. The equipment was budget-priced by potential vendors, wherever possible, based on the limited information available. A large percentage of the vendors' quotes were on a supply-and-erect basis, so an allowance of 25% of the total equipment costs was made for erection. The costs are based on December 1986 budget prices, and the accuracy cannot be considered to be anything less than 25% due to the limited information supplied.

Exclusions from the total equipment costs include:

- commissioning
- spares (running and commissioning)
- 'first fill' (initial catalysts and chemicals)

Only a summary of the estimated equipment costs for each process appears in this Appendix; a confidential annex to the Manual, with more detailed cost data, has been supplied to the Fellowship of Engineering.

6.1 Limestone with Forced Oxidation

	£
Total Equipment Erected Cost	38,377,000
plus 5% for Miscellaneous Process Equipment	1,918,850
Sub-Total	40,295,850
Excluding 25% allowance for erection:	
Total Equipment Cost	30,221,890

NB. No flue gas prescrubbing or gypsum de-watering facilities are included in the costs, although EPRI estimates indicate that their inclusion could increase capital costs by about 13%.

6.2 Chiyoda

	£
Total Equipment Erected Cost	43,122,230
Plus 5% for Miscellaneous Process Equipment	2,156,110
Sub-Total	45,278,340
Excluding 25% allowance for erection:	
Total Equipment Cost	33,958,760

6.3 Saarberg-Hölter

	£
Total Equipment Erected Cost	41,470,130
Plus 5% for Miscellaneous Process Equipment	2,073,510
Sub-Total	43,543,646
Excluding 25% allowance for erection:	
Total Equipment Cost	32,657,740

6.4 Lime Spray Drying

	£
Total Equipment Erected Cost	34,993,730
Plus 5% for Miscellaneous Process Equipment	1,749,690
Sub-Total	36,743,420
Excluding 25% allowance for erection:	
Total Equipment Cost	27,557,570

6.5 Wellman-Lord

	£
Total Equipment Erected Cost	66,691,700
Plus 5% for Miscellaneous Process Equipment	3,334,590
Sub-Total	70,026,290
Excluding 25% allowance for erection:	
Total Equipment Cost	52,519,720

6.6 Magnesia (MgO) Scrubbing

	£
Total Equipment Erected Cost	98,972,020
Plus 5% for Miscellaneous Process Equipment	4,948,600
Sub-Total	103,920,620
Excluding 25% allowance for erection:	
Total Equipment Cost	77,940,470

Index Appendix 4

Abatement, nitrogen oxides – 4
Abatement process – 164
Absorption oxidation process – 109, 121
Absorption reduction process – 121
Acidic emission – xv, xvi, 3, 5, 6
Active carbon adsorption process – xvii, xviii, xxi, 26, 50, 56, 137, 154
Active carbon selective catalytic reduction process – 163, 164
Airoil-Flaregas Ltd – vii
Alkali dry injection process – 29, 53
Alkali scrubbing process – 12, 32
Alkali scrubbing/spray drying process – 16, 35
Alliance process – 79, 82, 85, 165
Allied Chemical Co – 20, 21
Ammonia scrubbing process – 13, 21, 33
Andersen 2000 Inc – viii
Appraisal, process – 5, 7, 31, 55, 113, 121, 153, 162
Aqueous sodium carbonate process – 23, 47
Aroskraft, AB – viii
Asahi Chemical process – 145, 159
Ash, coal – 41
Axial-staged recirculating burner – xix, 95, 98

Babcock and Wilcox – 156
Babcock Power Ltd – vii
Babcock-Hitachi KK – 103
Babcock-Hitachi process – 61
Battelle Institute – viii, 35, 36, 39, 41, 42, 45, 49
Bed, catalyst – 125
Bed, moving catalyst – 116
Beijer Institute – vii
Bergbau-Forschung active carbon process – 84
Bergbau-Forschung GmbH – 50, 164
Berridge Environmental Laboratories Ltd – vii
BF-Mitsui active carbon combined abatement process – 164
Biased burner firing – 98, 101
Bischoff GmbH, G – vii
Bituminous coal – 50
Blowing, soot – 116, 124, 126
Boiler, coal-fired – 4, 42, 52, 60, 64, 159, 161
Boiler, gas-fired – 4, 117
Boiler, oil-fired – xvi, 4, 42, 49, 60, 64, 73, 76, 77, 82, 99, 114, 116, 117, 118, 119, 120, 126, 128, 159, 162, 164, 167
Boiler, oil-fired twin fire-tube – 5
Boiler, oil-fired water tube – 5, 57, 59, 63, 66, 122
Boiler, steam – 117
Boiler, tangentially fired – 101
Boiler, utility – 117, 120, 129, 131, 154
British Coal Corporation (Coal Research Establishment) – vii
British Gas – vii

British Gypsum – vii
Buckau-Walther Group – 16, 17
Buoyancy, plume – 61, 64, 68
Burner, axial-staged recirculating – xix, 95, 98
Burner, dual-register – xix, 95, 98
Burner, fuel-staged – 95, 99
Burner, limestone injection multistage (LIMB) – 55, 141, 156
Burner, low-NOx – xix, 99, 156
Burner, natural gas – 52
Burner, oil – 52
Burners out of service (BOOS) – 98, 100, 101

Calor – vii
Capital cost – xviii, 59, 62, 66, 70, 76, 81, 83, 86, 87, 88, 103, 128, 133, 167, 170
Catalyst – xx, 13, 27, 33, 52, 107, 108, 114, 115, 116, 117, 118, 119, 120, 122, 123, 124, 125, 126, 127, 128, 129, 139, 151, 152, 165
Catalyst, molecular sieve – 125
Catalyst bed – 125
Catalyst erosion – 116
Catalyst fouling – 116
Catalytic Inc – 21
Catalytic inc/IFP process – 44
Catalytic oxidation process – 28, 51, 141, 155
Catalytic reduction process – 141, 155
Cement industry – 71, 77
Central Electricity Generating Board – vii, 31, 32, 33, 56, 57
Central Electricity Research Laboratories – vii
Chelating compound – 145, 146, 159
Chem Systems – vii
Chemical oxygen demand (COD) – 38, 71
Chemico – 48, 82
Chemiebau Rheinluft – 50
Chiyoda Chemical Engineering & Construction Co Ltd – viii, 39
Chiyoda Thoroughbred 121 process – xvii, 38, 39, 40, 56, 59, 61, 63, 64, 66
Circulating fluidised bed combustion – 29
CITEPA – viii
Citrate process – 22, 45
Claus process – 79, 82, 85, 165
Coal, bituminous – 50
Coal, high-sulphur – 40, 47
Coal ash – 41
Coal gasification – 3
Coal-fired boiler – 41, 42, 52, 60, 64, 159, 161
Combined SO2-NOx abatement process – 137, 153, 162, 164
Combustion, circulating fluidised bed – 29
Combustion, fluidised bed – 51, 52
Combustion, low excess air – 98, 99, 101

213

Combustion, off-stoichiometric – 95, 97, 98, 99, 101
Combustion, reduced heat load – 97, 98, 99, 100
Combustion, staged – 97, 98, 99, 100, 101
Combustion Engineering – 39
Composition, flue gas – 5
Composition, oil – 5
CONCAWE – viii
Conoco Inc – 53
Conosox process – 22, 46
Copper oxide absorption process – 140, 154
Copper oxide process – 27, 51
Cost, capital – xviii, 59, 62, 66, 70, 76, 81, 83, 86, 87, 88, 103, 128, 133, 167, 170
Cost, installation – 87
Cost, operating – xvi, 6, 59, 63, 66, 70, 76, 81, 83, 87, 89, 103, 128, 133, 167, 170
Cost index – 89
Crude oil – 93

Davy McKee Corp – viii, 66, 77
DB Gas Cleaning Corp – viii
De Jong Coen BV – viii
Denitrification, flue gas – xix, 5, 6, 108
DeNOx process – 12, 102, 107, 121, 128, 130
Department of the Environment – vii, ix, xv, 3
Deposition, dust – 117
Desulphurisation, flue gas – xvi, xvii, xviii, 4, 5, 6, 11, 31, 164
Desulphurisation, oil – 3, 5
Deutsche Babcock Anlagen – vii
Direct absorption process – 109
Double alkali process – 15
Dowa process – 14, 34
Dry adsorption process – 109, 128
Dual alkali process – 15, 34
Dual-register burner – xix, 95, 98
Dunphy Oil and Gas Burners Ltd – vii
Dust deposition – 117
Dust erosion – 124
Dust plugging – 118, 124

Effect, thermal – 124
Efficiency, plant – 5
Efficiency, thermal – 5
Efficiency factor – xvi, xviii, xxi, 167
Electric Power Research Institute – viii
Eletrolytic Zinc Co of Australia Ltd – 31
Electron beam radiation process – 142, 157
Emission, acidic – xv, xvi, 3, 5, 6
Emission, particulates – xvi, xvii, 5, 16, 17, 18, 19, 21, 24, 25, 26, 41, 48, 70, 75, 79, 83, 132, 144, 158, 165, 167
Emission factor – xvi, xxi, 5, 7, 58, 62, 65, 69, 75, 80, 83, 86, 127, 128, 131, 166, 170
Energy and Environmental Research Corp – 156
EPRI – 34, 87, 88, 89
Equimolar absorption process – 110
Erosion, catalyst – 116
Erosion, dust – 124
ESTS – viii
ETSU – vii
Exxon Research Engineering Corp – 117, 119, 120, 130, 131

Factor, efficiency – xvi, xviii, xxi, 167
Factor, emission – xvi, xxi, 5, 7, 58, 62, 65, 69, 75, 80, 83, 86, 127, 128, 131, 166, 170
Fellowship of Engineering – vii, ix, xv, 3

Fertiliser industry – 142
Firing, biased burner – 98, 101
Firing system, tangential – xix, 95, 98
Fläkt Industri AB – viii, 77, 167
Fläkt-Boliden process – 20, 43
Fläkt-Hydro – 31, 32, 57
Fläkt-Niro process – 72, 73
Flue gas composition – 5
Flue gas denitrification – xix, 5, 6
Flue gas desulphurisation – xvi, xvii, xviii, 4, 5, 6, 11, 31, 164
Flue gas recirculation – 95, 97, 99, 101
Flue gas treatment – 107
Fluidised bed combustion – 51, 52
Fly-ash – 7, 18, 28, 37, 38, 54, 124, 141, 154
FMC – 35
Formation, plume – 116, 120
Foster Wheeler Development Corp – viii
Foster Wheeler Energy Corp – 84, 156
Foster Wheeler International Corp – viii, 50
Foster Wheeler Power Products Ltd – ix, 3
Foster Wheeler Resox process – 20, 21, 50, 79, 82, 85, 86, 165
Fouling, catalyst – 116
Fuel NOx – 93, 97, 98
Fuel-staged burner – 95, 99
Furnace, glass melting – 117

Gasification, coal – 3
Gas-fired boiler – 4, 117, 118, 120
General Electric Environmental Services Inc – viii, 39
Generator, steam – 117
Glass melting furnace – 117
Grid, injection – 119, 120, 129, 130

Halide – 3, 4, 11, 12, 16, 17, 18, 19, 20, 21, 22, 24, 26, 27, 110, 116, 139, 145, 165
Hamworthy Combustion Ltd – vii
Heavy oil – 99
High-dust system – 107, 123, 125
High-sulphur coal – 40, 47
High-sulphur oil – 117, 121, 122, 126
Hitachi – 51
Hitachi process – 27
Honeycomb catalyst reactor – 115, 116, 125
Hoy Associates (UK) Ltd – x, 3
Hydrated lime injection process – 28, 53

IHI limestone/gypsum process – xvii, 56, 59, 60, 66, 71, 151, 161
Incinerator, municipal waste – 117
Index, cost – 89
Industry, cement – 71, 77
Industry, fertiliser – 142
Industry, paper – 24
Industry, plasterboard – 14, 18, 62, 65
Industry, pulp – 24
Industry, wallboard – 71
Injection, overfire air – 97, 99, 101
Injection, sorbent – 156
Injection grid – 119, 120, 129, 130
Injector, wall – 118, 119, 120, 129, 130
Injector nozzle – 33, 41, 42, 116, 119, 130
Installation cost – 87
Institut Francais du Petrole – 21
International Energy Agency (Coal Research) – vii

International Flame Research Foundation – 156
Ishikawajima-Harima Heavy Industries Co Ltd (IHI) – viii, 31, 32, 39, 56, 57, 61, 62, 103
Ispra Mark 13A process – 23, 47

Johnson Matthey – vii
Joy-Niro – 73, 167

Kawasaki Heavy Industries – 103, 114, 121
Kawasaki process – 150, 161

Landfill – 62, 68, 72, 75, 76, 77, 167
Leachate – 12, 72, 76, 77
Light distillate oil – 93
Light oil – 99
Lignite – 50
Lime slurry scrubbing process – xvii, 18, 37, 56, 59, 63, 66, 72
Lime slurry scrubbing/spray dryer process – 40, 56, 72, 76
Lime spray dryer combined abatement process – 167
Lime spray dryer process – 144, 158
Limestone injection into multistage burners (LIMB) – 55, 141, 156
Limestone slurry scrubbing process – xvii, 18, 37, 41, 56, 59, 63, 66, 70, 72, 146, 159
Load swing – 100, 117, 118, 120, 124, 129
Lodge Cottrell – vii
Low excess air combustion – 98, 99, 101
Low-dust system – 123, 125
Low-NOx burner – xix, 99
Low-NOx coal burner – 156
Low-sulphur oil – 117, 126
Lurgi circulating fluidised bed lime absorber process – 29, 54
Lurgi Sulfacid process – 27
Lurgi (UK) Ltd – vii, 51

Magnesia slurry scrubbing process – xvii, 24, 48, 56, 81
Metra Consulting – vii
Mitsubishi Heavy Industries – 117, 117
Mitsui Miike Engineering Corp – viii
Mitsui Mining Co Ltd – 164
Molecular sieve catalyst – 125
Moretana calcium process – 152, 162
Moving catalyst bed – 116
Municipal waste incinerator – 117

National Swedish Environmental Protection Board – viii
Natural gas burner – 52
Niro Atomiser A/S – vii, 73, 76, 167
Nitrogen oxides abatement process – xxi, 4, 93, 107
Non-regenerable reagent process – xvii
Non-selective catalytic reduction process – 140, 155
Norsk Hydro – 31
NOx fuel – 93, 97, 98
NOx, prompt – 93
NOx, thermal – 93, 95, 97, 98
NOx abatement process – xiv, 49
NOx formation – 93
Noxso process – 157
Nozzle, injector – 16, 33, 41, 42, 119, 130
Nu-Way Ltd – vii

Off-stoichiometric combustion – 95, 97, 98, 101
Oil, crude – 93
Oil, heavy – 99

Oil, high-sulphur – 117, 121, 122, 126
Oil, light distillate – 93
Oil, light – 99
Oil, low-sulphur – 117, 126
Oil, residual – 93
Oil burner – 52
Oil composition – 5
Oil desulphurisation – 3, 5
Oil-fired boiler – xvi, 4, 42, 49, 60, 64, 73, 76, 77, 82, 99, 114, 116, 117, 118, 119, 120, 126, 128, 159, 162, 164
Oil-fired twin fire-tube boiler – 5
Oil-fired water tube boiler – 5, 57, 59, 63, 66, 122
Operating cost – xvi, 6, 59, 63, 66, 70, 76, 81, 83, 87, 89, 103, 128, 133, 167, 170
Overfire air injection – 97, 99, 101
Oxidation absorption process – 110, 121
Oxidation absorption reduction process – 120, 121
Oxidation/ammonia scrubbing process – 149, 160

Paper industry – 24
Particulates emission – xvi, xvii, 5, 16, 17, 18, 19, 21, 24, 25, 26, 41, 48, 70, 75, 79, 83, 132, 144, 158, 165, 167
Peabody Coal Co – 156
Peabody Holmes Ltd – vii
Peabody Process Systems – viii
Peat – 50
Pennwalt Ltd – vii
Plasterboard industry – 154, 18, 62, 65
Plate catalyst reactor – 115, 116, 125
Plugging, dust – 118, 124
Plume buoyancy – 61, 64, 68
Plume formation – 116, 120
Preheat, reduced air – 98, 100
Process, absorption oxidation – 109, 121
Process, absorption reduction – 121
Process, active carbon adsorption – xvii, xviii, xxi, 26, 50, 56, 137, 154
Process, active carbon selective catalytic reduction – 163, 164
Process, alkali dry injection – 29, 53
Process, alkali scrubbing – 12, 32
Process, alkali scrubbing/spray drying – 16, 35
Process, Alliance – 79, 82, 85, 165
Process, ammonia scrubbing – 13, 21, 33
Process, aqueous sodium carbonate – 23, 47
Process, Asahi Chemical – 145, 159
Process, Babcock-Hitachi – 61
Process, Bergbau-Forschung active carbon – 84
Process, BF-Mitsui active carbon combined abatement – 164
Process, Catalytic Inc/IFP – 44
Process, catalytic oxidation – 28, 52, 141, 155
Process, catalytic reduction – 141, 155
Process, Chiyoda Thoroughbred 121 – xvii, 38, 39, 40, 56, 59, 61, 63, 64, 66
Process, citrate – 22, 45
Process, Claus – 79, 82, 85, 165
Process, combined SO2-NOx abatement – 137, 163, 162, 164
Process, Conosox – 22, 46
Process, copper oxide absorption – 140, 154
Process, copper oxide – 27, 51
Process, deNOx – 102, 107, 121, 122, 128
Process, direct absorption – 109
Process, double alkali – 15
Process, Dowa – 14, 34

Process, dry adsorption – 109, 120
Process, dual alkali – 15, 34
Process, electron beam radiation – 142, 157
Process, equimolar absorption – 110
Process, Flakt-Boliden – 20, 43
Pricess, Flakt-Niro – 72, 73
Process, Foster Wheeler Resox – 20, 21, 50, 79, 85, 86, 165
Process, Hitachi – 27
Process, hydrated lime injection – 28, 53
Process, IHI limestone/gypsum – xvii, 56, 59, 60, 63, 66, 71, 151, 161
Process, Ispra Mark 13A – 23, 47
Process, Kawasaki – 150, 161
Process, lime slurry scrubbing/spray dryer – 40, 56, 72, 76
Process, lime slurry scrubbing – xvii, 18, 37, 56, 59, 63, 66, 72
Process, lime spray dryer absorption – xvii, xviii, xxi, xxii, 164
Process, lime spray dryer combined abatement – 167
Process, lime spray dryer – 144, 158
Process, limestone slurry scrubbing – xvii, 18, 37, 41, 56, 59, 63, 66, 70, 72, 146, 159
Process, Lurgi circulating fluidised bed lime absorber – 29, 54
Process, Lurgi Sulfacid – 27
Process, magnesia slurry scrubbing – xvii, 24, 48, 56, 81
Process, Moretana calcium – 152, 162
Process, nitrogen oxides abatement – xxi, 4, 93, 107
Process, non-regenerable reagent – xvii
Process, non-selective catalytic reduction – 140, 155
Process, NOx abatement – xix, 49
Process, Noxso – 157
Process, oxidation absorption reduction – 120, 121
Process, oxidation absorption – 110, 121
Process, oxidation/ammonia scrubbing – 149, 160
Process, Ralph M Parsons Co – 140, 155
Process, reburning – xix, 95, 99
Process, regenerable reagent – xvii
Process, Resox; see Foster Wheeler
Process, Saarberg-Holter – 56, 66, 67
Process, Saarberg-Holter-Lurgi lime/gypsum – xvii, 61, 63, 147
Process, sea water scrubbing – xvii, xviii, 11, 56, 57
Process, selective catalytic reduction – xx, 107, 113, 122, 128, 131
Process, selective non-catalytic reduction – xx, 108, 117, 122, 128, 131
Process, sodium carbonate adsorption – 143, 157
Process, sodium sulphite scrubbing – 145, 157
Process, sorbent direct injection – 30, 54
Process, spray dryer – 163
Process, Sulf-X – 25, 49, 147, 160
Process, sulphuric acid scrubbing – 14
Process, Sumimoto – 165
Process, thermal deNOx – 117, 128
Process, thiosorbic lime – 40
Process, Walther – 16, 33, 36, 149, 160
Process, Wellman-Lord – xvii, 19, 42, 56, 77, 78
Process, wet active carbon adsorption – 27, 51
Process, wet flue gas treatment – 120
Process appraisal – 5, 7, 31, 55, 113, 121, 153, 162
Prompt NOx – 93
Pullman-Kellogg – 39
Pulp industry – 24
Pulverised-coal boiler – 114

Radioactive material – 72
Ralph M Parsons Co process – 140, 155
Reactor, honeycomb catalyst – 115, 116, 125
Reactor, plate catalyst – 115, 116, 125
Reactor, ring catalyst – 125
Reburning process – xix, 95, 99
Recirculation, flue gas – 95, 97, 99, 101
Reduced air preheat – 98, 100
Reduced heat load combustion – 97, 98, 99, 100
Regenerable reagent process – xvii
Research-Cottrell – 38, 39
Residual oil – 93
Resox process; see Foster Wheeler
Rheinisch-Westfalisches Elektrizitatswerk – vii
Riley Stoker – 156
Ring catalyst reactor – 125
Rockwell International / 36, 156

Saacke Ltd – vii
Saarberg-Holter process – 56, 66, 67
Saarberg-Holter – 38, 39
Saarberg-Holter-Lurgi lime/gypsum process – xvii, 59, 61, 63, 147
Sea water scrubbing process – xvii, xviii, 11, 56, 57
Selective catalytic reduction process – xx, 107, 113, 122, 128, 131, 139
Selective non-catalytic reduction process – xx, 108, 117, 122, 128, 131
Shell – 51, 155
Shock, thermal – 117, 119, 127
Societee Foster Wheeler Francaise – viii
Sodium carbonate adsorption process – 143, 157
Sodium sulphite scrubbing process – 145, 159
Sommerlad (Consultant), R E – viii
Soot blowing – 116, 124, 126
Sorbent direct injection process – 30, 54
Sorbent injection – 156
Spray dryer process – 163
Staged combustion – 97, 98, 99, 100, 101
Steam boiler – 117
Steam generator – 117
Steinmuller GmbH, L &C – vii, 156
Stordy Combustion Engineering Ltd – vii
Sulf-X process – 25, 49, 147, 160
Sulphuric acid scrubbing process – 14
Sumitomo Heavy Industries – viii
Sumitomo process – 165
Swing, load – 100, 117, 118, 120, 124, 129

Tail gas system – 123, 124, 125, 126, 127
Tampella – viii
Tangential firing system – xix, 95, 98
Tangentially fired boiler – 101
Tennessee Valley Authority – viii
Thermal deNOx process – 117, 128
Thermal effect – 124
Thermal efficiency – 5
Thermal NOx – 93, 95, 97, 98
Thermal shock – 117, 119, 122
Thiosorbic lime process – 40
Thyssen Engineering – vii
Tipping, waste – 72, 77
Treatment, flue gas – 107
TRW Inc – 156

Uhde GmbH – 84, 164

Umweltbundesamt – vii
United Engineers & Constructors – 48, 49, 81, 82
US Department of Energy – 157, 158, 159
US Environmental Protection Agency – 154
Utility boiler – 117, 120, 129, 131, 154

VGB – vii

Wall injector – 118, 119, 120, 129, 130
Wallboard industry 71
Walther process – 16, 33, 36, 149, 160

Warren Spring Laboratories – vii, x, xv, 3, 4
Waste tipping – 72, 77
Wellman-Lord process – xvii, 19, 42, 56, 77, 78
Westvaco – 51
Wet active carbon adsorption process – 27, 51
Wet flue gas treatment process – 120
Wheelabrator-Frye – 36

York, Otto H – viii

Zink Co Ltd, John – vii